The Future of Helium as a Natural Resource

The purpose of this book is to inform the reader as to the future of helium: a most unusual substance, and a key natural resource with significant relevance for high technology. The frame of reference is international: helium is globalising.

Commercial helium is a by-product of the natural gas industry. The relationship of helium supply to natural gas is fundamental and has not been studied extensively in the past. The global natural gas industry is part-way through a profound change, with increasing emphasis on liquefied natural gas. This has major consequences for helium supply.

The authors turn their attention to helium supply and demand; most helium studies having previously been excessively focused on the supply side. Chapters study the potential impact of helium in a number of fields, including medical imaging and the nuclear industry, as well its influence in major national economies such as Russia and India.

William J. Nuttall is University Senior Lecturer in Technology Policy at the Judge Business School, University of Cambridge, UK.

Richard H. Clarke is a Cryogenic Process and Helium Specialist at Culham Centre for Fusion Energy, a UK-based research organisation developing fusion as a sustainable, long term energy source.

Bartek A. Glowacki is Reader in the Department of Materials Science and Metallurgy, University of Cambridge, UK.

Routledge explorations in environmental economics

Edited by Nick Hanley
University of Stirling, UK

1 **Greenhouse Economics**
Value and ethics
Clive L. Spash

2 **Oil Wealth and the Fate of Tropical Rainforests**
Sven Wunder

3 **The Economics of Climate Change**
Edited by Anthony D. Owen and Nick Hanley

4 **Alternatives for Environmental Valuation**
Edited by Michael Getzner, Clive Spash and Sigrid Stagl

5 **Environmental Sustainability**
A consumption approach
Raghbendra Jha and K.V. Bhanu Murthy

6 **Cost-Effective Control of Urban Smog**
The significance of the Chicago cap-and-trade approach
Richard F. Kosobud, Houston H. Stokes, Carol D. Tallarico and Brian L. Scott

7 **Ecological Economics and Industrial Ecology**
Jakub Kronenberg

8 **Environmental Economics, Experimental Methods**
Edited by Todd L. Cherry, Stephan Kroll and Jason F. Shogren

9 **Game Theory and Policy Making in Natural Resources and the Environment**
Edited by Ariel Dinar, José Albiac and Joaquín Sánchez-Soriano

10 **Arctic Oil and Gas**
Sustainability at risk?
Edited by Aslaug Mikkelsen and Oluf Langhelle

11 **Agrobiodiversity, Conservation and Economic Development**
Edited by Andreas Kontoleon, Unai Pascual and Melinda Smale

12 **Renewable Energy from Forest Resources in the United States**
Edited by Barry D. Solomon and Valeria A. Luzadis

13 **Modeling Environment-Improving Technological Innovations under Uncertainty**
Alexander A. Golub and Anil Markandya

14 **Economic Analysis of Land Use in Global Climate Change Policy**
Thomas Hertel, Steven Rose and Richard Tol

15 **Waste and Environmental Policy**
Massimiliano Mazzanti and Anna Montini

16 **Avoided Deforestation**
Prospects for mitigating climate change
Edited by Stefanie Engel and Charles Palmer

17 **The Use of Economic Valuation in Environmental Policy**
Phoebe Koundouri

18 **Benefits of Environmental Policy**
Klaus Dieter John and Dirk T.G. Rübbelke

19 **Biotechnology and Agricultural Development**
Robert Tripp

20 **Economic Growth and Environmental Regulation**
Tim Swanson and Tun Lin

21 **Environmental Amenities and Regional Economic Development**
Todd Cherry and Dan Rickman

22 **New Perspectives on Agri-Environmental Policies**
Stephen J. Goetz and Floor Brouwer

23 **The Cooperation Challenge of Economics and the Protection of Water Supplies**
A case study of the New York City watershed collaboration
Joan Hoffman

24 **The Taxation of Petroleum and Minerals: Principles, Problems and Practice**
Philip Daniel, Michael Keen and Charles McPherson

25 **Environmental Efficiency, Innovation and Economic Performance**
Massimiliano Mazzanti and Anna Montini

26 **Participation in Environmental Organizations**
Benno Torgler, Maria A. Garcia-Valiñas and Alison Macintyre

27 **Valuation of Regulating Services of Ecosystems**
Methodology and applications
Pushpam Kumar and Michael D. Wood

28 **Environmental Policies for Air Pollution and Climate Change in New Europe**
Caterina De Lucia

29 **Optimal Control of Age-Structured Populations in Economy, Demography and the Environment**
Raouf Boucekkine, Natali Hritonenko and Yuri Yatsenko

30 **Sustainable Energy**
Edited by Klaus D. John and Dirk Rubbelke

31 **Preference Data for Environmental Valuation**
Combining revealed and stated approaches
John Whitehead, Tim Haab and Ju-Chin Huang

32 **Ecosystem Services and Global Trade of Natural Resources**
Ecology, economics and policies
Edited by Thomas Koellner

33 **Permit Trading in Different Applications**
Edited by Bernd Hansjürgens, Ralf Antes and Marianne Strunz

34 **The Role of Science for Conservation**
Edited by Matthias Wolff and Mark Gardener

35 **The Future of Helium as a Natural Resource**
Edited by William J. Nuttall, Richard H. Clarke and Bartek A. Glowacki

The Future of Helium as a Natural Resource

Edited by William J. Nuttall,
Richard H. Clarke and
Bartek A. Glowacki

Routledge
Taylor & Francis Group
LONDON AND NEW YORK

First published 2012
by Routledge
2 Park Square, Milton Park, Abingdon, Oxon OX14 4RN

Simultaneously published in the USA and Canada
by Routledge
711 Third Avenue, New York, NY 10017

Routledge is an imprint of the Taylor & Francis Group, an informa business

First issued in paperback 2014

British Library Cataloguing in Publication Data
A catalogue record for this book is available from the British Library

Library of Congress Cataloging in Publication Data
A catalog record has been requested for this book

ISBN: 978-0-415-57697-0 (hbk)
ISBN: 978-1-138-77486-5 (pbk)

Typeset in Times New Roman
by Wearset Ltd, Boldon, Tyne and Wear

For Wheeler Sears

Contents

Figures and tables

Figures

Tables

Contributors

Rakesh K. Bhandari is Director of the Variable Energy Cyclotron Centre, Department of Atomic Energy, Government of India. He is a Distinguished Accelerator Scientist of National and International reputation. In addition, he is sincerely involved with several other research projects sponsored by the Department of Science & Technology, and the Ministry of Earth Sciences, Government of India. He is a prominent member of many international scientific forums such as the Accelerator Community, the International Vaccuum Society, and the Cryogenic Council.

Thomas W. Bradshaw is head of the Cryogenics and Magnetics Group at the Rutherford Appleton Laboratory. He completed his PhD at Lancaster University in 1979 on a nuclear adiabatic demagnetisation refrigerator for research on the properties of helium-3 at sub-mK temperatures. As a postdoctoral research assistant at Exeter University, he researched the transmission of phonons across solid–liquid helium interfaces, before joining the Rutherford Appleton Laboratory in 1983 to work on closed-cycle refrigerators in support of a range of satellite instruments. Recent projects include the building of a 4 K cooler for Planck Explorer (launched May 2009, for which he is a co-investigator), proximity cryogenics for the ATLAS detector on the Large Hadron Collider and the further development of a range of miniature cryocoolers for long-life space applications. Current projects include the development of a long-life closed-cycle 2 K cooler for ESA, the development of a superconducting magnet for the Muon Ionisation Cooling Experiment, development of superconducting undulators and the use of cryogenics for medical applications.

John W. Burgoyne leads the Magnet System Group in Oxford Instruments, responsible for the engineering, design and manufacture of superconducting magnet systems, of which a high proportion are custom-engineered to individual requirements. His technical background is in superconductivity and superconducting magnets, and he holds a BSc (Hons) in Applied Physics from the University of Strathclyde, and an MSc and PhD in Superconductivity and Cryogenics from the University of Southampton. Following his PhD, he performed postdoctoral work on high-temperature superconductors at Oxford University. Since joining Oxford Instruments in 1998, John has worked in business management, marketing, key account management and business development in several divisions of the company, including a year on secondment with Siemens AG Medical Solutions in Germany. In 2006 he achieved an MBA with Distinction from Warwick Business School, University of Warwick.

Zhiming Cai is currently working for AMEC UK Nuclear Limited as Senior Nuclear Safety Consultant. In 2005, Zhiming Cai entered the PhD programme after completing the MPhil course in Engineering for Sustainable Development at Cambridge University's Department of Engineering. His topic was the global helium market and resources, and the research was jointly sponsored by Linde-BOC Ltd and the UK Atomic Energy Authority. Before moving to Cambridge, he worked for seven years as a lecturer at an academic institution in China. In addition to an MPhil degree in Engineering for Sustainable Development from the University of Cambridge, he holds a Masters degree in Nuclear Reactor Engineering and Safety from Naval Engineering University at Wuhan City, China.

Roger Clare works for the Linde Group in its gas division, managing tonnage conceptual engineering for air separation plants in Asia Pacific. He has been providing specialist input for the helium resources project, examining the technology and cost of obtaining helium from air.

Richard H. Clarke is Cryogenics Specialist at CCFE (Culham Centre for Fusion Energy), a UK research organisation developing fusion as an alternative, long term, sustainable and clean energy source. He has developed a leading role in helium resource studies since joining CCFE as manager of the JET helium cryogenics plant. He was the industrial supervisor of Dr Zhiming Cai and is a Fellow of the Institution of Chemical Engineers. From 1995 to 2002 he was a principle heat exchanger specialist at the BOC group (now part of the Linde Group), designing and advising heat transfer solutions across the Group both in the UK and the US. While at the UK Atomic Energy Research Establishment, Harwell, from 1979–95 he negotiated and conducted contract compact heat exchanger (CHE) cryogenic, oil and gas research studies through HTFS (the Heat Transfer and Fluid Flow Service, now part of Aspentech) and he has many papers published on these topics.

Michael N. Cuthbert studied Chemical Physics at The University of Glasgow before his PhD at Imperial College in London. His thesis topic was "Dissipation in High Temperature Superconducting BiSSCO/Ag Tapes". Following this work on HTS materials using pulsed current transport techniques in low dc magnetic fields he went on to work on pulsed field susceptibility measurements on Low and High Temperature Superconductors as a PDRA at University of Bristol, UK. Since joining Oxford Instruments Cuthbert has worked in Japan and the US in various customer facing technical and commercial roles. He is currently Business Group Manager for the Ultra Low Temperature Group within the NanoScience Division of Oxford Instruments responsible for the engineering and manufacture of all the company's low temperature products operating down to 5 mK, for Quantum Computing and Low Temperature Applications.

Nisith K. Das is Head of the Helium Section at the Variable Energy Cyclotron Centre, Kolkata. The helium project is commissioned by the Indian Science and Technology Ministry for the Government of India, and taps indigenous sources of helium in India. The project is instrumental in ultra-low-temperature research and development activities.

Andrew Flower is the Principal Consultant at Flower LNG. As an independent consultant, he has advised clients in all parts of the world on issues such as strategy, markets, supply and demand, project structures and development, pricing, LNG shipping and contracts. He is a frequent speaker at major conferences and has had numerous articles published. He retired from BP in 2001, as Director of LNG, after 32 years of service, 22 of which were related to LNG. During that time he managed BP's LNG activities in Abu Dhabi, Nigeria, Qatar and Australia, and was involved in projects in Trinidad and Indonesia and feasibility studies for developments in Russia, Venezuela and Papua New Guinea. He has negotiated sales and purchase agreements with LNG buyers in Asia, Europe and the Americas.

Art Francis never retired from cryogenics. Throughout his career, until 1996, he was Helium Manager for Praxair Ltd USA, in charge of helium gas and liquid distribution. On formal retirement he set up his consultancy, trading as Cliffside Refiners, and continued to work for Praxair. He took charge of supplying liquid helium to the US research station at the South Pole for seven years, and helped to provide a continuous liquid supply lasting over the Antarctic winters from 2001. Following the USA 1996 Act to abandon the federal helium conservation programme, he wrote the 1998 manuscript for the book "Helium at the Crossroads", which was scheduled for publication as part of the OUP Cryogenic Monograph Series. In 1999 he was appointed manager of the Amarillo Liquid Helium Production Plant, and his book was never finished. However, the proposals in his manuscript, for an International Helium Management System, have been incorporated into this book *The Future of Helium*. He was concerned, right to the end of his life, about the US Bureau of Land Management's having to sell off the federal conserved helium stock. He died in 2007, aged nearly 81.

Bartek A. Glowacki is Belvedere Professor and University Reader in Applied Superconductivity. He takes an interdisciplinary approach to applied superconductivity and advanced hydrogen-oxygen production technologies for fuel and cryogenic purposes as well as the development of advanced cryocoolers. He is strongly engaged in an academic and industrial partnership to develop advanced superconductors for numerous diverse applications. He is currently establishing a unique hydrogen laboratory at the University of Cambridge. As the only UK expert in applied superconducting conductors he was invited by the United Kingdom Atomic Energy Authority to represent them in the EFDA initiative to design and develop MgB_2 and $YBa_2Cu_3O_7$ cables for the next DEMO fusion reactor. He is a World Energy Council member and European Energy Research Alliance representative. As Energy Expert for the Institute of Energy (Poland), he is responsible for policy governing the basic science programme.

Benjamin Hooker is an Accountancy and Finance graduate. He worked for Oceaneering International Inc., an oil service company in Aberdeen, before being transferred to Nigeria and Singapore, spending two years in each region. After returning to the UK, he was appointed Managing Director of Technotrak Limited, which designs, manufactures and retails safety equipment. Following the sale of Technotrak, he has consulted for a number of oil service companies and was until recently Director International of Newpoint Gas, a privately owned American manufacturer of modular gas treatment and processing skids responsible for the group's sales and marketing activities in Russia, FSU, UK, Europe and Africa. He is currently Managing Director of Eleuthera Capital Limited, a helium exploration and production company.

Nikolaos Kazantzis was educated at the Aristotle University of Thessaloniki (Greece), the University of Cambridge (UK) and the University of Michigan (USA), where he obtained his PhD in ChemicalEngineering. Currently he is a Professor of Chemical Engineering at Worcester Polytechnic Institute (WPI), Worcester, Massachusetts, USA, an Associate Researcher at the Electricity Policy Research Group and Senior Associate at Hughes Hall, Cambridge University, Cambridge, UK, as well as a member of the Technical Advisory Committee of the MKOC Process Safety Center, Texas A&M University, College Station, Texas. He is the recipient of the Career Award of the US National Science Foundation (NSF) and his research interests lie in the areas of energy systems analysis, monitoring and control in the presence of complexity, systems science, system dynamics and energy policy analysis, process safety and chemical risk assessment.

Shri C. Mallik is Project Director of the Superconducting Cyclotron Project and Head of Accelerator Physics Group, Variable Energy Cyclotron Centre, Department of Atomic Energy, Government of India. He has profound experience in Beam Dynamics and complete design and operation of Accelerator and related diagnostics.

Trevor Miller took an Honours Degree in Physics at Bristol University before joining Gestetner UK in Production Engineering. His involvement with Cryogenics began in 1985 when he was with the Oerlikon Leybold Group. Since then he has worked closely with many varied technological applications but MRI and helium conservation have been consistent themes throughout. He feels privileged to have been involved since the early days of the technology, playing his part in helping to develop it as an essential tool available in hospitals and laboratories around the world. Trevor Miller is currently Managing Director of Sumitomo (SHI) Cryogenics Europe

Ltd, a part of Sumitomo Heavy Industries and global market leaders in cryogenic cooling equipment for helium conservation and similar applications.

William J. Nuttall is a Senior Lecturer in Technology Policy at the Univeristy of Cambridge, UK. In 1987 he won a Fulbright Pat-Graduate Student Award to the Massachussetts Institute of Technology, USA to study for a PhD in Physics. On returning to the UK he worked as a scientist in universities and then for the Institute of Physics in science policy. He joined the teaching faculty of Cambridge University in 2002. He holds a shared post between Judge Business School, where he is based, and Cambridge University Engineering Department. He has a leading role in the Electricity Policy Research Group and in the recently established Cambridge Nuclear Energy Centre. He has been researching issues relating to helium resources since 2005. He is a Fellow of Hughes Hall Cambridge.

Joseph B. Peterson is the Assistant Field Manager of Helium Resources at the Amarillo Field Office, Bureau of Land Management, Department of the Interior, located in Amarillo, Texas. He has nearly 30 years of federal service and has been involved in the helium industry for over 15 years. He has authored several scientific publications in technical journals and his office is responsible for the Annual Helium Minerals Commodity Summary and Annual Minerals Industry Survey Report on Helium. He is a graduate of the University of Alabama with a BSc in Chemical Engineering.

Benjamin Reinoehl is the Owner and Principal of RMW Solutions. In this role he consults for industrial gas companies, government and research users of industrial gases. RMW Solutions support helium projects and the collection of worldwide helium industry production data. Ben has over 50 years of experience in the industrial gas and steel industries. Particularly of interest is his experience with helium recovery, purification, liquefaction and marketing at Air Products and Chemicals, Inc.

Ralph Scurlock is Emeritus Professor of Cryogenic Engineering, and formerly Director of the Institute of Cryogenics at the University of Southampton, UK. In 1999, Ralph Scurlock was selected by the Cryogenic Engineering Conference (CEC) for the Samuel C. Collins Award for his work on cryogenic technology. He was the first person from outside the USA to be awarded this prestigious prize. He is the author of many papers and several books, including *Low Temperature Behaviour of Solids* (1965), *History and Origins of Cryogenics* (1994), "Low Loss Dewars and Tanks" (2004) and *Low Loss Storage and Handling of Cryogenic Liquids. The Application of Cryogenic Fluid Dynamics* (2007).

Bo Sears has over 17 years of oil and gas exploration experience. Bo has been instrumental in the development of key helium exploration projects across North America. His experience in the exploration arts (geology, geochemistry, geophysics), coupled with his thorough understanding of helium generation and migration, has proved an invaluable tool in the identification of new helium resources. Bo graduated with a BSc from the University of Mississippi and an MBA from the University of Dallas.

Richard Stainsby is currently the Coordinator of the European Commission's gas-cooled fast reactor (GCFR) project and is the EC's Representative on the Generation IV GFR System Steering Committee. Richard, together with a number of his AMEC colleagues, act as consultants to the South African and United States nuclear regulators on the licensing of high-temperature reactors (HTRs) in those two countries. They are also active members of the European HTR network and the Materials and Components Sub-Project Leader within the EC's RAPHAEL project.

Adrian Thomas graduated in Physics from The University of Birmingham in 1985. He subsequently studied Reactor Physics before working for the UK Atomic Energy Authority. In 1988 he moved to Oxford Magnet Technology (now Siemens Magnet Technology), where for 20 years he has been responsible for the design of some of the most commercially successful superconducting magnets for use in Magnetic Resonance Imaging (MRI) systems.

Acronyms and abbreviations

(RE)BCO	HTS magnet material similar to YBCO where RE = yttrium, gadolinium, samarium … (rare earths)
^3He	Rarer isotope of helium (with one neutron), mostly primordial (Big Bang) in origin
^4He	Common isotope of helium (with two neutrons), mostly from radioactive decay of U and Th
A	Current (amps)
AC	Alternating current
AFM	Atomic force microscope
AGR	Advanced gas-cooled (nuclear fission) reactor
Al	Aluminium
APCI	Air Products and Chemicals Inc., Allentown, PA, USA (industrial gases and equipment, chemicals and processes)
Ar	Argon
ArcaTech	ArcaTech 100 market index (NYSE) (US listed companies engaged in computer hardware, software, semiconductors, telecommunications, electronics, aerospace and defence, health care equipment and biotechnology)
As	Arsenic
ASU	Air separation unit (industrial-scale cryogenic separation of components of air)
B	Magnetic induction (Tesla)
B//c	Where magnetic induction, B, is applied to a superconductor parallel to the crystallographic axis c
Ba	Barium
barg	Bar, gauge pressure (1 barg = 29.2 psia); bara, bar absolute
BBC	British Broadcasting Corporation, London
B_{c2}	Upper critical magnetic induction (Tesla)
Bcf	Billion (10^9) (standard) cubic feet
Bcm	Billion (10^9) cubic metres
BEA	Banque Extérieure d'Algérie, Algiers
Bi	Bismuth
Bi-2212	$Bi_2Sr_2CaCu_2O_x$ (an HTS superconductor)

Bi-2223	$(Bi,Pb)_2Sr_2Ca_2Cu_3O_x$ (an HTS superconductor)
B_{irr}	Irreversibility field (magnetic) at which superconductor does not show hysteresis
BLM	US Bureau of Land Management, administrative body that manages US federal helium reserves
BOC	BOC Group Plc, now part of the Linde Group, Munich, Germany (industrial gases and services)
BOM	US Bureau of Mines, agency responsible for US helium reserves before the Bureau of Land Management
BRIC	Acronym for Brazil, Russia, India, China, a grouping of emerging economies
B-T	Magnetic induction – temperature
Btu	British thermal unit (1,055 joules)
°C	Degrees Celsius
C	Carbon
C	Specific heat capacity (at constant pressure or volume)
Ca	Calcium
CAGR	Compound annual growth rate
CBM	Coalbed methane
CCFE	Culham Centre for Fusion Energy, Culham Science Centre, Abingdon, UK
CCR	Closed-cycle refrigerator
CD	Current drive, forms of additional heating used in MCF fusion to raise the plasma temperature
CERN	European Organization for Nuclear Research, Geneva
CH_4	Methane (the major component of natural gas)
CHEU	Crude helium enrichment unit, BLM's process plant that enhances helium recovery (Cliffside, TX, USA)
CHP	Combined heat and power, where more efficient utilisation is made of a primary energy source
CICC	Cable-in-conduit conductor, an arrangement of superconductor and normal conductors for high B uses
CISFHR	Committee (of NAS) on the impact of selling the federal helium reserves
cm	Centimetre
CNPC	China National Petroleum Corporation, Beijing, PR China
CO_2	Carbon dioxide
CPI	Consumer price index
CT	Computed tomography, a medical imaging technique
Cu	Copper
D	Deuterium, or ^2H, a naturally occurring heavy form of hydrogen (two neutrons)
DC	Direct current
DCFC	Direct carbon fuel cell
DCS	Distributed control system (process and power plants)

DEMO	Generic name of the first power-generating demonstration (MCF fusion) reactor
DNA	Deoxyribonucleic acid
DoI	US Department of the Interior, Washington, DC (an agency mandated to protect US natural resources and heritage)
DR	Dilution refrigerator (a "wet", sub-1 K refrigerator containing a ^3He-^4He mixture)
EIA	Energy Information Administration, Washington, DC (a US energy statistics/information centre)
EOR	Enhanced oil recovery (methods for increasing recovery of oil from a field)
Er	Erbium
Er_3Ni	Erbium-nickel alloy (has unusually high specific heat at LHe temperatures)
EU	European Union
FC	Fuel cell, chemical–electric energy converter
FCL	Fault current limiter, a device that prevents excessive current from flowing in a network or grid
Fe	Iron
FID	Final investment decision (in the context of LNG facilities)
FIR	Far infrared (spectroscopy)
FIS	Front Islamique du Salut (Islamic Salvation Front, Algeria)
FLIR	Forward-looking infrared (as in FLIR thermographic cameras)
FNL	Front de Libération Nationale (National Liberation Front, Algeria)
FSRU	Floating storage and regasification unit (for LNG)
FW	Flywheel, as used in energy storage devices (including superconducting versions)
G	Gas field size MMscf (or field production rate MMscf/day)
G	Giga (billion, 10^9)
GAN	Gaseous nitrogen
GATE	Major LNG import terminal in Rotterdam, the Netherlands (GATE = Gas Access To Europe)
GCFR	European gas-cooled fast (nuclear fission) reactor project
Gd	Gadolinium
GdBCO	Gadolinium-barium-copper-oxygen magnet (note: B means Ba and C means Cu in this context)
GDC	Gas Developments Corporation, Chicago, IL, USA
GDP	Gross domestic product
Ge	Germanium
GE	General Electric Company (a multinational energy, technology, consumer goods and finance conglomerate), Fairfield, CT, USA
GFR	Gas-cooled fast (nuclear fission) reactor
G-L	Gas-liquid (as in gas-liquid separator vessel)
GL1K, etc.	See GL1Z, etc. (LNG trains at Arzew, Algeria)
GL1Z, etc.	LNG process train (sub-unit) at Arzew, Algeria

GO	Gaseous oxygen
G-M	Gifford-MacMahon, a form of cryocooler (a "dry", cryogen-free or "dryogenic" cooler)
GMAW	Gas metal arc welding
Grade A	A recognised helium purity standard (cylinder gas having a purity of 99.996 per cent helium)
GTI	Gas Technology Institute, Chicago, IL, USA
GT-MHR	Gas turbine modular high temperature (nuclear fission) reactor
GWe	Gigawatt (10^9 watts) of electrical power output
GY	Billion years (gigayears, 10^9 years)
H_2	Hydrogen (gas, or H for elemental hydrogen)
H_2O	Water
H_2S	Hydrogen sulphide
HAV	Hybrid Air Vehicles, helium-filled surveillance and heavy lift vehicles (Hybrid Air Vehicles Ltd, Cranfield, UK)
He	Helium
Heliox	Breathing gas; typically ≤21 per cent O_2, >79 per cent He
HELOKA	High-pressure, high-temperature helium component test loop (Karlsruhe Institute of Technology, Germany)
HETL	Hybrid energy transfer line (a combined power and delivery line)
H_{irr}	Magnetic irreversibility field (μ_0 = vacuum permeability)
Hg	Mercury
HGMS	High-gradient magnetic separation or separator
HHV	Higher heating value, a definition of the energy content of a fuel (includes steam's latent heat)
HRVU	Helium resource value unit (a measure of the potential of a helium reserve)
HT	Heat transfer
HTR	High-temperature thermal (nuclear fission) reactors
HTR-PM	High-temperature – pebble modular (nuclear fission) reactor
HTS	High-transition-temperature superconductor (designed for use at 77 K or above)
HTSC	High-temperature superconducting cable
HTTR	High-temperature test (nuclear fission) reactor
Hz	Hertz, or frequency in cycles per second
ICE	Intercontinental Exchange, New York (commodities exchange)
ICEC	International Cryogenic Engineering Conference (biannual event combined with ICMC)
ICF	Inertially confined fusion, using laser compression of a D and T capsule
ICMC	International Cryogenic Materials Conference (see ICEC)
ICP	Indonesian crude price, an LNG pricing formula
IEEE	Institute of Electrical and Electronics Engineers, New York, USA
IFC	International Finance Corporation, Washington, DC, USA (part of the World Bank Group)

i-LH_2	Indirectly cooled by LH_2, using a sealed helium loop for heat transfer (Chapter 16)
IL	Intermediate liquid in an ASU
INSA	Institut National des Sciences Appliquées (INSA), Lyon, France
IPCC	International Panel on Climate Change, Geneva
IR	Infrared (spectroscopy)
ISO	International Organization for Standardization, Geneva
ITER	International Thermonuclear Experimental Reactor (MCF), Caderache, France
IVC	Inner vacuum chamber
J	Current density (of superconductor), A per mm^2 (J_c critical current density)
JCC	Japan Customs-cleared, an LNG pricing formula
JET	Joint European Torus, a tokamak MCF fusion device built at Culham Science Centre, Abingdon, UK
JT	Joule-Thompson effect (temperature change as gas flows through a JT valve in refrigerator)
K	Kelvin (=°C+273.15)
kA	Kiloamp
kg/h	Kilograms per hour
km	Kilometre
KPEI	Kryzhanovskij Power Engineering Institute, Moscow, Russia
Kr	Krypton
kV	Kilovolt
kW	Kilowatt
kWh	Kilowatt hour
l/h	Litres per hour
ln	Natural logarithm
La	Lanthanum
LAir	Liquid air (normal boiling point 78.6 K)
LAr	Liquid argon (normal boiling point 87.2 K)
LASER	Light amplification by stimulated emission of radiation
lbs	Pounds (1 lb = 0.454 kg)
LCD	Liquid crystal display
LED	Light emitting diode
LH_2	Liquid hydrogen (normal boiling point 20.3 K)
LHC	Large Hadron Collider (world's largest and highest-energy particle accelerator located at CERN)
LHe	Liquid helium (normal boiling point 4.2 K)
Li	Lithium
LIN	See LN_2
LLR	Leak learning rate (per cent per year reduction in leak rate or consumption of helium) (Chapter 15)
LN_2	Liquid nitrogen (normal boiling point 77.3 K)
LNe	Liquid neon (normal boiling point 27.0 K)

LNG	Liquefied natural gas (liquefies at approximately –161 °C)
LO_2	Liquid oxygen (normal boiling point 90.1 K)
LoS	Low-transition-temperature superconductor (see LTS)
LOX	see LO_2
LPG	Liquefied petroleum gas (primarily propane and butane)
LTS	Low-transition-temperature superconductor (5 K or below)
mbar	millibar
M	Magnitude of earthquake (Richter scale)
M	Mega (million, 10^6)
MASER	Microwave amplification by stimulated emission of radiation
Mcf	Thousand (standard) cubic feet (sometimes written as MCF)
MCF	Magnetically confined fusion, using a tokamak in which plasma is confined by a magnetic field
MCR	Multicomponent refrigerant, generic name of a process used for LNG liquefaction
MgB_2	Magnesium-diboride (an MTS superconductor)
MIT	Massachusetts Institute of Technology, Cambridge, MA, USA
mK	Millikelvin (10^{-3} K)
ml	Millilitre
MMBtu	Million Btu (see Btu)
MMcf	See MMscf
MMscf	Million (standard) cubic feet
mmtpa	Million metric tonnes per annum (conventional notation used in the LNG industry)
Mo	Molybdenum
mol.	Mole, amount of chemical substance (6.022×10^{23} atoms or molecules)
MP	Medium pressure
MPSC	Multi-pulse, stepwise cooling (a superconducting magnet conditioning technique)
MRI	Medical resonance imaging, a medical imaging technique
mtpa	Million tonnes per annum (see MMTPA)
MTS	Medium-temperature superconductor (nominally designed for use in the range 20–30 K)
MW	Megawatts (10^6 watt)
MWth	Megawatts (thermal, as distinct from electrical, basis)
MY	Million years
N	Newtons (metric unit of force)
N	North, as in north pole of a magnet
N_2	Nitrogen (gas)
N/cm^2	Pressure (newtons per cm^2)
NAS	US National Academy of Sciences, Washington, DC, USA
NASA	US National Aeronautics and Space Administration, Washington, DC, USA
NBP	National balancing point, a UK gas price virtual trading location

Nb_3Sn	Niobium-tin (an LTS superconductor)
NbTi	Niobium-titanium (an LTS superconductor)
NCNR	NIST centre for neutron research, Gaithersburg, MD, USA
Nd	Neodymium
NdFeB	Neodymium-iron-boron rare earth magnet
Ne	Neon
NG	Natural gas
Ni	Nickel
NIR	Near infrared (spectroscopy)
NIST	US National Institute of Standards and Technology, Gaithersburg, MD and Boulder, CO, USA
NISTEP	National Institute of Science and Technology Policy, Tokyo, Japan
Nm^3	Normal (temperature and pressure) cubic metres
Nm^3/h	Normal (temperature and pressure) cubic metres per hour
NMR	Nuclear magnetic resonance, a research imaging technique
NRC	US National Research Council, Washington, DC, USA
NSF	National Science Foundation, Arlington, VA, USA
NYMEX	New York Mercantile Exchange, New York, USA
NYSE	New York Stock Exchange, market index
O	Oxygen (O_2 oxygen gas)
OAO	Otkrytoe Aktsionernoe Obschestvo (Russian: Open Joint Stock Corporation)
OEM	Original equipment manufacturer
OGMS	Open gradient magnetic separation or separator (method of removing paramagnetic contaminants, e.g. from china clay)
OGP	International Association of Oil and Gas Producers, London, UK and Brussels, Belgium
OMB	Office of Management and Budget, a US fiscal oversight body, Washington, DC
OSTI	US Office of Scientific and Technical Information, Oak Ridge, TN, USA
P&I	Process and instrumentation diagram (for a process plant)
Pb	Lead
PbLi	Lead-lithium, a possible alloy for fusion (ICF or MCF) breeder blankets
PBMR	Pebble bed modular (nuclear fission) reactor
PEM	Proton exchange membrane or polymer electrolyte membrane (as used in fuel cells)
PEN	Refrigeration cycle that uses a propane-ethane-nitrogen refrigerant
PES	Power and Energy Society (part of IEEE)
PL	Product liquid in an ASU
PLC	Programmable logic controller
Po	Polonium

ppm	Parts per million, usually expressed volumetrically (see ppmv), e.g. 5.2 ppm of helium in atmosphere
ppmv	Parts per million (volume basis)
psi	Pounds per square inch
psia	Pounds per square inch absolute pressure (14.5 psia = 1 bar)
PSA	Pressure swing absorption (gas separation or purification method)
PSY	Plant start-up year (Chapter 15)
PTR	Pulse tube refrigerator
Q_{TH}	Thermal power (of MCF reactor) including CD power
R&D	Research and development
RE	Rare earth (elements), e.g. Y, Gd, Sm, etc. (see also (RE)BCO)
regen.	Regeneration (of a cryopump to release absorbed or condensed gases before pump becomes overloaded)
RF	Radio frequency
RHQT	Rapid heating, quenching and transforming (a superconductor enhancement technique)
RL	Rich liquid in an ASU
RoI	Return on investment, a performance measure used to compare investment opportunities
RoW	Rest of world
R/P	Natural resource reserve/production ratio
RTB	Room temperature bore (in a superconducting magnet)
s	Entropy (a measure of a unit of energy's availability to do work)
S	Sulphur
ScHe	Supercritical helium (used instead of LHe for improved coolant stability (i.e. to avoid quenching)
S/C	Superconducting
SD	System Dynamics, a methodology for modelling complex feedback systems over time
Si	Silicon
SI	Superinsulated
SLPM	Standard litres per minute
SMES	Superconducting magnetic energy storage, efficient short-duration DC energy storage system
SOEC	Solid oxide electrolyser cell, an SOFC in regenerative mode for the electrolysis of water
SOFC	Solid oxide fuel cell
Sn	Tin
SPM	Stabilised permanent magnet
SQUID	Superconducting quantum interference devices (highly sensitive device for measuring magnetic field)
Sr	Strontium
std cc/s	Cubic centimetres per second at standard conditions (pressure 14.696 psia, and 60 °F/15.6 °C or 70 °F/21.1 °C)
STP	Standard temperature and pressure (see page xxxix)

T	Tera (trillion, 10^{12})
T	Tesla, a unit of magnetic field (Earth's magnetic field ~$6 \cdot 10^{-5}$ T)
T	Tritium, or ^3H, an artificially produced heavy form of hydrogen (three neutrons, radioactive gas, 12.3 year half-life)
T_c	Superconducting transition temperature; reverts to normal conduction above T_c
Tcf	Trillion (10^{12}) (standard) cubic feet
TF	Toroidal field (TF coil: the principal magnetic coil used in MCF to confine the plasma)
Th	Thorium
Ti	Titanium
THz	Terahertz (10^{12} Hz)
Tl	Tellurium
TMI	Three Mile Island, Harrisburg, PA, USA (loss of coolant accident in fission reactor, 1979)
TTF	Dutch Title Transfer Facility, similar to UK's NBP
TV	Television (in the context of flatscreen plasma TVs)
T_λ	Lambda point: 2.18 K, the helium-I to helium-II transition temperature (helium-II is a superfluid)
U	Uranium
UHV	Ultra-high vacuum ($<10^{-9}$ mbar)
UISHR	See CISFHR
UKAEA	UK Atomic Energy Authority, Culham Science Centre, Abingdon, UK (parent body of CCFE)
UPS	Uninterruptable power supply
USBLM	See BLM
USBM	See BOM
USGS	US Geological Survey, Reston, VA, USA
UV	Ultraviolet (spectroscopy)
UV-Vis	Ultraviolet-visible (spectroscopy)
V	Vanadium
VHTR	Very high-temperature (nuclear fission) reactor
VNIGRI	Petroleum Scientific Research Institute of Geological Exploration, St Petersburg, Russia
VTI	Variable temperature insert (in superconducting magnet)
W	Tungsten
Wobbe	Wobbe Index, a measure of the interchangeability of fuel gases (=$HHV/SG^{0.5}$, where SG is gas specific gravity)
Xe	Xenon
Y	Yttrium
YBCO	$YBa_2Cu_3O_7$ (a HTS superconductor) (note: B means barium in this context)
z	Gas field concentration (per cent helium)

Preface

This book had its origins in 2004, during one of the first periods of surging worldwide demand for helium, particularly from Asia, combined with tightening supply. This left many helium users wondering how new helium sourcing would be developed if worldwide demand was to continue to grow at the then expected growth rate of more than 5 percent per year. And this concern was aggravated with the understanding of the legislated depletion of the US government's Amarillo, Texas crude helium reserve by 2015. While many knew that there were abundant sources of helium in the world, those sources were known to be geographically far-removed from the regions of increasing demand, and the sources were slow in being developed. Among the people concerned with all this were many scientists in fusion energy development, where helium was considered vital.

Richard Clarke, then the Plant Manager at the Joint European Torus Fusion Facility at the UK's Culham Science Centre, happened to be compiling a list of possible PhD study possibilities for British university PhD candidates. The list included "The Future of Helium", an appropriate topic for this time of concern about the future adequacy of helium supply. The same question happened to be considered by William Nuttall, a Senior Lecturer in Technology Policy at the Judge Business School of Cambridge University, and a PhD in Physics from MIT where future uses of helium have been in focus for many years. Clarke and Nuttall were acquainted and were each concerned with the long-term and reliable sourcing of helium from their work and interests in future energy development from nuclear fission and fusion. So a simple question shared by Clarke, Nuttall and others was, "What can be done so that the world does not run out of helium?"

Meanwhile, Nuttall was academic advisor to a young Cambridge student from China, Zhiming Cai, who needed to pick an interesting topic for his PhD study in System Dynamics at Cambridge. As system dynamics provided an advanced demand-modeling scheme for addressing complex demand–supply relationships, Nuttall, Clarke and Cai settled on framing Cai's PhD study and thesis around analyzing the future of helium demand and supply.

At this time Cambridge University was engaged in research on materials, which was important to advancing more efficient superconductivity, including improved performance at liquid helium temperatures. Bartek Glowacki, a Materials Science colleague of Nuttall and a Cambridge Reader in Applied Superconductivity Materials, became involved in the project. Cambridge University was close to the headquarters of BOC, a major UK-based supplier of helium to users around the world. So BOC was approached and indeed provided the other half of the funding for Cai's helium-modeling project. BOC's speaking partner to this project was Nick Ward from BOC's helium operations, who became an active participant in the study project as it expanded. (BOC is now part of Linde AG's worldwide industrial gases and engineering operations.)

With preparations complete, Cai's project started in October 2005 and was completed in early 2009. His work is referenced and described in several chapters of this book. As the project neared completion its participants decided to broaden the scope of Cai's project with a wider perspective on the future of helium demand and supply from other individuals with significant user, supplier and analytical interests in helium. A two-day conference was organized on the "Future for Helium" and was held at Hughes Hall, Cambridge University in March 2009. Cai briefed the 36 participants on his work, and several of the other participants provided their views on particular helium concerns and issues. As the breadth and depth of those papers were deemed to be useful to a much larger group of government and research and development people, users, suppliers and others, the project group decided to develop selected papers into chapters for this book, *The Future of Helium as a Natural Resource.*

At a personal level, I have been involved in helium since the beginning of my career in gases at Air Products in 1962. Since then there have been many projects, consulting assignments and articles in our *CryoGas International* journal, and recent committee participation with the US National Academy of Science analysis and report on "Selling the Nation's Helium Reserve", which operations were authorized under the "1996 Helium Privatization Act". My observations on this book include the following.

- The book provides an excellent analysis of the changing character of demand and supply during the last century and of the anticipated variety of complex factors influencing future helium demand. It provides a useful model to at least consider the forecasting complexities of the cause and effect of helium demand factors that will drive continuing investment in helium supply chain development and operations well into this century.

  An example of the complexity of helium demand forecasting can be seen in the potential effects of large increases in US and international shale gas developments. This will impact the future of LNG development and supply and those, in turn, affect on the availability of crude helium from various LNG demand and supply scenarios.

  Another example of the complexity of assessing demand is that while this book is correct in assuming increased supply of electric and thermal power from nuclear plant operations, particularly via nuclear fusion, is likely to create significant new demand for helium. The worldwide scare from Japan's March 2011 tsunami and Fukushima nuclear plant accidents has potentially changed at least the short-term view on nuclear fission power and, in turn, the future for helium in those developments and potentially for longer-term operations, such as fusion power. This is just another illustration of the vulnerability to the changing face of environmental versus energy concerns faced by a product like helium, and shows the way that strategies arise from the ebb and flow of popular and political opinion.

While one must be careful to estimate future demand for a product based on its demand history, it appears to me that there is high probability that demand for helium will continue to grow over the long run because of its unique properties and performance, in tandem with the growth of complex technology development around the world. But I must confess that few people, if any, have a clear picture as to how that development will occur.

- Chapter 17 includes some interesting ideas from the late Art Francis, whom I knew well and who was dedicated to "things helium". His ideas on the need for a worldwide strategic reserve for helium are useful and important. It is very likely that the world's helium users do not fully appreciate the importance of the US government's role in the historically critical and reliable supply of helium since 1920. Equally it is likely that they do not realize that crude helium was the source of much of the world's private sector refined pure gas and liquid until Poland's helium development and until private sector liquid helium became available from Exxon's private sector crude in 1987. More recently, the operations of the Bureau of Land Management (BLM) large crude helium reserve at the Cliffside, Texas have provided a critical volume of crude, being the first and last source of crude for pure gas and liquid production by the major Hugoton gasfield's private sector refiners. While that reserve position will continue until at least 2015, it will probably be depleted (or sequestered) by about 2020, depending on the US government's disposition decisions by 2015.

  The need for a private sector "spinning reserve" to provide product during periods of peak demand and plant outages and to replace BLM's reserve role is now becoming recognized. The planning, financing and reliable operations of such a spinning reserve will need to be factored into the economics of supply if the world's current and relatively reliable supply chain is to be maintained for the rest of this century and beyond.
- On another and related front, few users understand the basis of current supply scheme costs and end-user helium pricing, which has been reliant in the past on the US Bureau of Mines price for cylinder, tube trailer and liquid helium, and now upon the USBLM's pricing of crude. Now the BLM auction crude price is the primary basis for pricing global helium supply at the plant spigot, through a few major refiners in the world's helium supply chain. If and when the BLM ceases operations at the Cliffside reservoir, its current crude auction price system will cease to exist and the world's crude and refined players will have to establish a more "market-based" pricing system. This book will likely benefit that effort.

I have found that this book provides a valuable contribution to the understanding of helium demand, both current and anticipated, and trust that if you have current or future interests in helium you will benefit from reading it.

J. R. Campbell
Publisher, *CryoGas International*
Lexington, MA

Editor's acknowledgements

As editors we are very conscious of the large amounts of effort put into this book by many people and we are extremely grateful to all the authors of the chapters that make up this book. We suggest that each chapter is best regarded as representing the opinion of its authors; it is not necessarily the case that we concur with all our various chapter authors on all points. We do believe, however, that what they have to say will be of interest.

We are especially grateful to the many anonymous referees who kindly doubly peer-reviewed all the chapters other than the first and last. Their generous efforts greatly assisted this project. We are also most grateful to all those who attended the Future for Helium conference held at Hughes Hall, Cambridge in March 2009. That meeting is discussed further in Chapter 1.

John R. Campbell kindly prepared the preface for the book and provided us with much good assistance and advice. His contributions are very much appreciated. We are also most grateful to colleagues who provided crucial assistance during the development of the project. Special thanks are due to Philippa Bradshaw, who ably steered the project through its most difficult phase. We are very grateful to Jim Krane and Shelley Lockwood for assistance with copy-editing in various places, but of course any remaining errors are in no way their responsibility. We thank Camilla Burgess for assistance with proof editing.

We would like to record our thanks to Chad Heys and Holly Ponting for assistance with the graphics and figures. We are grateful to Derek Stork and David Ward for editorial advice concerning nuclear fusion applications.

Our interest in helium resources overlaps with our interest in the potential of liquid hydrogen to substitute for liquid helium in some cryogenic applications; these ideas are described in Chapter 16. Numerous people have helped us in various ways with such ideas. In particular we would like to record our thanks to: Dan Baciu, Leslie Bromberg, Bruno Coppi, Jeffrey Friedberg, Maura Garvey, P. M. Grant, Malcolm Haines, Richard Kennedy, K. T. McDonald, Joe Minervini, Michael O'Brien, A. Piotrowska-Hajnus, R. Tomov, Nick Ward and S. Yamada, among others. While we are grateful to them all for their assistance it should not be inferred that that they necessarily agree with our assessment of the issues. We are also most grateful to Gordana Najdanovic and Lisa Wears for administrative and managerial assistance in the early days of our developing ideas.

Numerous organizations have also been of great assistance at various stages of our interest in this topic. Their assistance is much appreciated but it should not be assumed that the organizations listed necessarily agree with the information or conclusions presented in this volume.

In particular we would like to thank:

BOC Ltd, part of the Linde Group
Cambridge Judge Business School
Cambridge-MIT Institute
Culham Centre for Fusion Energy
Engineering and Physical Sciences Research Council
Hughes Hall, Cambridge
India-UK Collaboration in Fuel Cells
Low Carbon University Energy Alliance
MIT Plasma Science and Fusion Center
Science and Technology Facilities Council
Siemens Magnet Technology UK
UK Energy Research Centre
University of Cambridge Department of Materials Science and Metallurgy

This work was also helpfully informed by the SUPERGEN 14 project "H-delivery" and the EU FP7 "EFECTS" Project.

Our project has attracted media interest over the years and we are especially grateful to *Gasworld* magazine and to *Physics World* for covering our activities. We would like to record our thanks to Routledge and the Taylor and Francis Group for publishing this book. Most especially we would like to thank Simon Holt at Routledge Economics for his support and advice throughout the development of this book. His help is very much appreciated.

Last, and most important, we would like to record our special thanks to Zhiming Cai. It was his University of Cambridge PhD project that lay at the heart of our growing interest in helium. Without his efforts none of the later work, including this book, would have come into being. We are most grateful to him. In this context, we record our thanks to David Martin for initiating the PhD project in early 2005. We trust that you will find this book to be of benefit and of interest. We apologize if, despite our efforts, any errors remain. If they do they are our responsibility and we would be grateful to learn of them.

W. J. Nuttall[1]
R. H. Clarke[2]
B. A. Glowacki[3]
January 2012

Notes

1 wjn21@cam.ac.uk
2 richard.clarke@icheme.org
3 bag10@cam.ac.uk

Unit conversions applicable to helium and natural gas

Metric		*Imperial*
1 standard cubic metre (1 sm^3)	=	35.315 standard cubic feet (scf)
(1 foot = 0.3048 metres by definition)		
1 normal cubic metre (1 Nm^3)	=	36.054 standard cubic feet (scf)
(at 1.013 bara [bar, absolute], 15.0 °C)[1]		(at 14.696 psia, 70.0 °F)[1]
27.736 Nm^3	=	1 Mscf (1,000 scf)
27,736 Nm^3	=	1 MMscf (1 million scf)
1 tonne helium	=	0.2133 MMscf (213.3 Mcf)
(density of helium 0.169 kg/m^3 at 1.013 bara, 15.0 °C)		
Liquid/gas equivalent	=	740 volume/volume
1 tonne natural gas	=	0.042–0.054 MMscf
(density of natural gas 0.66–0.85 kg/m^3 at 1.013 bara, 15.0 °C)		
Liquid/gas equivalent (1.013 bara, 15.0 °C)	~	600 volume/volume

1 Definitions of normal and standard may vary: these conditions have been adopted by the BLM. In Table 8.3 the temperature used is 0 °C.

1 Introduction

Richard H. Clarke, William J. Nuttall and Bartek A. Glowacki

In 2011 the world celebrated the centennial of an amazing discovery: superconductivity, where an electrical current can flow almost without resistance.[1] While seeking to prove Johannes van der Waals's thermodynamic theories, Heike Kamerlingh Onnes stumbled upon superconductivity in April 1911. It was the unexpected outcome of his team's meticulous efforts and ingenuity in being the first to liquefy helium gas. For that work he received the Nobel Prize for Physics in 1913 (van Delft, 2007). At the time, helium, which had only been discovered on Earth in 1895, was extremely scarce and was found only in a few rare earth minerals such as monazite and cleveite.

Today, helium gas is widely extracted as a by-product of natural gas production. It underpins many high technologies. Yet, just as superconductivity has finally come of age, many people are asking how long known helium supplies can be sustained. It is the purpose of this book to explore the global helium industry, helium's uses and possible alternatives in an attempt to answer the question: what is the future of helium as a natural resource?

Helium has a unique combination of properties: extreme nuclear and chemical inertness (a book on helium chemistry would be very thin), low density, the lowest boiling point of all, superfluidity and the highest ionisation potential, to name just a few. If it did not exist there would have been no magnetic resonance imaging (MRI) scanners in hospitals, a heavily retarded space and satellite programme, great difficulty in manufacturing optical fibres and computer chips (the backbones of the internet) and, of course, no party balloons (the immediate and only association the general public seems to make with helium). "Big physics" and CERN, the world's single largest user of helium, would be a much reduced affair.

As we write, it is true: there is a helium crisis; a helium-3 crisis (Cartwright, 2012). This is the isotope of helium that (on Earth) is a million times rarer than normal, radiogenic helium-4. Helium-3 is the decay product of radioactive tritium (heavy hydrogen) gas. Supplies of helium-3 have largely come from sources linked to US nuclear weapons. Historically, their warheads have made use of tritium gas, which decays with a 12.3-year half-life to helium-3. The United States had accumulated a small stockpile of helium-3 that has been heavily depleted since the 9/11 attacks. Most of it has been consumed in neutron detectors, a security measure found in many airports. Speciality applications of helium-3 in space cryogenic instruments, and vital down-hole oil and gas instruments have become very expensive or impossible, as supply remains severely restricted. Does the same fate await helium-4 users in a few years?

Reading the industrial gas industry magazines, and sometimes the tabloids too, one might think that is the case. "Up, up and away … where has all the helium gone?" (Garvey, 2007); "The coming helium shortage – where will you be when the world runs out of helium?" (Deakin, 2001); "Helium shortage hampers research and industry – if new sources of helium aren't developed, the world's supply of the gas will dwindle and prices will soar" (Kaplan, 2007); "Indispensable helium is routinely squandered" (Browne, 2008); "Goodbye helium, goodbye brainscans" (*The Oil Drum: Europe*, 2008); "Destroying science in a helium balloon" (Perelman, 2003); and "Scientists say Earth's helium reserves will 'run out in 25 years' (and party balloons should cost [$100])" (Firth, 2010), in which Nobel Laureate Professor Robert Richardson lamented that the US government was selling its stockpile "at a ridiculously low price". Soon afterwards the official US "open market" price went up by 15.8 per cent.

Feelings have been running high for a few years and with some reason. Unexpected disruption and delays in 2007–8 hampered the throughput of helium-producing process plants around the world by up to 10 per cent. This upset an already tight supply–demand balance, sparking anxiety that helium was indeed "running out". In the US, some users' supplies were cut off completely while others had theirs reduced.

The coverage also revealed some misconceptions about helium supplies:

- Although helium is the second most abundant element in the universe, not all helium-4 on Earth is radiogenic (deriving from natural radioactive decay) helium – about 7 per cent is primordial ("Big Bang") helium, as is essentially all the helium-3 (Tolstikhin, 1975).
- Around half of all helium molecules disturbed during natural gas production are vented to the atmosphere during burning (Elhelwa, 2008) – this dwarfs, but does not necessarily negate, helium recovery efforts by helium users (Chapters 2, 4, 5 and 6 give some insight into why).
- Helium supplies may well run low in the coming decades but effectively the Earth will never run out of helium – there are 3.8 billion tonnes of it in the atmosphere (nonetheless, how much can be extracted and at what price is debatable and is discussed in Chapter 8).
- There is not a helium-rich layer at the top of the atmosphere (Keesom, 1942).

The origin of the atmospheric equilibrium for helium has been explained clearly by a US helium policy review in 2000: "Atmospheric helium is in dynamic equilibrium between the gain of helium diffusing from Earth's crust (as a product of radioactive decay of elements such as uranium and thorium) and losses of helium into space" (National Academy of Sciences, 2000). The nature of the dynamic equilibrium underlying the apparently stable concentration of helium in the atmosphere remains somewhat controversial and is discussed in Chapter 8.

In a wider context, helium is just one of several "high-technology" elements where economic availability may be threatened in the twenty-first century and supply disruptions are anticipated (Davies, 2011). There are ten elements, including silver, that are on the "critical" list (Figure 1.1).

What distinguishes helium from all the other elements is its volatility after use, as it quickly dissipates into the dilute pool that is the atmosphere, where the helium concentration, arising from the dynamic equilibrium described above, remains 5.2204 ± 0.0041 ppm by volume (ppmv), virtually unchanged by anthropogenic releases (Holland and Emerson, 1987). Other elements exist mostly in the solid phase, although dispersion makes them increasingly hard to recover.

In May 2004 one of the editors, Richard Clarke, then Helium Plant Manager at the Joint European Torus (JET) fusion facility at Culham Science Centre (UK), was beginning to wonder just how sustainable it would be to extract helium from natural gas in order to feed a fusion power plant when he received a request from BBC Radio's Quentin Cooper to help with a programme called "Up, up and away" in a Radio 4 science series called "Connect". It was all about the supposed helium crisis and the questions were challenging.

What would fusion and other industries of the future be doing about the looming helium shortage? Would hydrogen be a good substitute? Would we mine the Moon (for deuterium/helium-3 fusion)? Or would deuterium/deuterium fusion be possible? It would take another book to answer those questions, but in a major study it was observed that "With serious difficulties in burning this fuel,

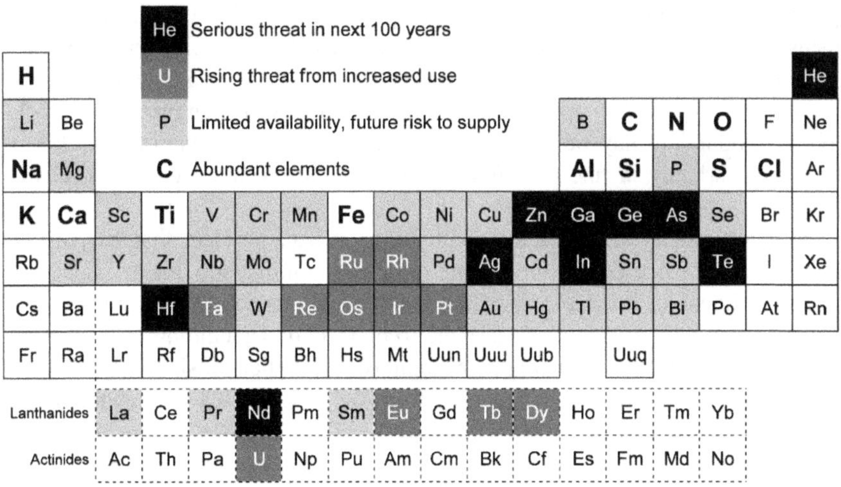

Figure 1.1 Endangered elements: helium is high on the list (courtesy: Chemistry Innovation Ltd).

it is unlikely that demand for deuterium/helium-3 fusion will drive a massive lunar mining effort" (Stott, 2005).

Under its Outreach programme, Culham explored the potential for a PhD study with the University of Cambridge, where there was interest from co-editors Dr William Nuttall at the Judge Business School and Dr Bartek Glowacki at the Department of Materials Science and Metallurgy. Respectively, they had two different stances: technology policy, and hydrogen cryomagnetics (alternatives to helium and low T_c magnets). It was also recognised that there was a need to bring reliable industrial insight to the project, and Nick Ward, Helium and Special Gases Manager at BOC Gases (now BOC-Linde) agreed to participate in the PhD project.

Each sponsor of the study needed to understand the helium market, but for different reasons and on different timescales. For BOC Gases, a medium-range perspective would help to ensure a secure supply for customers well into the future. The fusion researchers needed to know whether helium would remain readily available for long enough to be a "stepping stone" to future solutions that were either helium-free or helium-minimised. The timescales of such research dictate a need for the long view. However, one thing became clear from the start – and let us dispel an urban myth here – fusion is not the answer to any "helium crisis". A 1 gigawatt (1 GWe) fusion power plant might make around 400 kg of helium-4 in a year (Cai *et al.*, 2006). That would be insufficient to sustain a fusion plant's own needs for the future, even in the best possible scenario thus far foreseeable (this is discussed further in Chapter 15). Fortunately, as for other helium users, there were and are other technological solutions on the horizon.

Gradually, the PhD project evolved and an examination was made of current and future supply and demand using the Vensim® system dynamics (SD) market-modelling software. SD is a well-established methodology for capturing quantitative and qualitative aspects of complex systems where feedback mechanisms are important. The modelling results are discussed in Chapter 10. Vadim Udut, Head of OAO Geliymash, a Moscow-based helium research company that advises gas-export monopoly OAO Gazprom, in 2011 ventured that world helium production may peak at ~10.6 Bcf around 2030 (Shiryaevskaya, 2011). This is remarkably in line with PhD student Zhiming Cai's predictions (Cai *et al.*, 2010).

As Zhiming Cai investigated the literature, it became clear that only a few studies had considered helium's long-term availability, and the mathematics used was linear in the sense that there was no allowance for feedback paths. Among most helium users there seemed to be little understanding of the supply chain or of the factors that might profoundly affect their business or research activities. Helium was seen as coming from the "Yellow Pages", available on demand with just the price to be negotiated.

While most thinking related to under-supply, it was, however, imperative that an open-minded approach be taken. What if an abundant supply emerged, either through discovery or technological means? For example, Japan has a set of national goals, and developing a practical method of recovering helium from air was in fifty-first place on their list, to be achieved by 2017 (NISTEP, 2000). Subsequently, the target appears to have been dropped. Without clear insight into developments it becomes extremely hard to plan business or research efforts that might be important in the future. Examples include the following.

- Whether to use helium as a coolant in a new generation of small and large fission reactors that could play a large part (or not) in developing economies, or fuel-scarce countries, in meeting energy demand and environmental goals.
- A new generation of airships such as the Hybrid Air Vehicle (HAV) is crucially dependent on helium. A glut or dearth in supply and the arising price effects could profoundly affect such projects. Perhaps airships will become the workhorse of Canadian logging, oil and gas equipment transportation as the permafrost recedes in regions without any other infrastructure (Limb, 2009).
- Will a biofuelled jet aircraft, a hypersonic hydrogen plane or a luxury, helium-filled airship be the way for a select few to travel across the Atlantic in 2030? We note, however, the classic "What price speed?" paper (Gabrielli and von Kármán, 1950). The constraints described there are still much the same today and may well persist.

The interconnectivity between the oil and gas industry and helium is profound, as geology has intertwined the two. Carbon emission targets, if achieved, may constrain helium extraction from natural gas. Alternatively, if natural gas (with LNG from the Arctic perhaps) became the world's fuel of choice, due to its lower carbon intensity than oil or coal, this might open up more helium production opportunities. It is clear, then, that market-modelling progress can only be achieved through fundamental understanding of the underlying knowledge and economics of supply, demand and the feedback loops thus created. Helium is not unique in the economic sense, although there are quirks, such as it not yet being a commodity (it is not traded, except by long-term contracts among a limited group of market players), and that it is a by-product of a major commodity (natural gas). A linear approach to resource exploitation (years of supply = reserves/extraction rate, the R/P ratio) can provide misleading and possibly over-optimistic projections.

Fortunately, unlike fossil fuels, the waste product from using helium is not CO_2 but helium – indestructible helium. But if minerals like helium become scarcer, the energy intensity of sustaining their use or implementing substitutes may increase with time. In turn, this may reverse the energy efficiency improvements of recent decades and lead to an *increasing* energy use per unit of global GDP, with concomitant CO_2 implications. Add in the arising cost or price factors and this possibility becomes apparent when considering helium substitutes (Table 1.1).

Four years on, with the PhD completed, the project sponsors decided it was appropriate to publish some of the results and develop further understanding at an invitee-only conference held at the University of Cambridge in Spring 2009. It was simply entitled "The Future for Helium".

Table 1.1 Helium substitutes and their energy implications

Necessity	*User group*	*Substitute(s)*	*Reference*
"Have to have helium, need lots."	Fusion, "Big Physics", fission (fourth generation).	Extract from air?*	Chapters 8, 14, 15.
"Have to have helium-like properties, but have or will find ways to minimise use."	MRI scanners, diving (heliox mixtures), electronics manufacture, optical fibres, purge gas, inerting.	Helium recovery*, cryocoolers* (dryogenics)?	Chapters 9, 11, 12, 13, 16.
"Good to have helium, but if price or supply intermittency increase will stop using helium or find alternatives."	Leak detectors, airships (manned or unmanned), welding, party balloons.	Hydrogen gas, other gas mixtures, or stop.	* Increased energy intensity.

The conference's success was made possible through the enthusiastic participation of around 40 experts covering many aspects of the helium chain from across the world (Figure 1.2). It was one of the precursor events that spawned this book, with many original contributions that will enable readers to obtain an insight into an important resource not often considered. It is hoped that the book will appeal to students (especially postgraduates) in a wide range of disciplines, including physics, chemistry, engineering, energy, earth sciences, environmental sciences and resource economics.

In Chapter 10, Nuttall and colleagues seek to develop further some of the system dynamics ideas proposed by Zhiming Cai in his University of Cambridge PhD research. In particular, consideration is given to the system dynamics of the demand side of the market.

Some say that we can live without helium, and in the past we have. Yet, as we shall see, if the triumph of technology (that Lord Alec Broers identified as mankind's liberation in his 2005 BBC Reith Lectures (Broers, 2005)) matters, then so does this little molecule.

If we conclude that the supply of helium is going to grow scarcer, then the crux of the matter becomes the rate of adaptation which helium users are able or willing to make. Here, as is explored in Chapter 11, cost, anxiety or a fascination with technology (that is also relatively easy to use) drives change – in this case to "dryogenics". The desire to reduce anxiety may exaggerate the pre-existing role of "willingness to pay" underpinning the concept of "elasticity" in economics. Individuals will pay more than they might otherwise in the face of anxiety, at least until the drivers (supply shortage or supply intermittency) reduce. In a sense we argue that anxiety can cause a temporary adjustment to bounded rationality amongst consumers (Caplin, 2001), and these effects have been captured to some extent in the modelling work (Chapter 10).

Figure 1.2 Participants at the "Future for Helium Conference", Hughes Hall, University of Cambridge, 30–31 March 2009.

In other areas, such as helium's use in welding, ingrained behaviour, customs or standards may inhibit change. There may be a relationship here between "traditional" and "high-technology" helium applications. In the former, economic growth may be low, perhaps 3 per cent (or even negative in the West), so the latter (i.e. mainly cryogenics-based) might overtake and in time dominate the helium market. This was one of the reasons we have concentrated in this book on current and future cryogenic and advanced helium applications, although Chapter 9 describes the "traditional" uses too. The "high-technology" applications are less well known and may eventually comprise significant parts of the helium market. The effect is illustrated in Figure 1.3, where the "high-technology" market for helium could become dominant, as it appears to be growing at rate that seems to be reasonably well described by a specialist market index (the NYSE ArcaTech 100) where the average growth rate (pre-recession) was 11 per cent over 20 years. If it turns out that there is a shift of growth from West to East then, perhaps, a more global measure than the ArcaTech100 Index is needed. The overall market growth in 2011 was around 4 per cent, but there was a wide variation: in Asia, 8.5 per cent; in the US, 1.5 per cent (Garvey, 2011). However, it is also evident that some of the "high-technology" applications may have shorter life cycles than "traditional" uses, as was the case for the "dotcom" optic fibre boom in the late 1990s in which helium was extensively used.

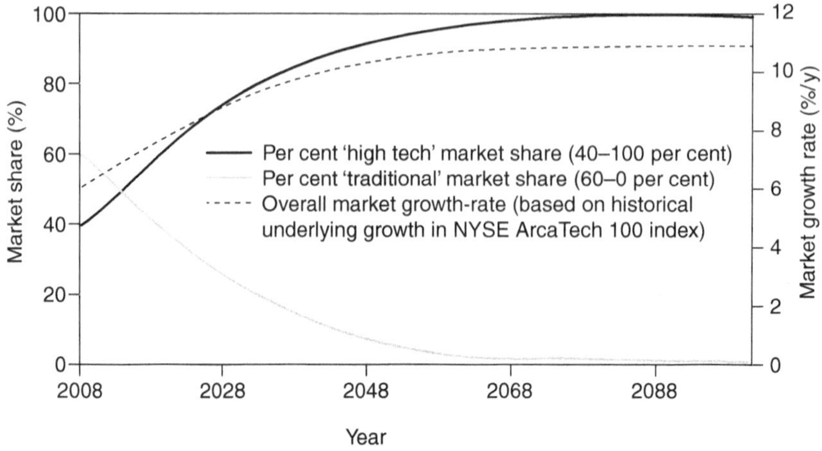

Figure 1.3 Possible change in balance of helium users if underlying growth rates persist (in the absence of feedback mechanisms).

The extremities of the helium market need examination too. At one end there is ongoing venting of helium molecules "unearthed" during man's activities (mainly natural gas extraction). On the other, there is now serious interest in dedicated exploration for helium where helium mining is the main goal as opposed to hydrocarbons. Let us first consider venting a bit further.

There are many reasons why venting occurs, some of which include: disincentives (mainly in the US) to develop resources because of past or present government policies; potential liabilities for retrospective royalties, taxes or lawsuits; potential disruption of the (main) revenue stream, i.e. from natural gas; lack of interest from the industrial gas companies in purchasing helium from independent operators. Outside the US, the issue is about getting access to the LNG condenser "purge gas" that contains helium. Here there are only a few windows of opportunity to bid and the barriers to market entry (pre-qualification exercises) effectively exclude all but the five or six industrial gas majors – the oligopoly of gases (Downie, 2002). Given the sheer volume of LNG exported, as discussed in Chapter 5, the wastage caused by delayed tenders is significant even though the helium content of LNG feed gas is just 0.015–0.05 per cent by volume, which is low by comparison with most US-sourced helium. The delays in accessing the raw material may be partly operational, political or because the owners want to be sure a helium demand needs to be satisfied. These were key factors in developing Algerian helium, the involved story of which is told in Chapter 4. Despite being the first large plant associated with LNG, the Algerian helium project started successfully. In some Middle East plants there have been problems with front-end purity control, but it is believed that those issues are now much better understood.

Nonetheless, a point the industry may need to watch is end-product purity. With the growth in precision and analytical uses of helium steadily rising, it is important that liquid helium can be sourced from any geography. Industry purity standards should be reviewed periodically, so that any liquid helium contaminants (such as ice, hydrogen and neon) do not interfere with the increasingly exact demands of these markets.

The LNG conundrum is complicated by ongoing developments in the natural gas industry. If unconventional or "shale" gas keeps developing (and certain environmental issues are addressed) it may start to suppress LNG demand: or, in a post-Fukushima world, maybe not (Platts, 2011). If the cost of LNG plants escalates due to skilled labour shortages or materials costs then this could significantly affect access to future helium sources (shale gas formations, such as the Marcellus in Pennsylvania, generally contain only meagre traces of helium). This is important, because LNG has already become the lead non-US route to helium and when the Qatar-2 plant comes online in 2013 it will be the world's largest single-train liquefier, capable of producing 20 tonnes of liquid helium per day (4.3 MMscf per day). This will make Qatar the world's dominant LNG-sourced, helium-exporting nation – provided it can sustain its current production rate of 77 million tonnes of LNG per year.

Further uncertainty is caused when the large volume of Russian helium is considered (Chapter 6). There are immense challenges in reaching this helium because of geology, infrastructure and climate. On the one hand, potential revenue and geography (the EU to the west and China to the east) encourage Russian gas development. In early 2011, the first shipment of liquid helium from Orenburg reached South Korea, a distance of 9,000 km, by train (Nekhai, 2011). Yet the Russian Central Congress has long recognised the value of conserving their country's helium for future generations in the twenty-first and twenty-second centuries (Yakutseni, 1995).

The questions of policy and governmental involvement that have dominated the US helium industry have fed into the thinking of the Indian government and its key research institutes. While the cultural approaches are profoundly different, it is the similarities that are striking. Chapter 7 explores the technological responses that India is making to become independent and self-sufficient in helium. Nisith Das *et al.* indicate that India's specialist helium plant developments are seen to be crucial for its independent nuclear power programme.

Helium has long been considered a "high-technology" gas (even before the term was coined). It was also a gas of strategic and military importance. This started 90 years ago when helium was commandeered for airships. In the early years of the twentieth century the demand was dominated by rigid (and semi-rigid) airships, or dirigibles as they were also known. Prominent among these early examples were the ill-fated large US airships, the Shenandoah, Akron and Macon, which used helium as a lifting gas. During and after the Second World War interest moved to smaller, non-rigid airships known as "blimps", and for a while these were a major military/strategic use for helium. This period has been expertly described by Clifford Seibel in his book *Helium, Child of the Sun*

(1968). This was followed by demands for rocketry (including the 65 tonnes of helium needed for each Apollo launch in the 1960s space race). A quandary was thus created: how to ensure a strategic supply, whether by government or by private industry operations. Understanding this is the objective of Chapter 2. Original archival documents have been accessed, together with extensive material written by the late Art Francis (in his unpublished manuscript "Helium at the crossroads" (Francis, 1996)). Francis was responsible for many aspects of the US Bureau of Land Management (BLM) helium operations leading to the construction of the Crude Helium Enrichment Unit (CHEU) vital to today's operation (and sell-off) of the Cliffside reserve in Amarillo, Texas.

Chapter 2 also covers the US National Research Council (UISHR, 2010) report investigating the impact of selling the (US) helium reserves. The price increase that followed was justified by the need to "enable a more rapid retirement of the BLM facilities debt to the US Treasury – a central goal of the 1996 Act [of Congress]" and "…to get a fair price for a scarce government resource" (Leslie Theiss, USBLM Amarillo Field Manager (USBLM, 2010)). This is just the latest episode in the helium saga. At the end of the book, in Chapter 17, Ralph Scurlock reflects on these issues and suggests policy options, one of which might be to establish a number of large-scale strategic (geological) reserves for crude helium around the world – perhaps commencing in Southern Europe. This might be a way around the Cliffside problem, where the objective now is not the conservation of helium but the raising of revenue.

An unintended side-effect of the Cliffside sale has been the loss of the "flywheel" mechanism, whereby reduced production at other plants worldwide (scheduled or unscheduled viz: the 2011 data in Chapter 3) used to be covered by excess capacity on the BLM pipeline. This has been a significant factor in the disturbances in the downstream market. The currently tight supply situation should be alleviated when the Qatar-2 plant starts up (Kornbluth, 2012). Ahead of that a 200 MMscf/y Matheson–Air Products joint venture at Big Piney (Wyoming), and a Linde 350 MMscf/y unit at Skikda, Algeria, are expected to come online. And, in the medium term, there are a number of interesting and potentially large-scale helium prospects in Central and Western Australia.

Returning to the exploration question, in a seminal paper geologist Steven J. Maione (Maione, 2004) anticipated the need for a dedicated helium exploration campaign – the deliberate search for helium-rich gas reservoirs. Perhaps there is a balance to strike here? There may indeed be medium and far-horizon opportunities for dedicated "helium mines" but for now those reserves are tight and secure.

In the short term, as Joe Peterson (Assistant Field Manager, Helium Operations at the US Bureau of Land Management) points out in Chapter 3, there is no reason to expect a crisis pre- or post-2015 when the Cliffside reserve in the US is scheduled for withdrawal. This is because, belatedly, there has been investment and development of US and other reserves that will ensure ongoing supply for most of this decade. As with oil and gas, one of the difficulties is knowing with any certainty what reserves exist. Explorers then have the challenge of developing prospects even if the geology looks promising. There may be significant difficulties in bringing helium to the market. These include separating helium from other gases such as nitrogen and CO_2, often with low value, in remote locations, or liquefying or trucking Grade-A helium to the industrial gas majors for on-selling to the market, if prices and terms can be agreed, perhaps through joint ventures.

There is still much debate about what constitutes an "economic reserve" of helium. In the 1970s, when helium was almost exclusively purified in absorbers (mole sieves), a feed containing >0.3 per cent helium was deemed necessary. Nowadays, with LNG-associated plants, a feed of just 0.04 per cent helium as is found in Qatar's North Field is considered satisfactory. This is because the feed from an LNG plant to a helium plant contains ~50 per cent helium; the helium is an incondensable component in LNG liquefaction. Earl Cook (Cook, 1979) captured the importance of field size and concentration and proposed the use of a helium resource value unit (HRVU) that equals $G.z^2$, where G is the field size or production rate and z is the helium concentration. Higher z values imply easier separation and more product for a given G. The HRVU helps prospectors to determine whether, or when, to produce a field. There was angst, back in the 1970s, about whether helium should be recovered for future generations. In the interim, the so-called "intergenerational equity" factor has played little role in helium policy. Technology and politics have dominated the thinking and Cook's paper highlights the need constantly to re-evaluate underlying assumptions. By 1970s standards, LNG would not be a route to helium, and neither would extraction from air. Yet by comparison, some of Algeria's natural gas contains 0.19 per cent – a rich resource that is being less exploited than the HRVU formula might suggest, for reasons that Chapter 4 makes clear.

Richard Clarke and Roger Clare discuss, in Chapter 8, the economics of extracting helium from air – *the backstop* – by using industrial, cryogenic air separation plants. New plants would be fitted with an additional distillation column to produce a 60/40 nitrogen/helium-neon stream. The mixture would then be compressed and trucked to specialist processing plants to make high-purity products, albeit at a flow of only around 2 per cent of today's demand. But the price might be around US$200 per Mcf, as compared to a 1980s estimate by Epple and Lave (1980) of US$1,600 (US$4,200 today).

Are such considerations "blue sky"? Possibly. But for fusion – the process of creating energy from fusing light isotopes of hydrogen together – it is the latter half of the century that matters, state Richard Clarke and Zhiming Cai in Chapter 15. Within this timeframe natural gas sources of helium look questionable to say the least. It would be easy, but unwise, for fusion to sidestep this issue given the long timeframes involved. It affects both magnetically confined (MCF) and inertially confined fusion (ICF). In any case, if the recent annual helium price inflation of 10 per cent continues, helium will be rather expensive in 2050. The real question here is the amount of energy required for fusion to obtain the helium it needs via air separation. Scaling current practices to a 1 GWe power plant would require up to a quarter of the output to be used to drive the associated air separation plants (assuming anyone would want to buy vast quantities of liquid oxygen or nitrogen) (Cai *et al.*, 2006). Thus helium supplies are a vulnerability for fusion, although there are already signs that the fusion community can and will become much more helium-efficient.

Bartek Glowacki, in his article on substituting hydrogen for helium (Chapter 16) picks up on these issues. Already promising ideas, such as 20 K liquid hydrogen cooling via a low-inventory helium loop, might substantially meet the needs of a fusion power plant cryogenics system. By finding safe ways to use hydrogen cryogenics, the potential benefits for superconducting magnet technology are significant as performance at 20 K (–253 Celsius) is so much better than at 77 K (–196 Celsius) where liquid nitrogen has often been proposed (Duchateau, 2012). A comprehensive examination is made of many aspects of the emerging hydrogen economy and the superconducting technology that supports it, and how helium use can be eliminated or reduced to a subsidiary role.

Returning, finally, to helium-specific gas exploration, this book is dedicated to the late Wheeler Sears, who will come to be seen, we believe, as a helium pioneer. Along with his son, Bo Sears at Inter-American Corporation, he developed novel techniques for finding and exploiting helium sources in the US mid-continent region.

Figure 1.4 shows one such find at Chupadera Mesa, New Mexico, near the Chupadera fault line (Sears, 2011). The field is helium rich but is associated with low-energy (low-BTU) gas, mainly CO_2. The challenge now is to bring this helium gas to market, and this requires full understanding of the technical, economic and policy choices and data. In principle it could contribute significantly to US helium needs.

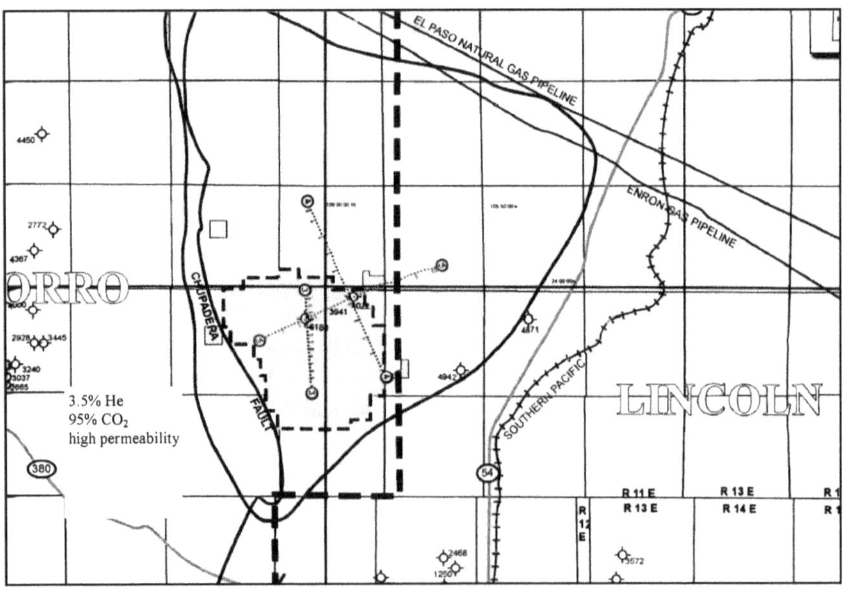

Figure 1.4 Helium-specific prospecting has produced results (Inter-American Corp.).

That, in essence, is what this book is about. Helping the reader to understand the myriad of ideas that inform helium-related policy decisions from wellhead to MRI machine, and from conventional resources to unconventional ones or helium substitutes. As we shall see, helium is a riddle of policy, molecules and machines.

Note

1 The attribute of zero resistance is not sufficient to define a superconductor; in addition a superconductor is a perfect diamagnet, giving rise to flux exclusion known as the Meissner effect. Frequently real superconductors exhibit a small residual resistivity even in the superconducting phase.

References

Broers, A. (2005) "The Triumph of Technology", BBC Reith Lectures 2005, BBC, London

Browne, M. (2008) "Indispensible Helium is Routinely Squandered", *New York Times* (Science), 14 May 2008

Cai, Z., Clarke, R. H., Nuttall, W. J., Ward, N. and Glowacki, B. A. (2006) "Is Helium From Air a Future Source for Helium Production?", *CryoPrague 2006* (ICMC 2006 – ICEC 21), Vol. 2, Icaris Ltd, Prague, pp. 219–22

Cai, Z., Clarke, R. H., Glowacki, B. A., Nuttall, W. J. and Ward, N. (2010) "Ongoing Ascent to the Helium Production Plateau – Insights from System Dynamics", *Resources Policy*, Vol. 35(2), June, pp. 77–89

Caplin, A. and Leahy, J. (2001) "Psychological Expected Utility Theory and Anticipatory Feelings", *Quarterly Journal of Economics*, Vol. 116, pp. 55–79

Cartwright, J. (2012) "Shortages Spur Race for Helium-3 Alternative", *Chemistry World* online, 12 January, http://www.rsc.org/chemistryworld/News/2012/January/helium-3-isotopes-shortage-alternatives-neutron-detectors.asp (accessed 13 January 2012)

Cook, E. (1979) "The Helium Question", *Science*, Vol. 206, 7 December, pp. 1141–7

Davies, E. (2011) "Critical Thinking: Endangered Elements", *Chemistry World*, January, pp. 50–4

Deakin, L. (2001) "The Coming Helium Shortage", *Chemical Innovation*, Vol. 31(6), June, pp. 43–4

van Delft, D. (2007) *Freezing Physics – Heike Kamerlingh Onnes and the Quest for Cold*, Koninklijke Nederlanse Akademie van Wetenschappen, Amsterdam

Downie, N. A. (2002) *Industrial Gases*, Kluwer Academic Publishers, New York

Duchateau, J.-L. (2012) "High Temperature Superconductors (HTS) in Fusion Technologies" in: *High Temperature Superconductors (HTS) for Energy Applications* (Edited by Z. Melhem), Woodhead Publishing, Cambridge, UK

Elhelwa, S. A. (US BLM engineer) (2008) Private Communications

Epple, D. and Lave, L. (1980) "Helium: Investments in the Future", *The Bell Journal of Economics*, Vol. 1(2), pp. 617–30

Firth, N. (2010) "Scientists say Earth's Helium Reserves will 'Run Out In 25 Years' (and Party Balloons Should Cost £65)", *Daily Mail* online, 23 August, London, www.dailymail.co.uk/sciencetech/article-1305386/Earths-helium-reserves-run-25-years.html (accessed February 2011)

Francis, A. (1996) "Helium at the Crossroads", unpublished manuscript

Gabrielli, G. and von Kármán, T. (1950) "What Price Speed? Specific Power Required for Propulsion of Vehicles", *Mechanical Engineering*, ASME, Vol. 72(10), pp. 775–81

Garvey, M. D. (2007) "Up, Up and Away … Where Has All The Helium Gone?", *Cryogas International*, October, pp. 22–6

Garvey, M. D. (2011) "Tight Supply Reins in the Worldwide Helium Market", *Cryogas*, October, pp. 30–2

Holland, P. W. and Emerson, D. E. (1987) "A Determination of the Helium 4 Content of Near-Surface Atmospheric Air within the Continental United States", *Journal of Geophysical Research*, Vol. 92(B12), pp. 12557–66

Kaplan, K. H. (2007) "Helium Shortage Hampers Research and Industry", *Physics Today*, Vol. 60(1), June, pp. 31–2

Keesom, W. H. (1942) *Helium*, Elsevier (Amsterdam, London, New York, Princeton)

Kornbluth, P. (2012) Private communication

Limb, D. (2009) "Airship Transportation System – Helium Handling and Storage", Fifth International Airships to the Arctic Symposium, Calgary, AB, 7–9 October, www.airshipstothearctic.com/docs/pr/19_David_Limb.pdf (accessed February 2011)

Maione, S. (2004) "Helium Exploration – a 21st Century Challenge", Houston Geological Society Online, 23 February 2004, www.hgs.org/en/art/162/ (accessed February 2011)

National Academy of Sciences (2000) *The Impact of Selling the Federal Helium Reserve*, National Academies Press, Washington DC

National Academy of Sciences (2010) *Selling the Nation's Helium Reserve*, National Academies Press, Washington, DC

Nekhai, O. (2011) "Russia Sells Helium to Asia-Pacific Region", *The Voice of Russia*, 1 February, http://english.ruvr.ru/2011/02/02/42453235.html (accessed February 2011)

NISTEP (National Institute of Science and Technology Policy, Japan) (2000) Sixth Survey Results in Resources and Energy, March, www.nistep.go.jp/achiev/ftx/eng/rep052e/pdf/3_07.pdf (accessed February 2011)

Perelman, M. P. (2003) *The Perverse Economy: The Impact of Markets on People and Nature*, Palgrave, New York

Platts (2011) "GDF Suez Official says Global LNG Market Tight in Wake of Fukushima", Santiago (Platts), 3 May

Sears, B. (2011) "Helium Exploration", www.helium-corp.com (accessed February 2011)

Seibel, C. W. (1968) *Helium, Child of the Sun*, The University Press of Kansas, Lawrence, KS

Shiryaevskaya, A. (2011) "Russia set to Top Helium Supply as U.S. Sells Reserve", Bloomberg (news), 9 February, www.bloomberg.com/news/2011–02–09/russia-aims-for-top-spot-in-helium-production-as-u-s-depletes-stockpiles.html (accessed May 2011)

Stott, P. E. (2005) "The Feasibility of using D-^{3}He and D-D Fusion Fuels", *Plasma Physics and Controlled Fusion*, Vol. 47, pp. 1305–38

The Oil Drum: Europe (2008) "Goodbye Helium, Goodbye Brainscans", blog posted by Rembrandt, 17 January, www.theoildrum.com/node/3484 (accessed February 2011)

Tolstikhin, I. (1975) "Helium Iisotopes in the Earth's Interior and the Atmosphere: A Degassing Model of the Earth", *Earth and Planetary Science Letters*, Vol. 26, pp. 88–96

USBLM (US Bureau of Land Management) (2010) "Bureau of Land Management Adjusts Price of Federal Crude Helium", Bureau of Land Management News Release, 20 August, www.blm.gov/nm/st/en/info/newsroom/2010/august/bureau_of_land_management.html (accessed May 2011)

Yakutseni, V. P. (1995) "The Status of Natural Helium Resources: Geopolitics of Production and Use", *Chemical and Petroleum Engineering*, Vol. 31(1–2), pp. 61–3

2 A history of the helium industry

Bo Sears

In the Spring of 1903, a small company called the Gas, Oil and Development Company drilled a well just off Main Street in the town of Dexter, Kansas in pursuit of oil and/or natural gas. Very quickly, a large gas reservoir estimated at 9 million cubic feet of gas per day was found at a shallow depth of 400 feet. Celebrations were planned and the excitement was rampant throughout the small Kansas town. Crowds gathered to witness (and hear) the huge gas flow and they anxiously waited to see the gas ignited, thus putting Dexter, Kansas on the map as a major resource. As the flame was placed in front of the gas stream, it was quickly *extinguished.*

Despite shutting the well in and attempting to ignite the gas again, it would simply not burn. News of this non-flammable gas traveled quickly within the State where it ultimately became known as "wind gas." Despite the fact that the dreams of a new Dexter were crushed, the news sparked the curiosity of the Kansas state geologist, Erasmus Haworth, who requested a gas sample. The sample was sent to the University of Kansas during the summer where David F. McFarland, a University of Kansas chemistry professor, performed a careful analysis. In the sample he noted a composition of 12.09 percent "inert residue."

The composition of this mysterious "inert residue" was later revealed by Dr E. H. S. Bailey, Head of the Chemistry Department at the University of Kansas, at the annual meeting of the American Chemical Society in New Orleans in 1906. Bailey revealed a helium composition of 1.84 percent (Cady and McFarland, 1906: 344).

Despite the rather large concentration of helium, the value of this helium would not be understood until years later. The helium present in this natural gas well, rather, was more of a laboratory curiosity. Immediately after receiving the results of the Dexter well, 44 gas wells in Kansas and Missouri were sampled and analyzed by the end of 1906, each of which contained helium albeit in smaller concentrations than the Dexter well (Cady and McFarland, 1903: 80). The results of this sampling also revealed a "remarkable regularity in the distribution of helium," showing that elevated helium concentrations appeared to be related to linear geological structures. The most important observation in this first natural gas sample survey was helium's relationship to nitrogen (helium reserves are most commonly found associated with high nitrogen concentrations).

Hamilton Perkins Cady (1874–1943) and David Ford McFarland (1878–1955), both of the University of Kansas, were the two chemists responsible for the analysis of these initial Kansas and Missouri natural gas wells, while members of the University Geological Society were responsible for the majority of the collection of samples (Cady and McFarland, 1907a: 1524). Cady and McFarland's observations in 1905 through 1907 would reveal some of the most important ideas on the relationship between helium accumulation and geological events. The results of their labors showed that helium was available in ample supplies, and in the summer of 1906 they stated, "we feel that we have here a very unusual opportunity for obtaining helium in practically unlimited quantities." (Cady and McFarland, 1906: 344).

Both Cady and McFarland were also cognizant of Ernest Rutherford and William Soddy's earlier proof of helium generation through the decay of radioactive elements (principally radium as per their studies). At this embryonic stage of helium gas sampling, it was too early to know exactly why helium was present in natural gas. In addition to helium found in these natural gases, Cady and McFarland in 1907 found that the Dexter gas well contained both argon and neon, which represented a new discovery in natural gas (Cady and McFarland, 1907b).

After Cady and McFarland made these significant discoveries of helium in natural gas, further helium work was curtailed in the coming years because there was no commercial need for the gas. After their unsuccessful attempts to liquefy

Figure 2.1 Professor Hamilton Cady in the laboratory where helium was discovered in natural gas in 1905 (courtesy: Spencer Research Library, University of Kansas).

Figure 2.2 Professor David Ford McFarland (credit: Department of Materials Science and Engineering, Pennsylvania State University, USA).

helium,[1] they went on to pursue their other duties as chemistry professors at the University of Kansas. All that was left of the helium gas collected was placed in three small bottles submerged in mercury, to prevent leakage, and placed on the top shelf of a laboratory at the University. Collecting dust, these bottles were simply labeled, "He 1905" (Seibel, 1968: 1).

The helium idea

As the Great War was firmly established by midsummer of 1914, one particular event and one publication in this year would provide the catalyst for a new helium "industry" to be born. Prior to 1914, helium was of little use beyond applications of experimental nature. Despite the fact that helium could now be found in apparently unlimited quantities, it was simply not required for any commercial application.

All of this changed in 1914, when a German book, entitled *Die angewandte Chemie in der Luftfahrt* (Applied Chemistry in Aviation) by Dr Géza Austerweil, was published. A section in Chapter 1 of this text, "Helium als Ballonfüllgas" (Helium as a Balloon-Filling Gas), specifically mentions the advantages of using non-flammable helium as a replacement to hydrogen for filling military balloons. Austerweil went on to mention, however, that the probability of obtaining enough helium to fill a balloon was unlikely because all of the world's known helium was held in extremely small quantities in the laboratory of Professor Kamerlingh Onnes, who had been the first to liquefy helium six years earlier

(Austerweil, 1914: 8). Austerweil clearly had no knowledge of the helium discovered in United States natural gases in 1906.

Another event in 1914 had a much more profound effect on the history of the helium industry. During that year, a German Zeppelin was attacked and pierced by many incendiary bullets but would not catch fire, leading many to believe that Germany had found a non-flammable gas (Moore, 1926: 203). If, as was suspected, Germany had somehow been able to obtain helium for military use, it would have a supreme advantage in the air over the Allies. (It would later become obvious that Germany did not possess any helium and in fact this particular dirigible was filled with hydrogen.)

One English scientist who caught wind of the German airship that could not be brought down (and possibly Austerweil's book) was Sir Richard Threlfall, who, fearing that helium had somehow been obtained by the Germans, immediately wrote to the British Admiralty voicing his concerns. Upon receipt of this letter, the British Admiralty authorized Threlfall (who would later be joined by Sir William Ramsay) to report on possible locations where helium could be extracted. Having not yet learned of the Dexter well and subsequent publications in the United States, Threlfall and Ramsay analyzed the gases emitted from coalmines and firedamps. They were able to locate two sources of helium in coalmines in Anzin, France and Frankenolz, Germany (near Luxembourg). The amounts expelled from these two locations, unfortunately, were not voluminous enough to fill military balloons adequately. The amount of helium liberated from the Anzin mine, for example, amounted to just over 141,000 cubic feet of helium *per year* (Moore, 1926: 204).

By December 1914, Threlfall finally learned of Cady and McFarland's work in the United States and by February 1915 had reported these US discoveries to the British Admiralty's Air Department. Realizing for the first time that helium could potentially be extracted from natural gas in unlimited quantities, the British Admiralty quickly assigned a small sum of money to investigate helium from natural gas within the British Empire, principally Canada.

The United States, not yet a participant in the war, learned in February 1915 of the British attempts to find helium for use in military airships in a letter from Sir William Ramsay addressed to Richard Bishop Moore (1871–1931) (Moore had formerly worked with Ramsay in the United Kingdom when studying the rare gases of the atmosphere). In this letter Ramsay writes:

> I have investigated blowers – that is, coal-damp rush of gas – for helium for our Government. There does not appear to be anything in the English blowers, but I am getting samples from Canada and the United States. The idea is to use helium for airships.
>
> (Seibel, 1968: 23)

After receiving this letter, R. B. Moore put it aside because the US was not yet involved in the war. In addition, President Woodrow Wilson forbade any war preparations as a part of his strict policy of neutrality.

Nevertheless, the British moved forward with haste on their Canadian project and assigned a sum of money in 1915 to Professor J. C. McLennan of the University of Toronto to experiment on the possibility of using helium for use in airships. McLennan was asked to determine the helium content of natural gases within the British Empire and to conceive a way to process helium from this gas. In relative short order, McLennan discovered several areas within eastern Canada that possessed helium concentrations ranging from 0.1 percent to 0.33 percent, and negotiated a contract with L'Air Liquide Company (now Air Liquide) to manufacture an experimental helium plant to process this natural gas. The site chosen for this plant was in the town of Hamilton, Ontario and the gas supply was to come from Western Ontario gases via pipeline (McLennan, 1920).

The L'Air Liquide plant in Hamilton was a modified air separation unit, which utilized cold temperatures in order to separate oxygen from the air (air is nearly 20 percent oxygen). This plant represented the first functional helium plant in the world but it was not without its problems. During the early years of its useful life, the problem of removing the heavier hydrocarbons found in the Western Ontario gas proved to be quite difficult (McLennan, 1920). Through trial and error and numerous modifications, the primary purpose of separating helium from natural gas was finally achieved by the middle of 1918. Toward the end of its operation in Hamilton it was able to achieve a helium purity of 87 percent, with the residual gas primarily composed of nitrogen. However, just as progress on purification was being made, the volume of gas from the Western Ontario fields began to decline substantially and the plant was subsequently moved to Calgary to process gas from the Bow Island Field.[2]

Meanwhile in the United States, still not in the war although the likelihood of its entry appearing imminent, a young chemistry student at the University of Kansas was seeking an advanced degree and sought the help of Professor Cady for an idea on his thesis. Cady had suggested that Seibel re-examine the previous work done on helium and the other rare gases of the Kansas natural gases. Seibel had no interest in undertaking this study but as this advice had come from such a distinguished chemist, Seibel reluctantly agreed and began his studies.

Seibel finished this thesis in early 1917 and was scheduled to present his work at the fifty-fourth annual meeting of the American Chemical Society, to be held in Kansas City in April, only days after the United States entered the war. Seibel reluctantly read aloud from his paper, "The Rare Gases of Natural Gas," and when he had finished expressed regret and stated, "I'm sorry, gentlemen, but there is no practical application for this information" (Cady and McFarland, 1906). Seibel's statement made reference to the point that, at a time when most Americans were focusing on war, his paper posed no practical benefit to the war effort.

Dr Richard Bishop Moore, who was keenly interested in Seibel's thesis, was present at this meeting and immediately responded to Seibel's last statement by producing his letter from Sir William Ramsay of February 28, 1915, and firmly stated that a practical application did indeed exist. Moore highlighted the British interest in helium for use in wartime airships, and reasoned that it would be possible to extract and process helium for American military use at a reasonable cost.

As the dialogue progressed, the mere idea of sequestering enough helium to fill an airship seemed incomprehensible to Seibel as he held in his had nearly the entire United States supply of the gas, which amounted to less than 1 cubic foot (Seibel, 1968: 23). The most obvious problem as perceived by Seibel was that the United States currently possessed only 1 cubic foot of helium, while a blimp would require nearly 100,000 cubic feet. As Seibel had recently sold 1 cubic foot of helium for approximately US$2,500, the cost to fill a blimp would be more than US$200 million.

Completely unfazed by Seibel's points, Moore began a dialogue with Dr Charles Lathrop Parsons, chief chemist of the Bureau of Mines, who was also in attendance at the meeting. Commenting on the superior benefits of using helium instead of hydrogen for airships, Moore asked Parsons, who was traveling back to Washington DC after the meeting, to pass on this information to his superiors at the Bureau of Mines and also the War Department.

The ensuing discussions held in April and May at the Bureau of Mines in Washington were very well received, especially by the Bureau's head of War Gas Investigations, George Arthur Burrell. Before the war, Burrell had examined numerous samples of natural gas wells in Texas, particularly the fully developed Petrolia Gas Field near the town of Wichita Falls, Texas. The gas from Petrolia had contained 20 percent to 30 percent nitrogen but was good enough to be used as fuel and pipelined down to Dallas and Fort Worth, Texas in large volumes. Having learned of helium's association with nitrogen, Burrell speculated that it was possible that gas from the Petrolia field might contain helium due to its nitrogen content. Burrell sent a gas sample to Dr Cady at the University of Kansas, where it was revealed to contain roughly 1 percent helium. Burrell immediately informed his colleague and chief metallurgist at the Bureau of Mines, Frederick Gardner Cottrell, to join him and R. B. Moore in discussing the use of this Petrolia gas for helium recovery. The Petrolia field, thus, became the likely place to begin should the project proceed.

Progress on the helium question gained steam on May 12, 1917 when Burrell wrote a letter to Major Charles deForest Chandler, who was in charge of the Balloon Division of the Aviation Section of the Signal Corps in the United States Army (Moore, 1926: 204). In this letter, Burrell had asked an opinion of Chandler on whether helium would prove to be advantageous enough to warrant production of the gas if it could be recovered inexpensively. Intrigued with the idea, Chandler would meet both Moore and Burrell in June to discuss its feasibility.

Before the start of the war in 1914, the US Army had already appropriated US$250,000 for aviation purposes but the majority of these funds were for airplanes. After the war had begun, the threat of airships was a growing problem for the United States. In addition, because of Woodrow Wilson's policy of neutrality, little focus had been placed on developing American airships, creating a tremendous disadvantage. By March 1915, an appropriation of US$1 million was awarded to the Navy, part of which was for its first airship (designed for hydrogen). Leading up to the United States' entry into the war, J. C. Hunsaker of the US Navy stated, "it seemed evident that a new weapon had appeared and that the U.S. should consider its possibilities carefully" (Althoff, 2004: xviii).

Now that the United States was in the war, interest in airships, especially those that could be filled with helium, was a high priority. The meeting in June 1917 between Moore, Burrell and Chandler would constitute the birth of the helium industry.

An industry born

The meeting between Moore, Cottrell and Chandler was designed to gauge further the potential interest of the Army and to determine whether it would be willing to contribute monies for a pilot helium project. Chandler became interested at once.

Interest in helium continued to expand as Chandler informed G. O. Carter of the Navy's Bureau of Steam Engineering, who immediately became interested on behalf of the Navy. Both Carter and Chandler agreed to move forward with a helium program as quickly as possible. Enthusiasm for helium would also reach the Assistant Secretary of the Navy, Franklin D. Roosevelt, who would later say, "with the fire risk eliminated the rigid airship or Zeppelin will hence be one of the most powerful weapons known" (Sherburne Rogers, 1920: 6).

Immediately as a result of these communications with Moore, Burrell, Chandler and Carter, a recommendation was made by Frederick Cottrell to contact Fred H. Norton of the Jeffries-Norton Corporation to begin discussions on the processing of helium. Norton, a Massachusetts Institute of Technology graduate and a very well-respected engineer of international reputation, was called to Washington on June 4, 1917 to discuss an experimental process where helium could be extracted from natural gas, which had never before been achieved in the United States. Cottrell, who was introduced to the Norton process by Bureau of Mines Director Van H. Manning in 1916, was of the opinion that Norton might be the engineer who could tackle this issue with relative low cost. Norton had developed the drawings for a "new process for air separation that embodied some novel and striking features" (Manning, 1919: 77) to produce oxygen inexpensively for use in blast furnaces, in which the Bureau of Mines had an interest. The oxygen in Norton's process was extracted using low temperatures like other air separation units, but his design was supposed to achieve the results with reduced compression costs, as these represented the primary cost of extraction.

Norton estimated that it would cost US$28,000 to build a plant that could process 5,000 cubic feet of helium per day (because this plant was experimental in nature, only very small volumes of gas were required to test the process). The request for US$28,000 was granted and soon Norton was making designs to build the first experimental helium plant and negotiating a contract with the Lone Star Gas Company to withdraw a small portion of the Petrolia gas from the pipeline in Fort Worth, Texas.

In the meantime, conferences were held in Washington DC in July 1917 to hold open discussions on all matters pertaining to the production of helium but also addressing many scientific questions concerning helium permeability through balloon fabric, solubility of helium in liquid methane and liquid nitrogen, and combustibility studies on mixtures of helium and hydrogen (it was learned that a 10 percent component of hydrogen in helium would not ignite). More importantly, a relationship between the United States and Great Britain was recommended in order to aid the Allied forces, resulting in the dispatch of Captain R. B. Owens of the US Army's Signal Corp to the United Kingdom to discuss cooperation between the two countries. All attendees of these conferences were in agreement to move forward on all fronts.

Not willing to risk the future of the entire United States industry on the experiment of one processing company, R. B. Moore recommended that two other companies well established in air separation also be considered. Moore had recommended to Frederick Cottrell that both Linde Air Products[3] and the Air Reduction Company be given a chance to compete for processing rights, to which he readily agreed (Moore, 1926: 204). By late July 1917, the joint Army and Navy Airship Boards recommended an increased appropriation of US$100,000 (split evenly between Army and Navy) to be divvied between these two companies and one other, which withdrew shortly afterwards. It was also recommended at this time that helium operations be consolidated and run by the US Bureau of Mines, with George Burrell leading operations. Dr Hamilton Cady and his assistant, Dr Clifford Seibel, would become the Bureau of Mines' consulting chemists in charge of the majority of helium analytical and research work.

As negotiations were proceeding with the well-known Linde Air Products and the Air Reduction Company, representatives from the Jeffries-Norton Company had already negotiated a verbal contract with the Lone Star Gas Company for processing rights and had chosen a location for their experimental helium processing plant in Fort Worth. The Norton representatives were quickly moving forward until all momentum came to a screeching halt in October 1917, when G. O. Carter prohibited any Navy funds to be diverted to the Jeffries-Norton Company. G. O. Carter, the Navy representative in Washington, was also a former Linde Air Products employee who had some knowledge of low-temperature processing. The Norton process, according to Carter, was an unsound technology undeserving of any Navy appropriation. The Army would ultimately side with Carter and leave Norton out of any Army appropriation as well (Seibel, 1968: 32).

Less than one week before the Norton process was excluded from any funds, the British Admiralty had informed the United States of their helium requirements for the war. The amount requested was enormous, 100 million cubic feet at once and 1 million cubic feet per week thereafter (Manning, 1919: 77). The British believed this amount was entirely feasible, primarily because of their belief in the Norton process. Through the convincing of Captain R. B. Owens on his mission to the United Kingdom after the helium conferences, he assured the British authorities that the Norton process would prove to be theoretically sound, thus assuring the Allies a low-cost method of helium extraction. Now that the

Norton process was excluded from any Army or Navy appropriation, this left the British Admiralty and the US Bureau of Mines somewhat embarrassed because of their unwavering belief in the Norton process.

Despite the stalling of the Norton fiasco, progress by the Bureau of Mines remained swift and an increased appropriation of US$500,000 (Seibel, 1968: 34) (a portion of which was supposed to be allocated to the Norton process) was approved for the two remaining processing companies. Both the Linde Air Products and Air Reduction contracts were signed in November 1917 and construction on both facilities began immediately. During construction and as a precautionary war measure, both plants became known as "argon" plants for added secrecy.

Luckily for the Jeffries-Norton Company, they would get a new life in early 1918 when, as a result of their process being reinvestigated by the National Research Council, were again deemed worthy of a chance to compete. On January 14 the National Research Council stated that

> the committee is unanimously of the opinion that the Norton process … is scientifically sound, that it should accomplish the desired result, and that every part of it seems to have been conceived in the light of a clear understanding of the problem and of the means which good engineering would suggest as conducive to economy.
>
> (Seibel, 1968: 34)

During the time it took to have the Norton process reinstated, the original plant site negotiated by Norton was assigned to both Linde Air Products and Air Reduction. Having no plant site at the end of the pipeline in Fort Worth, Norton decided to install his plant at the Petrolia field.

The three experimental helium plants were placed under the control of R. B. Moore in June 1918. A brief description of these plants is as follows.

1. The Linde Air Products plant (Argon Plant No. 1) in Fort Worth, which had a capacity of 5,000 cubic feet of gas per day, was completed in March 1918 at a cost of US$245,000. This plant utilized the Joule-Thomson effect, which is brought about by the rapid expansion of compressed gas through a small orifice (which is why an aerosol can becomes cold during spraying). As the incoming gas is compressed to 2,000 pounds per square inch (psi) and subsequently cooled by water, carbon dioxide and previously processed liquid gases, it is then expanded and thus cooled as it returns to normal pressure. This process is then repeated several times until the gases fall below their boiling points. The main drawback to this process was the enormous gas compression requirements, which necessitated large amounts of energy, thus increasing processing costs. Although this plant was the most expensive to operate, the plant was able to produce 70 percent pure helium initially, then 92 percent purity after reprocessing (Manning, 1919).
2. The Air Reduction plant (Argon Plant No. 2), with nearly identical capacity as the Linde Air Products one, began operations on May 1, 1918 and cost

US$135,000 to manufacture. Air Reduction used a "Claude" cycle (a process designed by Georges Claude of L'Air Liquide), which also utilized the Joule-Thomson effect. This system had much lower compression requirements, resulting in lower operating costs. The main difference between this plant and the Linde Air Products one was the use of an expansion "engine" that utilized the expanded gas to enhance further cooling effects, thus lowering compression requirements. Despite lower operating costs, the Air Products plant was only able to produce a maximum of 70 percent helium purity. In order to achieve higher-purity helium to send overseas, this 70 percent helium gas stream was reprocessed in the Linde Air Products plant, where 92 percent purity could be attained (Manning, 1919).

3 Construction on the Norton plant (Argon Plant No. 3) didn't begin until early April 1918 and was not completed until October 1, 1918. Manufacturing costs were just over US$148,000, with a capacity of 30,000 cubic feet per day (Manning, 1919). This process, which was the most heavily touted due to perceived low operating costs, was designed to operate with very low compression requirements. This process utilized three expansion engines instead of one, like that of the Claude process. It was believed that the use of three expansion engines that would negate heavy compression requirements and would be much more efficient. Despite repeated failures over the course of just under three years, the Norton process continued to draw support from Bureau officials. By April 3, 1919, the Norton process was only able to produce 20 percent helium and was finally sold as scrap two months later. The Norton process would ultimately cost more than the Linde Air Products and Air Reduction plants combined.

By June 1918 it was clear that Linde Air Products (known today as Praxair) had the superior process, and by August it was determined by the Army and Navy that a large, full-scale production plant was going to be required to meet the escalating demand for helium. In addition to new plant requirements, it was also deemed necessary to find new helium-bearing gas fields because the mature Petrolia field was rapidly declining. The Bureau of Mines then began an exhaustive search for helium reserves, beginning in June under the leadership of G. Sherburne Rogers of the United States Geological Survey.

The work carried out by G. Sherburne Rogers was initially to identify new helium reserves as quickly as possible, to aid the war effort. Natural gas wells from across the country were analyzed only for their helium content (and no other gas) due to the rush on the war. Throughout this initial research, the maximum helium content discovered was that of Petrolia field in Clay County, Texas.

Rogers' publication would become the most important document of helium resources for the government many years after its release. An extremely proficient geologist, Rogers delicately analyzed the geology of particular areas in order to hypothesize helium's probable origin. Although Rogers' work had been underway since the start of the experimental plants, it would not be released until 1920. It was also Rogers who predicted that the Petrolia field was never going to last at the rate in which it was currently being depleted, thus alerting government officials to negotiate with Lone Star Gas Company to reduce its production rate.

Operations on both experimental plants (Linde Air Products and Air Reduction) would continue throughout the duration of World War I, and by July it was well established that large-scale helium processing was going to be a reality. Both experimental plants had produced fair quantities of gas. Two small shipments of helium for experimental use were sent to France and Britain early in the year, and just prior to the signing of the November 1918 armistice ending the war, a much larger consignment of 145,000 cubic feet (US House of Representatives, 1922) of 92 percent helium, carried in 750 cylinders, was shipped to the New Orleans docks destined for France. The armistice was signed before this helium ever had a chance to aid the war effort and was shipped back to Fort Worth, where it would be stored and later used to fill the first US helium-filled dirigible, the Navy's C-7.

On October 22, 1918 Linde Air Products was awarded the contract for the first full-scale production plant in the United States, which was designed to produce 30,000 cubic feet of helium per day from the Petrolia field. Negotiations had already been under way to lease the Petrolia gas from the Lone Star Gas Company and construction of a new pipeline from the gas field to the new Linde Air Products facility was quickly under construction in Fort Worth. The armistice ending World War I would come into force on November 11, 1918, just 20 days after the contract with Linde Air Products was signed.

Although the war was now over and there was no longer a need for large volumes of helium gas, the Aircraft Board of the US Army made a review of the helium situation and determined that continuing the helium program was in the country's best interest.

Although the Norton process had been in operation since October, it had still been unable to produce any helium of significant quality. Regardless, Bureau of Mines engineers firmly believed that helium processing costs could be substantially reduced, and adamantly lobbied for Norton's continued efforts. In addition to this Norton lifeline, it was later recommended by Van Manning, Director of the Bureau of Mines, to continue operations of all three experimental plants so that possible cost-reduction refinements could be explored.

Two days after Christmas 1918, the Secretary of the Navy, Josephus Daniels, wrote to Van Manning of the Bureau of Mines expressing his intent to discontinue all government helium experimental work except for the Norton process (apparently the notion of producing ultra-inexpensive helium was too tempting, even for upper government officials). Daniels' plan was adopted and both experimental plants were ordered to be shut down on January 23, 1919. Manning persuaded Daniels to keep the Air Reduction Plant in operation a bit longer, at the company's expense, to test a potential improvement. Despite this extension the Air Reduction plant was never able to make any sort of meaningful improvements and was finally released on April 1, 1919 to the Bureau of Yards and Docks (Navy). Parts of this plant would be used in the first Bureau of Mines purification plant in Langley field, Virginia (Seibel, 1968: 48).

Post-World War I

Before World War I, helium was selling for US$1,700 per cubic foot (US$1,700,000 per thousand cubic feet), simply because so little of the product existed. No more than 100 cubic feet had ever been sequestered (House of Representatives, 1920). The two experimental plants in Fort Worth had been able to produce an unimaginable amount of helium just three years earlier and at a cost of roughly US$100 per thousand cubic feet.

Despite the end of the war in late 1918, the United States moved forward on the helium program so as not to be caught short again should another conflict arise. Plans on the Linde Air Products production plant would continue without interruption until it finally came on stream in April 1921, under the US Navy's authority. The first four months of operations produced just over 260,000 cubic feet of helium at a cost of US$480 per thousand cubic feet, but would later drop to US$174 per thousand cubic feet (Seibel, 1968: 50). During the fabrication of the first production plant, it was recommended by the Army, Navy and Interior Department to develop a Cryogenic Research Laboratory in Washington DC, in order to gain a firmer understanding of low-temperature natural gas processing to extract helium. On May 21, 1921, just one month after the production plant began operations, the Cryogenic Research Laboratory was dedicated by Madame Marie Curie, who happened to be in Washington DC during a six-month visit.

The Cryogenic Research Laboratory would prove to be a very successful program for the advancement of helium research, under the leadership of R. B. Moore and Clifford Seibel. The primary purpose of this research was to understand the science of helium within the processing of natural gas, in order to achieve cost reductions. The success of lower-cost separation depended upon various factors of helium experimentation. According to Clifford Seibel,

> it was anticipated that with data obtainable by means of such a laboratory, more efficient plants could be designed for the future. In due course, workers in that research laboratory, established in Washington, D.C., supplied the answers to many questions dealing with such factors as specific heats, phase equilibria, solubility of helium in the liquid components of natural gas, behavior of metals at low temperatures, heat exchange, insulating materials, removal of carbon dioxide from the natural gas, special valves, analytical recorders, and methods of obtaining the necessary refrigeration.
>
> (Seibel, 1938: 849)

In addition to this research, it was also recommended by R. B. Moore that underground storage for helium should be tested. Between the years 1918 to 1920, one company had proved the success of underground natural gas storage in order to maintain a reserve for high-demand winter months (Westcott, 1920: 240). It would be R. B. Moore's incredible insight that would ultimately lead to the usage of Cliffside field for helium storage decades later.

Another result of the Cryogenic Research Laboratory was the creation of railroad helium purification cars, which would enable balloons and dirigibles to remove contaminated air from their gas envelopes. It had been learned through extensive study at this point that while helium would permeate balloon fabric into the atmosphere, atmospheric air would also seep into the balloon envelope. Thus, repurification of air-contaminated balloon envelopes was a problem that needed prompt attention. In total, three purification railroad cars were built by 1925, and these proved to be successful and inexpensive to operate.

The only known use for helium after the war was for balloons and dirigibles, so repurification was still a very high priority for the United States government, especially as they were to receive three new dirigibles: one to be manufactured in the United States (ZR-1; *USS Shenandoah*), one from the United Kingdom (ZR-2), and one from Germany (ZR-3; *USS Los Angeles*) as a result of war reparations. Each of these vessels would require between 1 million and 2 million cubic feet of helium. With heightened helium activity, it was recommended in a US House of Representatives Hearing on January 22, 1920 to prohibit the export of any helium gas. A large, US$3 million hanger at the Navy Base in Lakehurst, New Jersey was already under construction in 1920 in order to facilitate the manufacturing of the ZR-1 (later *USS Shenandoah*) and house all three airships.

In 1923, the Bureau of Mines would lose the head of the Cryogenic Research Laboratory, Richard Bishop Moore, who accepted a private industry position at the Dorr Company in New York. Moore would be replaced by Dr Samuel Colville Lind, a former student of Marie Curie in the Sorbonne, France, who was the leading US authority on radioactivity. During both Moore and Lind's leadership at the Research Laboratory, experimentation for new uses for helium had been actively pursued. Trials for helium use in the treatment of tuberculosis and electric lightbulbs had been carried out, among others, but with no success.

The Helium Act of 1925

The Helium Conservation Act of March 3, 1925 would represent the first United States law involving helium production since the government became interested in the product in 1917. The purpose of this Act was to establish a helium production and sales program under the direction of one entity, the Bureau of Mines. The need for a law requiring conservation was a foremost concern because it was estimated that 500 million cubic feet of helium was being wasted annually as a result of non-extraction of particular helium-rich natural gases (US House of Representatives, 1924). The Bureau of Mines would now be responsible for all government helium production, including "the purchase, lease, or condemnation of land, conservation of helium gas, construction and operation of helium plants, conducting experimental work, and the leasing of helium under certain conditions" (Seibel, 1968: 73). The Fort Worth production plant, which had previously been under the Navy's control, would be transferred to the Bureau of Mines under the guidance of Clifford W. Seibel. After this first Helium Act of

1925, the Bureau of Mines (and later the Bureau of Land Management) would continue to manage the federal helium program until 1996.

The year 1925 was also when the first "other" use of helium would be discovered. Helium was found to be beneficial as a replacement to nitrogen in deep-sea diving as a way to prevent caisson disease (the bends) among Navy divers. Using helium enabled divers to stay underwater longer and ascend to the surface more quickly.

With the Bureau of Mines in full control of the US helium industry, no time was wasted in continuing the search for new reserves and improving laboratory work to facilitate processing. As a result of the second helium survey (1919–33), three new helium-rich areas were discovered, one of which would become the most important reserve in United States helium history, the Cliffside field near Amarillo, Texas. The two others would be the only helium-rich areas located on government lands; all other helium-rich areas would come from private sources. The first, located in Emery County, Utah, would become Helium Reserve No.1 by executive order on March 21, 1924 and the second, located in Grand County, Utah, would become Helium Reserve No. 2 by executive order on June 26, 1933 (Seibel and Kennedy, 1934: 759).

The US Bureau of Mines would finally meet competition from private industry in 1927 when two plants were erected by the Girdler Corporation of Louisville, Kentucky – at Dexter, Kansas and Thatcher, Colorado. Shallow helium-bearing gas at Model Dome in Thatcher, northern Las Animas County, Colorado was discovered in March 1927 to contain more than 7 percent helium and 81 percent nitrogen (Dobbin, 1933: 1057). Due to the passing of the Helium Act of 1927, very little helium could be used outside of government entities, which posed a problem for privately funded scientists experimenting with new uses for helium such as asthma treatment, which was discovered in 1926. These two private plants would go on to produce about 8 million cubic feet of helium from October 1927 until February 1930 (Seibel and Kennedy, 1934). Later, as a result of the Helium Act of 1937, the government would negotiate to purchase both plants, which would be dismantled by 1944.

A new era in Amarillo

By the time the Cliffside field near Amarillo, Texas was discovered in 1924, the gas field at Petrolia was nearing the end of its useful life and by 1927 it became clear that the United States could no longer count on Petrolia to meet any future helium requirements. As a result of the fact that helium supplies were believed to be scarce, the Helium Act of March 3, 1927 was established to prohibit the sale of helium to foreign countries and non-governmental domestic use. Until the Bureau of Mines could find another helium-producing field that would satisfactorily meet all government requirements, the Helium Act of 1927 was to stay in effect. After ten years the *Hindenburg* disaster at Lakehurst (see below) would help provide a catalyst for change in policy towards foreign helium distribution.

Although the nearby Nocona field in Montague County, Texas was given some consideration to supply the Fort Worth plant in order to continue operations, the Fort Worth plant was ultimately shut down and dismantled in early 1929 when the Amarillo plant became operational (this is discussed later). During its life of eight years, the plant produced about 480 million cubic feet of helium and towards the end of its life could produce this gas for only US$34 per thousand cubic feet.

The Cliffside field, located in Potter County, Texas was situated on a structure that covered roughly 50,000 acres and contained, on average, 1.75 percent helium. This newly discovered field was producing gas at 700 psi (pounds per square inch), and very little gas had been produced since its discovery. After careful geological review, it was estimated that the Cliffside field contained enough helium to last the US government 100 years. Negotiations by the Bureau of Mines to purchase this field from its owners (and the two companies who leased these properties) began in 1926, and by 1927 an operating contract was reached covering 20,000 acres, with an option to acquire the remaining 30,000 acres. The drilling for helium gas began in February 1928 followed by the construction of a pipeline from the new plant site to the new wells in July 1928. Construction on the Amarillo helium plant, located 7.5 miles west of the town, began in August 1928, and the first shipment of 200,000 cubic feet of helium was hauled to Langley, Virginia on May 6, 1929. The helium was transported in one of two railroad helium tank cars that had been designed by the Bureau of Mines (*Industrial and Engineering Chemistry*, 1929).

Now that the US Navy only had one dirigible in operation (*USS Los Angeles*) and the Amarillo helium plant was in full operation, helium was no longer a scarce commodity as it had been during the passage of the Helium Act of 1927. By December 1929, construction would begin on the largest US dirigible in Akron, Ohio by the Goodyear Zeppelin Corporation. The ZRS-4, later to be christened the *USS Akron*, would require 6,500,000 cubic feet of helium. The fact that an airship this large was ever commissioned confirmed the United States' belief that the helium supply was here to stay. The Cliffside field and the Amarillo plant could produce more than enough helium.

The early 1930s would be a bittersweet period for the US Bureau of Mines. The Amarillo helium plant was able to produce a quantity of helium inconceivable only years earlier. The four wells supplying the Amarillo plant were flowing at a volume of 30 million cubic feet of gas per day, resulting in over 450,000 cubic feet of helium per day (Seibel and Kennedy, 1934: 759). While operating seamlessly, the Amarillo plant and its staff would be dealt one blow after another, with the loss of the *USS Akron* on April 4, 1933, which killed 73 men, and later her sister ship, the *USS Macon*, which killed two men, on February 26, 1935. Government demand for helium ended abruptly after these two tragic losses. The loss of life on the *Akron* dealt a critical blow to the continuation of the dirigible program, but after the *Macon* tragedy there was little room in the Navy for dirigibles, despite efforts by the Navy's Captain Charles E. Rosendahl to continue their service. Therefore, the largest user of helium in the world would only require small volumes to fill smaller, non-rigid blimps and observation balloons. After the loss of the *Macon* in early 1935, the Amarillo helium plant reduced personnel and operated only intermittently for the rest of the year.

The government had more helium supply than demand after the loss of the *Macon* and it was decided to "lease" helium gas to private companies in order to keep the Amarillo plant functioning. Although the idea of a helium lease appears unreasonable due to the easily mobile helium atom, as per the Helium Act of 1927 no government helium could be sold to non-government entities. Therefore a leasing agreement was reached, and in 1936 President Franklin D. Roosevelt approved a lease to the Goodyear Zeppelin Corporation for commercial airship use. According to the Minerals Yearbook of 1937, "the purpose of leasing the government helium was to foster the development of commercial lighter than aircraft and to encourage the training of airship pilots" (Seibel and Kennedy, 1937). In addition to airship use, helium was supplied to the US public health service for use in the treatment of asthma and other respiratory ailments, which were aided by the use of helium. The success of helium used in respiratory illnesses soon represented a heavily growing demand.

By early 1937, it was clear that a change needed to be made to the Helium Act of 1927. New uses for helium were developing quickly and the US Army and Navy did not require anywhere near the output capacity of the Amarillo helium plant. Although met with criticism by government officials who still believed that US government helium should remain a military asset, the tension became unbearable after the *Hindenburg* disaster at Lakehurst, New Jersey on May 6, 1937 (Bain, 1997). Had the Germans had access to helium, the ship might have been filled with the non-flammable gas, thus saving many more lives. Nevertheless, the incident provided the impetus to amend the Helium Act of March 3, 1927.

The Helium Act of September 1, 1937 was approved four months after the *Hindenburg* disaster, authorizing the sale of helium gas not needed by the government. The bill which led to this Act was altered extensively during its time in the House of Representatives and Senate but its passing would ultimately lead to an explosion in helium usage across many scientific and commercial industries. The passage of this Act would also allow non-hostile foreign governments to purchase helium for their own commercial use.

The first government to apply for nearly 18 million cubic feet of helium was that of Nazi Germany, which needed helium more than ever in order to reassure the German public that airship travel could be safe. Germany's helium requirements were designed to fill the commercial airship, the LZ-130, which was then under construction.

Permission to export this helium to Germany was granted in November 1937 and the German freighter, the *Dessau*, was sent to the port of Houston, Texas to collect the gas in mid-January 1938. The helium, however, would never make it onto the German vessel because of just one man in President Roosevelt's Cabinet, Harold LeClair Ickes (1874–1952), US Secretary of the Interior. Although President Roosevelt felt an obligation to agree the sale due to the previous authorization, Ickes, who was deeply suspicious of Germany, held firm and refused any helium to leave American shores. Ickes would meet aggressive opposition in his pursuit to prohibit the sale.

Despite numerous hard-fought attempts by members of the US government to carry out the sale, including a push by President Roosevelt who lacked the legal authority to bypass the Secretary's verdict, Ickes simply would not budge, saying, "if helium has no military importance, why is a Zeppelin authorized for our own Navy?" (Stein, 1963: 53). The allotment of 18 million cubic feet to Germany expired on November 1, 1938, and the *Dessau*, which had been docked in Houston since the beginning of the year, was sent back to Germany empty-handed in mid-December. Harold Ickes would later feel vindicated after Germany invaded Poland in September 1939.

During the helium controversy of 1938, Bureau of Mines operations continued to supply helium to US private parties, selling more than 800,000 cubic feet through 20 contracts. In order to promote the private use of helium, the Bureau sold to pre-paying customers at slightly higher than production cost. Slightly lower prices were given to medical and scientific users. By the end of the year, the Bureau once again controlled a complete monopoly of the helium industry when it acquired the Dexter, Kansas and Thatcher, Colorado properties from the Girdler Corporation on November 3, 1938; today a quarter of the world's helium is sourced here.

During the year 1938, most of the mere 117,000 cubic feet destined for public use was to be used in medical facilities for the treatment of asthma. The method of using helium mixed with oxygen for asthma (and other respiratory ailments) was developed by Dr Alvan L. Barach, who had achieved great success with its use.

World War II

During the US fiscal year 1939 (July 1, 1938 to June 30, 1939), the private sector demand for helium increased to over 1 million cubic feet, representing nearly 30 contracts, all of which were supplied by the Amarillo helium plant. Although the majority of this private helium went to the Goodyear Tire & Rubber Company for the inflation of blimps, over a quarter went to medical and scientific purposes. In this year, the federal government was supplied with over 5.2 million cubic feet of helium gas (Seibel and Kennedy, 1940: 1104).

Helium received accolades for its use in deep-sea diving when its importance was revealed during the rescue of the Navy submarine *Squalus* (SS-192) in late fall 1939 (Seibel and Kennedy, 1940: 1104). During a test dive on May 23, 1939, the *Squalus* suffered a valve failure, became partially filled with water and sank 240 feet to the ocean floor. Mounting a large rescue effort, the Navy replaced nitrogen with helium to allow for the extended diving required for the deep waters. Four divers would earn the Medal of Honor for the rescue of 32 Navy sailors and one civilian. Helium was the only gas that allowed these divers to stay at these depths for extended periods of time without the occurrence of nitrogen narcosis (the bends).

After Germany invaded Poland in September 1939 and before the US involvement in the war, US military requirements for helium finally began to see an increase surpassing the capacity of the Amarillo helium plant. Helium's importance to the national defense program was noted in mid-1941, leading to an appropriation for the expansion of the Amarillo helium plant and finding additional helium resources. The Navy (and other government agencies) had estimated that helium requirements might reach 50 million cubic feet per year, thus requiring a rapid expansion of current helium operations. The expansion of the Amarillo helium plant would provide a maximum of 36 million cubic feet of helium per year, far less than the estimated requirements. Another plant was needed to make up for this pending shortage.

It did not take long for the Bureau of Mines to identify the next government helium source. Congress had appropriated an additional US$1.25 million in 1941 for the construction of a new government plant to be located near the small town of Masterson, Texas (the Exell plant), which would process gas from the southwestern portion of the Panhandle field that contained, on average, 1 percent helium. The capacity of this plant, which was to be completed by 1943, would supply an additional 24 million cubic feet of helium per year. Throughout 1941, private demand for helium had exceeded 2.7 million cubic feet, and perhaps its greatest "new" use was in magnesium welding, which provided far superior welds for airplane manufacturing. The use of helium for welding provided an "inert atmosphere," prohibiting the possible contamination of atmospheric air within the weld.

Four months after the United States entered the war in December 1941, the government determined that their helium requirements would reach upwards of 130 million cubic feet per year, a 160 percent increase over the course of a year. By June of the following year, they increased their estimated requirements to 230 million cubic feet per year. The combined output of both the Amarillo and Exell plants would be only 60 million cubic feet per year, and the Exell plant would not even finish construction until 1943. Clearly, more helium resources and plants were needed.

A lucky break occurred in the early part of the war when the Rattlesnake field near Shiprock, New Mexico (Navajo Reservation) was discovered to contain an enormous amount of non-combustible, high-pressure gas containing over 7.5 percent helium. The highest concentration of helium ever found before this had only approached 2 percent. The Continental Oil Company drilled this well to 6,950 feet when the non-combustible gas was found and the decision was made to plug the well. After the gas analysis was received on July 1, 1942, a contract was made with the Bureau of Mines to take financial responsibility over the field and associated leases. Despite the fact that Continental had spent over US$100,000 in drilling the well, the Bureau of Mines obtained the well and ownership of 7,800 leased acres for only $1 (Mex, 1947). The Rattlesnake field was considered a tremendous discovery considering the enormous demand for wartime helium.

Less than ten months after breaking ground on the new plant site, the Exell plant began operations on March 13, 1943, barely in time to meet the growing military demand. The gas supplying this plant was gathered from the Channing area (Moore, Potter, Oldham and Hartley Counties, Texas) of the Panhandle field, which contained an average helium concentration of 1 percent. The month of July 1943 would see production from both plants reaching 12 million cubic feet, which was beyond their rated capacities of 8 million cubic feet per month. Although still not enough to meet military demand, appropriations for three more plants were authorized for Otis and Cunningham, Kansas and Shiprock, New Mexico to process helium-rich gas from these areas. Construction began immediately in early 1943.

The addition of these three additional plants provided a boost to United States helium production immediately after their successive launches. The first of the three plants to be put on production was the Otis, Kansas plant, which was brought online in October 1943. The Cunningham plant went on production in January 1944, and the Shiprock plant in March 1944. With each of these new plants on production, over 137 million cubic feet was produced in the fiscal year 1944 (July 1, 1944 to June 30, 1944), which amounted to a 220 percent increase since 1939.

When World War II came to an end in 1945, the US military had been furnished with enough helium to meet all wartime demand thanks to the rushed work of the US Bureau of Mines. Helium's role in World War II proved to be an extremely important resource. In its lighter-than-air properties, helium's use was best stated by Rear Admiral Charles E. Rosendahl, "In all, blimps escorted 89,000 surface craft in World War II without a single loss to enemy submarines. Of these, 50,000 were in areas where U-boats were known to be present at the time" (Rosendahl, 1938–41). The other uses for helium during the war, as previously mentioned, would far exceed its use as a lifting medium in the years to come.

Production explosion

By the end of the war, demand for helium had dropped precipitously, thus leading to the redundancy of all plants except the Exell plant, which alone continued to produce. After only 18 days of operation, beginning in March 1944, the Shiprock plant was placed on standby status pending further notice due to the slackened demand. The Cunningham, Kansas plant would be dismantled in July 1945 because the gas field supplying it was inadequate to justify further operation. The Otis plant in Kansas was placed on standby the following month.

As all plants had been operating toward the end of the war when helium demand was slackening, the Bureau of Mines began a helium conservation program in January 1945, injecting surplus helium into the Cliffside field. Immediately before the Cunningham and Otis plants had been removed from production, over 20 million cubic feet had been injected into Cliffside from January to June 30, 1945. Just as Richard B. Moore had envisioned years earlier, a one-of-a-kind underground storage facility was created.

Although only the Exell plant remained in operation throughout the end of the 1940s, helium continued to be conserved in the Cliffside structure. By the end of the calendar year 1947, over 69 million cubic feet had been injected underground. More helium was available to supply non-government users, to whom 16 million cubic feet was supplied in 1948.

A technological breakthrough occurred in 1949 with the usage of activated charcoal at liquid nitrogen temperatures to achieve 99.95 percent (Grade-A) pure helium, the highest grade of helium ever produced. Prior to 1949, only purities of 98.3 percent had been achieved. Nearly half of the Exell plant's production in 1949 was Grade-A helium. Helium of this purity was extremely effective in creating a pure atmosphere for electric arc welding, where minute impurities could damage the integrity of a weld.

Demand for helium began to increase during the close of 1949 and early 1950, but this was adequately met by the operation of the single Exell plant. By the middle of 1950, however, demand for helium soared nearly 50 percent, primarily because of the production of Grade-A helium and its desirability for welding. The supply shock prompted the reactivation of the Amarillo plant by August of 1950 to bridge the gap before the Otis, Kansas plant could be ready for reactivation. The Otis plant would begin operation again in March 1951, while the Shiprock (Rattlesnake field), New Mexico plant would remain in standby condition, despite government increases due to the Korean War conflict.

The Exell, Otis and Amarillo plants continued to operate until further escalation in demand required the reactivation of the Shiprock (Rattlesnake field) plant in 1953. With all four plants in operation in 1954, over 190 million cubic feet of helium was produced during the calendar year, while also beginning a withdrawal of some conservation helium from Cliffside field. During this year another important source of helium-bearing gas, containing nearly 2 percent helium, was noticed by the Bureau of Mines.

The Keyes field of Cimarron County, Oklahoma, which was mostly controlled by the Colorado Interstate Gas Company of Colorado Springs, contained an enormous supply of gas and a feasibility study was quickly authorized in August 1954 to determine its potential. The primary problem with the Keyes field was its high nitrogen content, which required removal if the natural gas was to be supplied to consumers. Therefore, a joint venture would be required to clean up two valuable products in the Keyes gas stream, methane and helium.

During the summer of 1955, the Rattlesnake field would ultimately fail due to the encroachment of water during gas production (Royston and Wheeler, 1956: 444), representing one of the largest disappointments in the United States helium program. In its initial discovery, Bureau of Mines engineers believed that the Rattlesnake field contained roughly 800 million cubic feet of helium. When it was placed on standby status in March 1944 after only 18 days of operation, the Navajo Reservation stopped receiving royalty payments from the production of helium on its land. After lengthy negotiations, the Navajo Reservation and the Department of the Interior reached a settlement whereby the government agreed to pre-pay a royalty based on total calculated reserves, so that the plant and wells could be placed in standby until needed. Thus the Rattlesnake field, which was placed on full production in 1953 until its end in mid-1955, produced a mere 42 million cubic feet of helium, far below the calculated reserve of 800 million cubic feet. The government helium plant at Shiprock would briefly begin operations again in July 1955 with the addition of a privately owned gas well in the nearby Hogback field (New Mexico).

Helium demand in 1955, although met by the four producing government helium plants, escalated dramatically. It was recognized by the Bureau of Mines that massive volumes of helium were being wasted through the production of gas that was not connected to any government plant. The demand for Grade-A helium soon reached a critical stage, resulting in the appropriation of US$6 million for the expansion of the Exell plant, increasing its capacity by another 150 million cubic feet of helium per year, which was expected to come on stream by April 1957.

Continued escalating demand for the next few years required Cliffside storage withdrawals in order to cushion the inadequate production. In 1957, 291 million cubic feet had been produced, while another 22 million cubic feet had been withdrawn from Cliffside leaving only 24 million cubic feet left in the storage field. Exell's expansion came online two months later than planned, on June 2, 1957. In addition, a new source of helium-bearing gas was found near Otis, Kansas, which helped to alleviate some of the supply issues. Despite the increased production, demand far outweighed supply, leaving the Bureau of Mines to institute an "informal" allocation system allowing "essential purpose" users first call on the product. A helium triage unit was thus established. By 1958, the Macy's Thanksgiving parade in New York had no helium to inflate their balloons, which had to be filled with air.

As negotiations between the Colorado Interstate Gas Company and the US government became tenable by both parties, a contract was signed in April 1958 for the exclusive rights to process helium out of the natural gas owned by the Company. An appropriation of US$18 million was delivered in August 1958 for the construction of a new plant to be located at Keyes, Oklahoma (Figure 2.3). The Fluor Company of Los Angeles was awarded the contract and construction on the US$12 million plant began by the end of the year.

One month after the contract with the Colorado Interstate Gas Company for the Keyes gas field was signed (April 1958), a conservation policy was proposed in order to manage the current supply issues that the Bureau of Mines was facing. An estimated 4 billion cubic feet of helium was being lost annually (Lipper, 1959: 474), especially from the giant Hugoton gas field, which spanned from the Texas panhandle all the way into southwestern Kansas. Although the helium content in this gas field was 1 percent or lower, the gas was marketable enough to burn, thus wasting precious helium, for example, every time a stove was turned on to boil water. The conservation policy was designed to encourage the participation of private industry to construct and manage new helium plants to process helium that would normally be wasted. By August, the conservation bill was introduced to Congress to amend the Helium Act of 1937. It was hoped that such an Act would conserve over 50 billion cubic feet of the precious commodity.

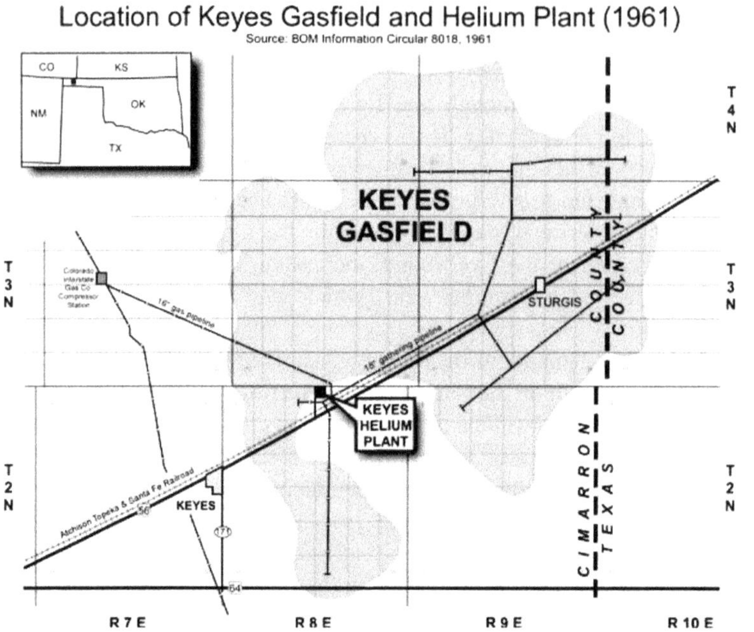

Figure 2.3 The start-up of the Keyes Helium Plant, in 1959, ended a severe helium shortage.

The severe helium shortage came to an abrupt end when the Keyes plant became operational in August 1959, only nine months after breaking ground. Supply became so abundant, in fact, that the Bureau was able to store 108 million cubic feet at Cliffside by the end of the 1959 fiscal year (June 30, 1960). In addition, the Shiprock, New Mexico plant resumed operations in July and a new large field at Pinta, Arizona was discovered to contain 8 percent helium (and 90 percent nitrogen) from a shallow formation (780–1,200 ft). The operator of this Pinta Dome field, Kerr-McGee Oil Industries, Inc., began construction on the first private plant since the Girdler plants were acquired in 1937. The additional Arizona plant and reserves, however, were not considered to have a sufficient impact on forecasted helium demands.

The Helium Act of 1960

The helium conservation bill, despite two years in Congress, was approved on September 12, 1960 by Congress and President Dwight D. Eisenhower and became effective on March 1, 1961. Private participation in the new conservation program was requested in October 1960, resulting in proposals from 14 different companies. The terms of the new Act would enable the government to purchase helium (under contract) from private crude helium plants and store it for future use in the government's Cliffside field. Further helium refinements to Grade-A purity would be processed in the government's Exell or Keyes plant.

The 1960 Act allowed the Department of the Interior to borrow (from the US Treasury) up to US$47.5 million per year to purchase helium from private sources based on 22-year take-or-pay contracts. Using a fixed price, the conservation program was designed to be fully self-sufficient and self-liquidating to cover all operating costs, helium cost and interest on the Treasury loans. It was estimated that the Bureau of Mines could stockpile 52 billion cubic of feet of helium by 1985, thus solving any long-term helium demand issues.

The plan, at first glance, was a very simple one. After the crude helium was purchased by the Bureau of Mines, it would enter Cliffside field for storage. Some would be processed by government helium plants to produce Grade-A helium. In order fully to service the Treasury debt, the Bureau's wholesale price of helium was raised from US$19 (for federal customers) to US$35 per thousand cubic feet. This price increase, established to pay off the helium debt in 25 years, essentially doubled the price any user would pay for helium. The Bureau would also maintain a near government monopoly on helium production because the contracted crude helium refiners on the pipeline could not legally sell any commercial helium to end users. However, any private plant not under contract with the Bureau of Mines could sell to any third party. The Bureau's primary (and flawed) assumptions with the passing of this Act were:

1 All US helium produced would be purchased and sold by the federal government.
2 Both federal and private demand would continue to escalate.

The first five plants were awarded to four different companies in mid-1961, with construction beginning immediately.

1 **Northern Helex Company** (a subsidiary of Northern Natural Gas Company) was awarded a contract on August 15, 1961, to be located in Bushton, Kansas.
2 **Cities Service Helex, Inc.** (a subsidiary of Cities Service Company) was awarded a contract on August 22, 1961, to be located in Ulysses, Kansas.
3 **National Helium Corporation** (a subsidiary of Panhandle Eastern Pipe Line Company and National Distillers & Chemical Corporation) was awarded a contract on September 13, 1961, to be located in Liberal, Kansas.
4 **Phillips Petroleum Company** was awarded contracts for two plants on November 13, 1961, to be located in Dumas, Texas and Sherman County, Texas.

In order to connect these private plants to the Cliffside field where crude helium would be stored, the Bureau of Mines contracted for a 450-mile pipeline worth US$8 million in September 1961, which was expected to become functional by July 1962. The plants along this pipeline would first extract helium from the natural gas stream, then return the helium-free natural gas to the companies to sell to their customers.

Northern Helex at Bushton, Kansas and Phillips at Sherman County, Texas were the first two conservation plants to come on stream in December 1962, delivering over 2 million cubic feet of crude helium to the government pipeline before the end of the year. The three other plants would come on stream in 1963, the largest being National Helium's Liberal, Kansas plant, which was designed to produce 5.8 million cubic feet of helium per day. The price that the Bureau of Mines paid for their crude helium product ranged from US$10.30 to US$11.78 per thousand cubic feet (Kinney, 1963: 54).

By November 1962, the private Kerr-McGee plant in Pinta Dome, Arizona came online, representing the only company who did not have any contract with the federal government over the sales of helium. The helium produced by Kerr-McGee would ultimately find its way to non-government users on the West Coast of the United States.

Another resource outside the United States, representing the first non-US helium resource to be revealed outside US borders, was a non-combustible gas found near the town of Swift Current, Saskatchewan. The discovery well (B. A. Wilhelm #1–9), drilled in October 1958, revealed helium content of 1.9 percent, with the residual being primarily nitrogen (96.6 percent). A small plant with annual capacity of 12 million cubic feet of gas was built by Canadian Helium Ltd (a company formed by British Oxygen Company and L'air Liquide), which came on stream in December 1963. It was believed that another Canadian discovery by Texaco in the Wood Mountain area of Saskatchewan province (1.38 percent helium), coupled with the Swift Current reserves, would satisfy all Canadian demand through the year 2000 (*Oilweek*, 1965: 14). The Swift Current plant would shut down permanently in 1977, after 14 years in operation. The Wood Mountain helium would never be produced.

The increased United States helium activity due to the 1960 Act resulted in enormous volumes of crude helium to be injected into Cliffside field. Although government demand was increasing rapidly, especially due to the nation's space race, the US$35 per thousand cubic feet charged by the Bureau of Mines was high enough to create private competition, which could produce and sell Grade-A helium at lower cost. By 1964, construction was under way on a new private plant in Otis, Kansas that would produce 1–2 percent helium gas from the Reichel field in Rush County, Kansas. The plant, owned by Kansas Refined Helium Company, was designed to produce 180 million cubic feet of helium per year. Later, in 1966, another private plant in Elkhart, Kansas was built and operated by Alamo Chemical Company and Gardner Cryogenics to process gas from the Greenwood field in southeastern Colorado/southwestern Kansas.

By 1967, three private plants (Navajo, Otis and Elkhart) capable of producing Grade-A helium (gaseous and liquid) were selling helium to commercial users at a significant discount to the Bureau's price of US$35 per thousand cubic feet. Under the terms of the 1960 Helium Act, all government agencies were obligated to purchase helium from the Bureau of Mines at the established US$35. Government demand during the late 1960s began to wane due to cutbacks related to the Vietnam War. In addition, the January 1967 fire in the *Apollo 1* at Cape Kennedy (later Cape Canaveral), which claimed the lives of three astronauts, sharply reduced NASA's helium requirements for the remainder of the year. Virtually all of the helium supplied to non-government users was purchased from the three private helium plants. The Swift Current, Saskatchewan plant was also enlarged to a capacity of 36 million cubic feet per year to meet Canadian demand (some of this gas was exported to Japan and other Asian countries (Thomason, 1967)).

All five of the government-owned helium plants were in operation at the beginning of 1968 but due to dwindling reserves and associated high costs, the government Otis, Kansas plant ceased operations in April 1968 and ownership of the Shiprock, New Mexico plant was transferred to the Navajo tribe in July 1968. During the year, four more private plants would come on stream making a total of 16 plants, producing both crude and Grade-A helium. Cities Service Cryogenics built a crude helium facility in Scott City, Kansas, which utilized a private pipeline to carry crude helium to the company's facility in Ulysses, Kansas, where it was further refined to Grade-A helium (Cities Service installed a purification and liquefaction facility in Ulysses during the year). The other three private plants, two in Arizona and one in Amarillo, Texas, were also in operation by the end of the year. With seven private plants in operation in 1968, the likelihood of any purchaser (besides government agencies) buying expensive US$35 helium from the Bureau of Mines was unlikely. The established US$35 price mandated by the 1960 Act did not take into account advances in technology that allowed private companies the opportunity to offer a better price.

By the end of 1969, it was readily apparent that helium sales were continually declining from the peaks that had occurred in 1966. Although exports had increased to roughly 60 million cubic feet, operations at four private plants would cease by the end of 1969. All three private Arizona plants and the new Linde Air Products Amarillo plant would either be abandoned or non-operating. The following year, the government Amarillo helium plant, which had been in operation since 1929, was closed on April 15, 1970. A portion of the facility was still used for loading smaller helium canisters for distribution.

Termination of contracts

The early 1970s would see a dramatic increase in United States helium exports to foreign countries. A 130 percent increase in exports occurred from 1970 to 1971. Helium demand in foreign countries would ultimately result in foreign helium extraction units for the first time since Canada's entry in 1963. A helium-bearing field in Poland was discovered (0.4 percent helium) and a contract to design a plant was signed by the end of 1971 (the Odalanow Poland plant came on stream in 1977). The same year another plant near Paris, France was producing 7 million cubic feet per year (Koelling, 1971).

Despite the aggressive foreign demand for helium, the United States, which was suffering from an economic slowdown in the early 1970s, would continue to see declines in domestic helium sales. In addition, a lowering of the borrowing allotments by the Treasury to the Department of the Interior resulted in delayed payments to the conservation plants (which would spark a lawsuit by Northern Helex in December 1970 for breach of contract). Without the large federal demand for helium, the Bureau of Mines helium program debt burden grew to US$210 million by the end of the fiscal year 1970. The forecasted revenues in 1961 were roughly US$120 million short in 1970 due to the presence of lower-cost private helium producers and lower government demand. Private producers were able to sell Grade-A helium to private end users for about US$21, which was US$14 lower (per thousand cubic feet) than the government sales price. It became clear that the Department of the Interior could not continue with the conservation program established in 1960.

On January 26, 1971, the Secretary of the Interior made a determination that the conservation program was no longer necessary to meet the objectives of the US helium program. All government helium requirements had been fulfilled by government plants, which meant that the primary purpose of the 1960 Act had been satisfied. The 1960 Act, which read, "a sustained supply of helium which, together with supplies available were expected to become available otherwise, will be *sufficient to provide for essential government activities*." A total of 28.5 billion cubic feet of crude helium had been stored in Cliffside since the conservation program began, thus allowing the interpretation of the Act being "sufficient to provide for essential government activities." The four conservation companies were notified that their contracts would terminate on March 28, 1971.

The termination of the conservation contracts would ignite a flurry of lawsuits, both by the four companies directly affected by the termination and by landowners who would no longer be receiving royalty payments for helium extracted. The damages estimated by the conservation companies were roughly US$375 million, for which they felt the government should be held responsible. Despite two years of legal wrangling, the Bureau of Mines discontinued accepting helium on November 12, 1973, thus officially terminating the conservation contracts. This termination would result in the unfortunate loss of 2.2 billion cubic feet of helium per year, either through venting or through waste in the burning of fuel gas. In order to prevent the wastage of precious helium gas, the Bureau of Mines in 1975 offered the use of Cliffside field, for a storage fee, for private storage of helium gas. Northern Helex would be the first to take advantage of this arrangement in September 1975, storing about 600 million cubic feet of helium per year.

From the mid-1970s until the end of the decade, the United States would no longer maintain the status of sole world exporter of helium. Although the helium content was quite low, new production was established near Paris, France and the new facility in Odolanow, Poland became operational in 1977. Another small plant in Orenburg, Russia would come online in the same year. As US exports continued to grow throughout the late 1960s and 1970s, economic extraction of low concentration helium became feasible due to the quantities of gas being processed. Production at the Swift Current, Saskatchewan plant would come to an end in 1977 due to reserve depletion.

In the United States, the Kerr-McGee Navajo, Arizona plant would be shut down in 1976 because of diminished reserves, but a helium resource was identified in the newly discovered Dineh-bi-Keyah field in Apache County, Arizona, which was estimated to contain 1.3 billion cubic feet of recoverable helium. In addition, a large (world's largest in 1979) helium purification plant developed by Union Carbide Corporation was brought on production in 1979 in Bushton, Kansas to process crude helium from the Northern Helex Company. While tremendous advances were being made by private helium producers, the helium debt incurred by the federal government by July 1, 1976 had reached a staggering US$412 million (and interest of US$24 million), to be serviced by a mere government net income of $1 million from helium sales.

Near the end of the decade, the most important American helium industry event to occur since the Hugoton, Kansas production was the discovery of the Tip Top field in Sublette County, Wyoming by Mobil Oil. The field had actually been discovered in 1962 but had been shut in due to the poor composition of the gas (66 percent CO_2, 21 percent methane, and 5 percent hydrogen sulfide (H_2S)). The helium content, a relatively high 0.6 percent, was (on average) higher than the giant Hugoton field and the estimated recoverable helium in this field was over 42 billion cubic feet of gas. As a result of this discovery, Exxon (later ExxonMobil), the developer of the field and manufacturer of the processing plant, would become the largest single domestic producer in the United States since production began in 1986; today a quarter of the world's helium is sourced here.

Mobil began drilling the Tip Top field in 1978 for methane recovery but extensive processing was required to make the gas marketable. First the hydrogen sulfide had to be removed, then CO_2 and finally nitrogen, before any methane sales could be monetized. At the time, there was no market in the area for CO_2, which is now used for enhanced oil recovery, meaning that all non-combustible gases were vented into the atmosphere during processing. Negotiations between Exxon and the federal government (the federal government was the owner of most of the Exxon leases in Wyoming) began in the early 1980s to pursue the extraction of helium.

The Minerals Leasing Act of 1920 required that all helium found on federal lands must be sold to the United States government at a stipulated price. This 1920 Act was written shortly after World War I when helium was still considered of great military importance. Section 181 of Chapter 3A, Subchapter 1 reads:

> The United States reserves the ownership of and the right to extract helium from all gas produced from lands leased or otherwise granted under the provisions of this chapter, under such rules and regulations as shall be prescribed by the Secretary of the Interior: Provided further, That in the extraction of helium from gas produced from such lands it shall be so extracted as to cause no substantial delay in the delivery of gas produced from the well to the purchaser thereof.

Therefore, no incentive existed to produce helium from federal lands unless the price paid for the gas was high enough to warrant extraction. After years of negotiations between Exxon and the Bureau of Mines, a contract was established and Exxon began construction of a helium extraction and purification plant at their Shute Creek facility in Lincoln County, Wyoming in the early 1980s. Helium production at Shute Creek began in October 1986 with the capacity to produce 1.5 million cubic feet of helium per day, since extended to 4 million.

By the end of the 1980s, government involvement in the helium industry was beginning to appear irrelevant and a push began to privatize the industry. Although many factors played a part in the termination of government involvement in helium, the fact that private usage far exceeded government usage must have played a large role. On January 28, 1986, when the Space Shuttle Challenger disaster occurred, NASA demands dropped precipitously, meaning even fewer government demands. Therefore by 1987, the Office of Management and Budget (OMB) established the notion that government should no longer have any role in helium production. The main asset in question was the stalwart Bureau of Mines' Exell plant, which had been in operation since 1943. In 1988, however, the House of Representatives Appropriations Committee prohibited the sale of the plant. A further due diligence study of the government's helium assets and operations would need to be addressed. By the late 1980s, the prices that the Bureau of Mines received for Grade-A helium and those received by private companies were approaching parity, around US$37 per thousand cubic feet.

Helium Privatization Act of 1996

At the beginning of the 1990s, it became clear that change was needed in the United States helium industry. Proposed legislation to amend the 1960 Helium Act allowing government agencies to purchase helium from private producers was recognized. Clearly, much debate was needed in order to tackle the privatization issue, not least to find a reasonable, non-competitive government agreement with energy giants like Exxon and the major industrial gas companies who were firmly established in the private helium industry. If the government were to reduce the price of government-sold helium in order to clear its helium inventory, then a likely public relations campaign against the government would ensue.

In 1995, a bill was introduced to Congress highlighting various points to be addressed (paraphrased from Peterson, 1994):

1. Discontinue Bureau of Mines helium sales and allow government purchases of privately produced helium.
2. Dispose of government helium assets.
3. Establish an orderly sales program of Cliffside reserves.
4. Maintain operations of Cliffside and associated 425-mile government pipeline.
5. Continue leasing of federal lands where helium-bearing reserves exist.
6. Eliminate the helium debt.

By the next year, President Clinton signed into law on October 9, 1996 the Helium Privatization Act of 1996. This law mandated that the US government terminate the production and sale of Grade-A helium by 1998, dispose of all its helium-related assets, and sell the helium reserve held in Cliffside field by 2015 (except for 600 million cubic feet ordered to be maintained by the government). In March 1998, the last federal helium plant, the Exell, was shut down as mandated by the Act and was later sold to a private buyer.

Approximately 30 billion cubic feet of helium stored in the Cliffside field was ordered to be sold on a straight-line basis no later than January 1, 2015, and in such a manner as to cause "minimum market disruption" (National Research Council, 2000). The price to be charged for the Cliffside helium was formulated in order to pay down the US helium debt completely by 2015. The formula derived for pricing this helium was simply the total helium reserves stored in Cliffside divided by remaining helium debt (and then adjusted for the Consumer Price Index, CPI). In 2000, this price was calculated to be US$43 per thousand cubic feet, while privately produced helium was then selling for US$32 per thousand cubic feet. Due to this price differential, it was believed that any helium sold from Cliffside would represent a supply of last resort. That is, if an industrial gas company could buy helium cheaper than from the US Government, it would.

Shortly after the passing of this Act, a feasibility study was released by the National Academies Press titled *The Impact of Selling the Federal Helium Reserve* highlighting current government role and any potential harmful consequences that might arise as a result of this Act. During the time of this study, the helium industry was relatively stable and it was not believed that Cliffside withdrawals would exceed estimated projections. In 1998, roughly 98 percent of all domestic helium demand was met by private industry, while the remaining 2 percent was supplied by the government. Of this 2 percent supplied by the government, most was allocated to the Department of Defense, Department of Energy, and NASA (Hayes, 1998: 32). At this time, it was widely believed that enough helium existed from private sources that withdrawals would not begin from Cliffside until much later than anticipated. All domestic demand and a portion of international demand was adequately met by US helium producers and the remaining foreign demand was supplied by plants operating in Poland, Russia and Algeria (the Algeria plant located in Arzew was brought online in the mid-1990s).

Almost immediately after the 2000 study was released, foreign demand began to increase in 2000 and 2001, by 38 percent and 16 percent respectively (Pacheco, 2000 and 2001), signaling a growing trend in international consumption (especially in Asia). The 2000 study had grossly miscalculated foreign demand, most likely due to inadequate data. This growing demand led to the establishment of a new helium facility to be constructed in Skikda, Algeria, and another in Ras Laffan, Qatar in the mid-2000s. Start-up problems with both of these facilities led to lengthy delays prior to any helium production, resulting in increased US export.

Throughout the 2000s and prior to the economic recession in 2008–9, helium demand continued to escalate, especially overseas, resulting in annual price increases from helium suppliers. Approximately one-half of US domestic demand and nearly one-third of global demand were supplied by the government's Cliffside field. As no new sources of helium had been found during the 2000s, industrial gas companies had to rely on this government supply to meet growing demand. The helium from Cliffside was no longer a resource of last resort but rather a primary resource.

Immediately prior to the economic recession of 2008–9, the helium supply issue was becoming critical. Some users were faced with supply disruptions, while others became too exposed to the ever-increasing price of helium. Many small, government-funded research projects were severely impacted by both supply and pricing issues, resulting in shutdowns in some instances. In nearly all cases the price of helium represented a major constituent in the government grant, thus limiting additional research.

The supply and pricing issues of the late 2000s, coupled with the enormous drawdown of the Cliffside field, prompted the US National Research Council to form the *Committee for Understanding the Impact of Selling the Helium Reserve*, which released its results in early 2010. The primary thrust of this committee was to see if the 1996 Act "has had any adverse effect on U.S. scientific, technical, biomedical, and national security users of helium." The committee unilaterally agreed that the 1996 Act was having a severe negative impact on US users of helium and was not in the best interests of the country or US taxpayers.

The 2010 committee outlined three concerns and established recommendations for each in order to help steer the US helium industry back to a state of normalcy. If no changes were made to the 1996 Act, it was likely that the US would become a net importer of helium for the first time in history. If the US were to rely on foreign helium, this would severely affect nearly all US users of helium in that prices would simply become out of reach for many domestic users.

The first concern of the 2010 study was Cliffside helium pricing. Currently, in 2012, the price for Cliffside helium (crude helium) is US$75.75 per million cubic feet. This price, however, is not calculated on any fundamental market-based activity. Rather, the price was established by the 1996 Act in a way that would eliminate the helium debt (debt created by the Helium Act of 1960) by 2015. The price was simply the remaining helium reserves left in the Cliffside field divided by the total helium debt and adjusted for the Consumer Price Index (CPI). This minimally established price, the 2010 study maintained, was not in the best interests of US taxpayers. If a market-driven price were established, the debt could be eliminated much sooner than 2015. Moreover, this artificially low price might prohibit future helium development (i.e. looking for new sources), which would further cause price and supply disruptions.

Secondly, the 2010 committee dealt with the Cliffside reserve itself and its drawdown. The 1996 Act ordered that all helium be sold on a straight-line basis so that the entire reserve (minus the minimum required by the US government) is completely sold. This straight-line method was deemed to be wasteful by the 2010 committee, in that it did not represent prudent reservoir maintenance. The committee believed that if the present reservoir management were to continue, it might pose a severe adverse effect on helium supply in the future.

Last, the committee specifically addressed smaller-scale (especially government grant) research facilities that had been severely affected by helium pricing and supply disruptions in the recent past. The committee suggested that these smaller research facilities be protected from US helium instability and should be guaranteed supply at economic prices. If Cliffside were able to reserve more helium than mandated by the 1996 Act, adequate helium resources would be ensured for smaller facilities that had been impacted the most.

The 2010 study was a starting point required to address the growing helium problem. No new US helium sources have been found in decades and the reserves left in Cliffside are not infinite. Clearly, the US cannot rely on Cliffside helium within the next decade without facing certain supply disruptions in the future.

Notes

1 The Dutch physicist, Heike Kamerlingh Onnes, was the first to liquefy helium in 1908 in his laboratory at the University of Leiden.

2 By the time the Calgary gas was processed for helium, the war had ended. During its time in Calgary, the plant produced nearly 60,000 cubic feet of gas ranging from 60 percent to 90 percent pure helium.

3 Linde Air Products is now known as Praxair and is not affiliated with the Linde Group (Munich, Germany) or Air Products (Allentown, PA, USA).

References

Althoff, W. F. (2004) *USS Los Angeles: The Navy's Venerable Airship and Aviation Technology*, Potomac Books, Washington, DC

American Chemical Society, "The Discovery of Helium in Natural Gas" http://portal.acs.org/portal/acs/corg/content?_nfpb=true&_pageLabel=PP_ARTICLEMAIN&node_id=928&content_id=CTP_006343&use_sec=true&sec_url_var=region1

Austerweil, G. (1914) *Die angewandte Chemie in der Luftfahrt*, Druck und Verlag von R. Oldenbourg, München und Berlin

Bain, A. (1997) "The *Hindenburg* Disaster: A Compelling Theory of Probable Cause and Effect", Proceedings of the National Hydrogen Association, Eighth Annual Hydrogen Meeting, Alexandria, VA, March 11–13: 125–8

Cady, H. P. and McFarland, D. F. (1903) "Helium in Kansas Natural Gas," *Transactions of the Kansas Academy of Science*, Vol. 20, December: 80

Cady, H. P. and McFarland, D. F. (1906) "Helium in Natural Gas," *Science*, New Series, Vol. 24(611), September 14: 344

Cady, H. P. and McFarland, D. F. (1907a) "The Occurrence of Helium in Natural Gas and the Composition of Natural Gas," *The Journal of the American Chemical Society*, Vol. 29(2), November: 1524

Cady, H. P. and McFarland, D. F. (1907b) "Neon and Argon in Natural Gas," *Transactions of the Kansas Academy of Science*, Vol. 21, December: 64–5

Dobbin, C. E. (1933) "Geology of Natural Gases Rich in Helium, Nitrogen, Carbon Dioxide, and Hydrogen Sulphide," US Geological Survey

Hayes, D. V. (1998) "Helium", *Minerals Yearbook 1998*, US Geological Survey

Industrial and Engineering Chemistry (1929) "Helium Plant Makes Initial Shipment," *Industrial and Engineering Chemistry*, Vol. 21(6): 524

Kinney, G. (1963) "World's Biggest Helium Plant Opens," *Oil and Gas Journal*, September 30

Koelling, G. W. (1971) "Helium", *Minerals Yearbook 1971*, US Geological Survey

Lipper, H. W. (1959) "Helium", *Minerals Yearbook 1959*, US Geological Survey, p. 474

Manning, V. H. (1919) "Petroleum Investigations and Production of Helium," Bureau of Mines Bulletin 178c, Government Printing Office, Washington, DC

McLennan, J. C. (1920) "Report on Some Sources of Helium in the British Empire," Bulletin No. 31, Canada Department of Mines

Mex, N. (1947) "Helium-Bearing Gas Lands in the Navajo Indian Reservation," Eightieth Congress, First Session, Document No. 212, US House of Representatives, 1947

Moore, R. B. (1926) "Commercial Production of Radium, Mesothorium, and Helium," *Industrial and Engineering Chemistry*, Vol. 18(2)

National Research Council (2000) *Impact of Selling the Federal Helium Reserve*, National Academies Press, Washington, DC

Oilweek (1965) "Canadian Helium to Lose Present Markets in Europe," *Oilweek*, November 1: 14

Pacheco, N. (2000) "Helium", *Minerals Yearbook 2000*, USBLM, Amarillo, TX

Pacheco, N. (2001) "Helium", *Minerals Yearbook 2001*, USBLM, Amarillo, TX

Peterson, J. B. (1994) "Helium", *Minerals Yearbook 1994*, USBLM, Amarillo, TX

Rosendahl, Captain C. E. (1938–41) Personal Files, James Doolittle Section, Eugene McDermott Library, University of Texas at Dallas

Royston M. K. and Wheeler, H. P. (1956) "Helium", *Minerals Yearbook 1956*, US Geological Survey

Seibel, C. W. (1938) "Production of Helium at Amarillo," *Industrial and Engineering Chemistry*, Vol. 30(8): 848

Seibel, C. W. (1968) *Helium, Child of the Sun*, University Press of Kansas, Lawrence, KS

Seibel, C. W. and Kennedy, H. S. (1934) "Helium", *Minerals Yearbook 1934*, US Geological Survey

Seibel, C. W. and Kennedy, H. S. (1937) "Helium", *Minerals Yearbook 1937*, US Geological Survey

Seibel, C. W. and Kennedy, H. S. (1940) "Helium", *Minerals Yearbook 1940*, US Geological Survey

Seibel, C. W. and Kennedy, H. S. (1956) "Helium", *Minerals Yearbook 1956*, US Geological Survey

Seibel, C. W. and Kennedy, H. S. (1959) "Helium", *Minerals Yearbook 1959*, US Geological Survey

Sherburne Rogers, G. (1920) *Helium-Bearing Natural Gas*, Government Printing Office, Washington, DC

Stein, H. (ed.) (1963) *American Civil-Military Decisions: A Book of Case Studies*, University of Alabama Press, Birmingham, AL

Thomason, E. M. (1967) "Helium", *Minerals Yearbook 1967*, US Geological Survey

US House of Representatives (1920) "Exportation of Helium Gas," Hearing before the Committee on Interstate and Foreign Commerce of the House of Representatives, Sixty-Sixth Congress, Second Session on HR 12376, May 21, Government Printing Office, Washington, DC

US House of Representatives (1922) "Helium Gas, Committee on Public Lands," Washington, DC, December 5

US House of Representatives (1924) "Conservation of Helium Gas," Hearings before Committee on Military Affairs, Sixty-Eighth Congress, First session, HR 5722

Westcott, H. P. (1920) *Storage of Natural Gas in the Gas Sand, Handbook of Natural Gas*, 3rd edn, Metric Metal Works, Erie, PA

3 The US federal helium reserve

Joseph B. Peterson

Introduction

The US federal helium conservation program was initiated in 1960 after enactment of the 1960 Helium Act Amendments. The program was administered by the US Department of Interior's Bureau of Mines until 1998. In 1998, administration of the program was transferred to the Bureau of Land Management (BLM). The purpose of the program was to provide an adequate supply of helium to meet growing demand from US federal government helium users and was spurred by increasing applications of helium in the aerospace and medical industries. The Act enabled the federal government to construct a 450-mile pipeline, upgrade helium extraction facilities, and purchase helium from private companies located in the Panhandle-Hugoton gas fields of Kansas, Oklahoma and Texas. By 1975, nearly 40 billion cubic feet of helium was placed in storage at Cliffside field located near Amarillo, Texas (Clarke, 1976).

US helium conservation program

By the end of World War II, the US federal government had identified several strategic uses of helium and developed a research program to increase domestic supplies. Figure 3.1 describes early results of the program to identify and quantify helium resources in the US. Engineers used reservoir characteristics to determine the most economically efficient strategy for helium recovery. Fields that contained 0.3 percent helium or greater in a gas stream were considered economic for helium recovery. As shown in Figure 3.1, several fields were identified in the US that met the criteria. The greatest activity was in northern Texas, Kansas and Oklahoma. Other notable areas of helium recovery activity, dominated by private sector players, include western Wyoming, Utah, Colorado, New Mexico and Arizona.

Figure 3.2 depicts the results of the conservation program. It shows the volume of helium placed in storage from 1960–72. During the program, private industry built five helium extraction plants that were connected to the helium conservation pipeline. These plants provided the helium that was sold to the federal government and was stored in Cliffside field. In 1972, Congress determined that an adequate amount of helium had been stored to meet future US requirements and purchases of helium were stopped. In the mid 1970s, the concept of the "flywheel" approach to managing helium developed. Helium is a co-product of natural gas production, but has opposite peak demand periods. Natural gas peak demand occurs in winter, as gas is needed to heat homes, run factories and generate electricity; however, peak demand for the helium industry occurs during the summer. Therefore, in the early 1970s, commercial storage of private helium in Cliffside field began. Helium was stored during the winter (low demand) and removed from storage and utilized during the summer (high demand), giving rise to the "flywheel" concept. Finally, Figure 3.2 also shows helium redelivered/removed from the reserve in accordance with the 1996 Helium Privatization Act, which required the federal government to cease from participating in the helium refining business and to offer for sale, to the public, helium from the reserve. Since 2003, the amount of helium in the reserve has decreased, due in part to open market sales and increased demand for helium sourced from outside the US.

USBLM facilities

At the heart of the USBLM facilities lies the Bush Dome storage reservoir in Cliffside field. Bush Dome is a natural geological formation (dolomite) with a dome of impermeable rock (anhydrite) overlaying the dolomite and bounded on two sides (north and east) by an intersection with the water table and a pinch point on the other side. This reservoir and a 450-mile pipeline from Amarillo to Bushton, Kansas form the primary means by which the United States is able to store and redeliver helium to its customers along the pipeline from Kansas, Oklahoma and Texas.

In 2004 the BLM began operation of the Crude Helium Enrichment Unit (CHEU), which was designed to help manage helium in Cliffside field. The CHEU upgrades helium from the field by removing hydrocarbons from the gas stream and enriching helium for delivery to the crude helium conservation pipeline. The CHEU can process 21.4 MMcf/d of gas, with an average helium concentration of 29.4 percent. The products are: a stream of roughly 80 percent helium, that is sent directly to the helium conservation pipeline; natural gas that is feed into a natural gas pipeline; and natural gas liquids, which are sold by contract.

US helium resources and production

The BLM latest report on helium resources in the United States indicates that as much as 581 billion cubic feet of helium may be available for recovery, as shown in Table 3.3. Compared to the 2002 report this represents a 24 percent increase in US helium resources and is due to more areas of gas exploration being developed over the past five years. Table 3.2 shows annual sales and exports of grade-A helium in the United States. Exports have increased steadily from 2004–8, reflecting a continuing demand for helium worldwide; while domestic (US) demand for helium decreased during the same period by 23 percent.

This level of export particularly reflects strong growth in microelectronics and fibre optics in East Asia and growing demand from Europe in several sectors. The situation has also internationalized on the supply side, with six non-US private helium refiners operating in 2009. The US industry, however, remains the largest and strongest supplier nation in the world with ten private refining plants and ten private production facilities, in addition to the USBLM facilities at Cliffside. Some key data relating to the helium industry are presented in the Appendix to this chapter.

Global helium plants

Even though the US currently supplies more than 74 percent of the world's helium needs, the global helium market continues to see increased demand and other countries have stepped up to meet this demand. Over the past three years non-US helium production has averaged 1.5 Bcf/year. Future plans include plant capacity increases that will boast production over the 3.5 Bcf levels by 2015. As helium prices increase, other sources of helium throughout the world become more likely candidates for development. The list below provides a global overview of where helium plants are located:

- USA – 17 private helium plants on BLM helium pipeline;
- USA – five private helium plants independent of helium pipeline;
- Algeria – two private helium plants (Arzew and Skikda);
- China – one private helium plant (Sichuan);
- Russia – one helium plant (Orenburg);
- Poland – one private helium plant (Odolanow);
- Qatar – one private helium plant.

In summary, global production is roughly 25 percent non-US originated and US production is holding roughly constant. The trend is that the proportion of non-US sourced helium in the global industry is growing.

National Academy of Sciences study

The BLM requested that the National Academy of Sciences update the 2000 report, "Impact of Selling the Helium Reserves", which looked into the potential impacts of the 1996 Helium Privatization Act on industrial and scientific communities. The new report, "Selling the Nation's Helium Reserve", was completed in April of 2010 and provided several recommendations (NAS, 2010). Some specific recommendations for immediate improvements to the federal helium program include: changing the methods for pricing the helium sold from the reserve; committing additional resources to managing the physical facilities at the reserve; and providing assistance to small-scale scientists/users of helium. The Academy also included more general recommendations for how best to meet the nation's current and future helium needs. They included: (1) collecting and making available the information needed to manage the helium reserve and to formulate future helium policy; and (2) initiate strategies to develop a comprehensive long-term program for meeting the nation's helium needs. The BLM has taken these recommendations under advisement and has begun to implement them where possible. For example, the BLM has increased the open market price for crude helium and is developing a strategy to price helium more effectively, more in line with the market price. The BLM is making more information available to the general public concerning helium resources in the United States. It is also responding to requests from the scientific community concerning taking advantage of the in-kind helium regulations, where users obtain access to federal helium through federal agency contractors (as opposed to commercially available helium). For more information, see the NAS report (NAS, 2010).

Concluding observations

While the Cliffside facility has been responsible for supplying 35 percent of helium demand in recent years, there is no reason to infer that a crisis looms for the moment when the resource ceases to be available. In 2007 there was a sense that the BLM might be depleted as a source of helium for the global market by 2015, but during late 2008 and early 2009, because of the economic downturn, it started to become increasingly clear that the reality would be rather different, and less problematic. The recent NAS report (2010) further illustrates why any fear of a helium-related crisis in 2015 is misplaced. However, concern from the scientific community over losing the federal helium reserve as a long-term source of helium raises additional questions. In early 2009, private industry did not purchase all of the helium available from the reserve due to the economic downturn. The technical life of the BLM facility is longer than the dates implied by a 2015 rampdown. Depending on the Actions of Congress, it is possible to foresee an operational facility for eight to ten, or more, years ahead.

Currently, a supply crunch gas emerged since 2011 that may last well into 2013, even though new helium production capacity is coming to market. More production capacity is needed – but this is effectively independent of the policy for the BLM in 2015.

Consistent with my sense of optimism I should report that the sense of an impending and enduring crisis post-2015 is unwarranted. Beneficial investment in helium infrastructure over the past three years and expected investment over the next three years should compensate for any constraints arising from the BLM system. Within the US it is important to stress that new private capacity is coming into play. New reserves are being accessed in the Four Corners area and in southeastern Wyoming. Outside of the US, similar developments are occurring through expansion of existing facilities.

While there is surely ample helium to address core demands, it may become necessary for policymakers to address issues concerning access to helium, pricing, availability to meet the needs of the scientific research communities and other topics addressed in the recent NAS report (NAS, 2010).

Appendix

Key trends in helium (data from the US Bureau of Land Management):

Table 3.1 US Government sales of Grade-A helium, Bcf

	Domestic	*Export*	*Total*
2004	3.10	1.57	4.67
2005	3.11	1.83	4.94
2006	2.93	2.07	5.00
2007	2.64	2.32	4.96
2008	2.37	2.40	4.77
2009	1.69	2.63	4.33
2010	1.89	2.66	4.56
2011	1.83	2.96	4.79

Table 3.2 US-identified helium resources in Bcf @ 14.65 psia and 60 °F, as of January 2007

	2002	*2006*
Probable	117	189
Possible	149	210
Speculative	103	182

Table 3.3 Global helium resources in Bcf, estimated as of January 2007

United States	581
Qatar	364
Algeria	303
Russia	241
Canada	72
Poland, China and others	152

Table 3.4 Global helium production, Bcf

	US	*Rest of the world*	*Total*
2004	4.67	0.87	5.54
2005	4.94	0.95	5.89
2006	5.00	1.02	6.02
2007	4.96	1.26	6.22
2008	4.77	1.60	6.37
2009	4.33	1.44	5.77
2010	4.56	1.51	6.07
2011	4.79	1.43	6.22

Table 3.5 Foreign (non-US) helium demand, Bcf

Year	*US exports*	*Foreign*	*Total*
2004	1.57	0.87	2.44
2005	1.83	0.95	2.78
2006	2.07	1.02	3.09
2007	2.32	1.26	3.58
2008	2.40	1.60	4.00
2009	2.63	1.44	4.07
2010	2.64	1.52	4.16

Table 3.6 Global helium production, Bcf (expected versus actual)

Year	*Expected*	*Actual*
2004	5.54	5.54
2005	5.89	5.89
2006	6.20	6.02
2007	6.43	6.22
2008	6.45	6.38
2009	6.00	5.77
2010	6.20	6.07
2011	6.3	6.22

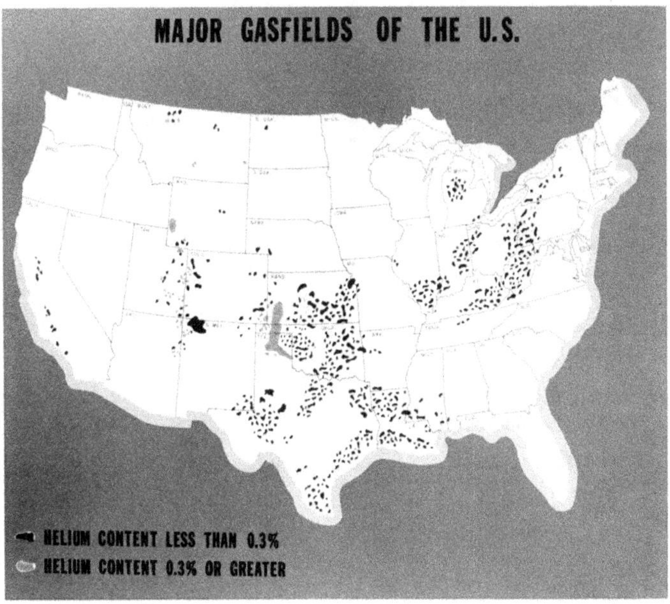

Figure 3.1 Distribution of helium gas reserves in the continental United States (source: US Bureau of Land Management, 1998).

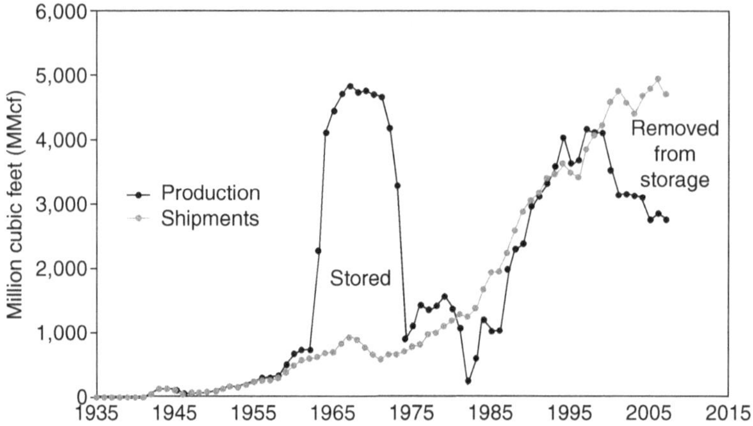

Figure 3.2 Historical production and sales of helium from the US federal helium reserve. Production is shown as dashed line and sales are shown as solid line (source: US Bureau of Land Management, 2009; compiled by Steven J. Maione).

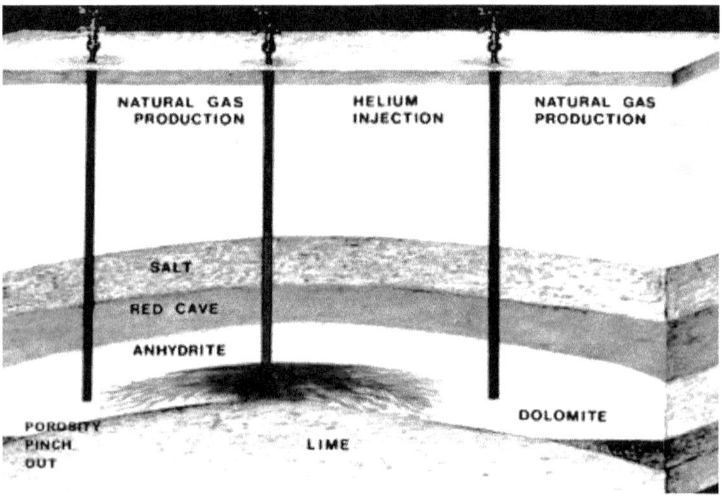

Figure 3.3 Bush Dome Helium Reservoir in the Cliffside field, north Texas, USA (source: US Department of the Interior, 2009).

References

Clarke, T.G. (1976) "Helium", *Minerals Yearbook 1976*, http://minerals.usgs.gov/minerals/pubs/usbmmyb.html (accessed March 5, 2012)

National Academy of Sciences (NAS) (2010) *Selling the Nation's Helium Reserve*, National Academies Press, Washington, DC, www.nap.edu/catalog/12844.html (accessed December 22, 2010)

4 Helium in Algeria

Pioneering work for the first large helium plant sourced by tail gases from an LNG facility

Benjamin Reinoehl

Prologue

This case study illustrates the dynamics involved in building relationships between existing global helium industries seeking new sources of supply from within the national oil and gas industries of Middle Eastern nations. The saga of helium in Algeria is as much about crossing business practices and cultures as it is about the technical aspects of extracting helium from liquefied natural gas (LNG) that, as this first experience shows, were challenging enough in themselves.

The initial Air Products approach to Sonatrach, the Algerian state-owned oil and gas company, to recover and refine helium from the tail gases of Sonatrach's proposed LNG projects, was defined after a visit in February 1977 to the Institute of Gas Technology (IGT) in Chicago. An IGT subsidiary, Gas Developments Corporation (GDC), had prepared an eight-volume confidential study for Sonatrach, reporting on all aspects of helium. Since Air Products had been working with IGT on LNG studies, it was aware of the Sonatrach study but did not have access to the confidential document.

In December 1976, Air Products' President received a letter from the Manager of the Gas Department at Sonatrach, suggesting a meeting to discuss a potential helium project in Arzew, Algeria. Air Products' designated helium team comprised its Manager – Industrial Products (J.B. Reinoehl), and its European Sales Manager for Helium. The team responded to Sonatrach by letter and collateral printed materials in February 1977. Sonatrach reacted positively to the information, and a meeting in March 1977 was scheduled in Algiers. At this time, Sonatrach had already been operating the first LNG plant in the world for over a decade. Sonatrach's three-train La Camel LNG Plant (GL4Z) at Arzew started up in 1964 and was built by the consortium Technip and Pritchard International Corporation. Air Products' expertise in helium was well known, given that it was a major technical and equipment contributor to the US Bureau of Land Management's (USBLM) 1961 helium conservation program authorized by the US Congress.

At the Algiers meeting, Sonatrach stated that it was evaluating the feasibility of helium extraction from its natural gas plants, but had not made any decisions regarding marketing arrangements, and questioned Air Products' representatives regarding worldwide helium consumption and supply patterns. The approach was fairly pessimistic toward helium, most likely due to the GDC study and Sonatrach's estimate that current world production capacity exceeded demand at that time. Sonatrach believed that the supply situation would become even less attractive by installing more capacity at Arzew when an under-construction Polish helium plant at Odolanow was placed on stream. The Polish plant, built by Petrocarbon and having a design capacity of 150 million cubic feet (MMcf) per annum, was to come on stream in 1977. Sonatrach's options were completely open at this date with regard to the type of participation it would require for a helium project, though it seemed certain that a commitment for liquid helium production at a defined price from Air Products or other industrial gas companies would be necessary in order to proceed with a project. Sonatrach would expect a "take or pay" arrangement to assure a return on its investment, and some degree of capital participation might also be desired. Air Products stated that the crude helium production from the 1961 US Helium Conservation Program plants was expected to decline through late 1980s and 1990s, and a new Algerian supply would be a very promising project. The European helium market at the time was supplied by the United States. New US helium applications were forecasted, which could consume and possibly exceed the current US helium capacity. These new applications were fiber optics manufacturing, atmospheres for chip manufacturing for electronic devices, magnetic resonance imaging (MRI), and, potentially, diving gases for North Sea oil and gas activity. Price discussions were next on the meeting agenda, and with Sonatrach's knowledge from the GDC study, Air Products stated that its pricing analysis seemed reasonable.

Air Products recommended that Sonatrach consider recovery of helium from the tail gases of the LNG-2 (GL2Z) project then being negotiated. Air Products provided copies of material balances for the La Camel plant (GL4Z), LNG-1 (GL1Z) facilities and the LNG-2 project to Sonatrach's Head of LNG Services, who was also in attendance at the meeting. He pointed out that the sale of liquid helium to Air Products and purchase of the LNG-2 facility helium plant were to be separate negotiations. Air Products further requested Arzew electricity values to help later cost evaluations, and also agreed to produce a forecast of worldwide supply and demand for the 1980s and 1990s. At the close of the meeting, Sonatrach suggested the parties meet again in a month in Washington, DC.

Air Products provided Sonatrach with a helium review prior to the April meeting detailing the worldwide helium market and sources, and an outline draft for a helium supply agreement. Air Products also included information on a preliminary investment for a helium purifier and liquefier, plus energy horsepower required to produce 150 MMcf per annum of liquid helium at Arzew.

The Air Products helium team met with Sonatrach in Washington, DC in May 1977 to proceed with discussions on the helium project. Sonatrach reported that it intended to proceed with the helium recovery project at the LNG-2 (GL2Z) facility and requested a bid from Air Products for a purifier/liquefier of

150 MMcf/annum (approximately 700 liters per hour). Sonatrach management preferred multiple customers, rather than a single customer, for liquid helium production. Air Products' earlier March letter clearly stated its position on this matter, which was "the desire to purchase all the Sonatrach liquid helium." Sonatrach stated that "the single company concept is not attractive to Sonatrach," but left the door open by asking Air Products to prepare a capability proposal to convince Sonatrach management that Air Products should be permitted to purchase and market the entire liquid helium production. If Sonatrach were to consider a single partner, the criteria would be: (1) overall technical capability; (2) willingness to be flexible; (3) growth situation; (4) success should benefit both Sonatrach and partner; and (5) partner's marketing ability.

Sonatrach felt that it could be competitive in the US market since it owned large LNG ocean tankers that could piggyback liquid containers to the US coast without charge. Air Products felt that while the idea had merit, LNG shipments would not always coincide with helium customer demand and the storage of liquid helium is very expensive. Sonatrach also had concern about the reliability of the US helium market due to the possibility of US government intervention. Air Products believed that US government intervention was possible, but planned to market the majority of the Sonatrach production in Europe. The Washington, DC meeting concluded with the understanding that Sonatrach would set aside time for this topic at the coming August 1977 LNG Conference #5 in Düsseldorf, Germany, and requested additional copies of Air Products' helium capabilities and a more definitive draft of a liquid helium supply contract, further stating that it planned to make a decision on its helium partnering during the calendar year 1977.

The next serious discussion about the helium project was at the 1977 LNG Conference #5 in Düsseldorf, where the Manager of Sonatrach's Gas Department was the keynote speaker. Sonatrach and Air Products met and discussed helium recovery, purification and liquefaction from LNG tail gases at Arzew, Algeria and Air Products' capabilities. Sonatrach now seemed favorably inclined to the idea of aligning with only one major industrial gas company. Air Products informed Sonatrach that M. W. Kellogg and Air Products planned to bid jointly for the total helium plant, with Air Products providing the major hardware and operating support while Kellogg would provide, and be responsible for, project management and plant installation. Later, Kellogg and Air Products did work together as a team for the LNG-2 project after it was awarded, but there was not a helium project as part of the award. At the conclusion of the Conference, Sonatrach requested Air Products to prepare a document discussing the worldwide helium status and visit its headquarters in Algiers for further discussions. No date was set for a meeting.

Liquefied natural gas

Air Products' first LNG success occurred around 1968–70 with Esso (now Exxon) at an LNG project in Marsa el Brega, Libya, utilizing coiled-wound

Figure 4.1 MCR™ Cryogenic Heat Exchanger for LNG-1, Arzew, Algeria, being loaded onto an ocean-going vessel at the Port of New York, USA.

exchangers in Air Products' multicomponent refrigerant process systems (MCR). Air Products developed the heat exchanger technology and manufacturing expertise from helium recovery equipment for plants supplying crude helium to the US government's 1961 Helium Conservation Program. Sonatrach purchased 12 C3MR™ exchangers for the LNG trains from Air Products for Arzew. Six trains were for its LNG-1 (GL1Z) facility in 1977, and six more in 1981 were for its LNG-2 (GL2Z) facility. For the trains at the GL2Z facility, Air Products installed a flash drum and core on the main exchanger of each of the trains for helium recovery, even though a helium project had not yet been agreed upon. These small two units, attached to the main exchanger, would flash and separate the helium with some nitrogen from the tail gases of each train making helium recovery more efficient.

All LNG trains were later revamped to upgrade their capacity, LNG-2 in 1996 by Air Products and M. W. Kellogg, and LNG-1 in 1997 by Air Products and Bechtel. Tractebel, a Belgium firm, was hired by Sonatrach to supervise the engineering work done.

Continuing negotiations for a helium project at Arzew, Algeria

Air Products and Gardner Cryogenics received letters from Sonatrach in June 1980 under the signature of its Director of Gas Exports, inquiring about each company's interest in a helium project at Arzew. (Air Products had purchased Gardner Cryogenics in April 1980 from US companies Carpenter Technology and Phillips Petroleum. Air Products' Reinoehl led the purchasing team.) Air Products replied that Gardner was now part of Air Products and the July response expressed both continued interest in a helium project at LNG-2 at Arzew and discussed the previous meetings in 1977. Air Products also offered to meet with Sonatrach in Washington. Sonatrach responded in September 1980 with an invitation for Air Products to come to Algiers, but Air Products' work schedules delayed a meeting with Sonatrach until early 1981. Air Products did offer a meeting in Paris with its new European Helium Manager.

Air Products' next effort to move the helium project forward was initiated with a telex to the Secretary General of the Ministry of Energy and Petrochemical Industries of Algeria. The contact telex referenced an article by R. Hammoutene (1983) and mentioned Air Products' interest in a helium project at Arzew and the previous meetings and discussions with Sonatrach management. It suggested a meeting with Air Products' Vice President Marketing, Europe (who spoke fluent French) and Reinoehl as the Air Products helium team.

Sonatrach's *Directeur de la Transformation des Hydrocarbures* offered to meet in Algiers in November 1983. The Assistant to the Director General and the *Chef de Projet* were the key people in a total attendance of ten Sonatrach personnel at the meeting in Algiers. Air Products' presentation started the meeting with a video film entitled "Helium – We Do It All," plus a number of 35 mm slides. The remainder of the meeting of over two hours was conducted in French and English and covered technical and commercial subjects. From those discussions at the meeting, it was clear to Air Products that Sonatrach planned to include in its five-year plan the Arzew helium project that was being developed for 1984. Sonatrach stated that it planned to get its government's go-ahead authority for a helium project with an on-stream date of 1986. Sonatrach intended to invite separately Air Products, L'Air Liquide and an unknown third party to the negotiations. The Hassi R'Mel field was the major source of native gas supply for Sonatrach's LNG facilities. Air Products agreed to provide a helium project outline and offered Sonatrach a visit to Air Products' new 350 MMcf per year helium plant, on steam in mid-1982, and located at the top of the Texas Panhandle next to the Sherhan native gas processing plant of Phillips Petroleum. Sonatrach was still considering a 150 MMcf per year plant, which now seemed small because the worldwide helium market was by this time about 1.4 Bcf per year, and had grown at a rate of 8.45 percent per year over the prior ten years. Before the intended visit to Air Products, Sonatrach was asked to provide energy rates, total annual cost per employee, and the approximate cost per trip to move liquid helium in cryogenic ISO containers from Arzew to the docks for shipment to Europe. Sonatrach personnel agreed to visit Air Products' headquarters in December 1983, but the trip was delayed until February 1984. Ahead of the meeting, Air Products prepared a Sonatrach/Air Products Helium Project outline covering all aspects of the project:

1 helium facility (description/capacity/total investment);
2 operating agreement draft (technical support/maintenance back-up/operator training);
3 liquid helium supply agreement draft (initial prices/a high percentage take or pay);
4 finance;
5 tentative project schedule.

The Sonatrach contingent arrived at JFK Airport on February 19, 1984 for three days of meetings. Day one meetings were held at Air Products' headquarters in Allentown, Pennsylvania and included reviews of helium processing, operations and distribution. Later the group toured Gardner Cryogenics' manufacturing shop for liquid helium cryogenic ISO containers. Then the group traveled to Amarillo, Texas for dinner with the General Manager of the US Bureau of Mines (BoM) Amarillo, Texas Helium Field Office. On day two, Sonatrach and Air Products personnel visited with the BoM at its offices in Amarillo for a helium presentation and later a tour of the US government's nearby Cliffside field, crude helium storage and pipeline operations. The BoM visit was followed by a tour of Air Products' Panhandle helium plant with a later tour and presentation at its Liberal, Kansas distribution terminal. The group returned to Allentown for the day three meeting and discussed the economics of a helium project at Arzew, an operating support agreement, financing the helium project and a liquid helium supply agreement outline. The meeting concluded with a summary of the three-day visit. Sonatrach's representatives were very pleased with the visit and were convinced that Air Products could provide the best helium plant and technical support were Sonatrach to enter into the helium business, stating that there were still two Sonatrach internal hurdles: (1) including the helium project in the Sonatrach five-year plan; and (2) the Algerian government must agree with the selection of an American company. Sonatrach said the latter hurdle was somewhat outside of its control since there existed very hard feelings within the Algerian government over the previous Trunkline LNG take-or-pay situation. That situation had been created by the US government's setting of maximum prices for LNG delivered into the United States, but the specific LNG contract between Sonatrach and Trunkline LNG had negotiated prices higher than maximum price allowed by the US government. Ultimately, the dispute was resolved in the World Court in favor of Sonatrach. Sonatrach believed the best plan was to have Air Products sell Sonatrach the helium plant, provide technical support and sign at least a 15-year product supply agreement. The agreement would be used to finance the investment via the American banks. The Sonatrach team left Allentown and proposed to meet again in Algeria in April 1984.

In April 1984, as requested, Air Products sent Sonatrach copies of the charts used by the BoM at its presentation for Sonatrach the previous February and suggested agreement on a date for the next meeting. However, prior to any meeting related to the helium contracts, Air Products requested approval for a separate Air Products' project and operations personnel visit to Arzew in May 1984. Approval was granted and an Air Products team (Production Manager, European Operations Director and a Project Manager) made the visit.

Also in May, Air Products personnel met with Sonatrach in Algiers. The meeting covered a liquid helium purchase agreement and the major points for the purchase and installation of a liquid helium plant. Air Products proposed that: (1) Sonatrach buy the helium plant and Air Products sign a long-term purchase agreement with an 80–85 percent take-or-pay provision; or (2) the parties could enter into a long-term joint venture arrangement. Sonatrach was in favor of a joint venture limited to helium only provided it included a strongly worded take or pay clause and was in accordance with a new 1982 Algerian law allowing joint ventures. Sonatrach was to make its final decision when it had completed an economic comparison of both alternatives. Surprisingly, Sonatrach asked that Air Products agree to a 32-month delay in the helium project until the completion of LNG-2 – plus all other LNG-related work and the new harbor at Arzew. Air Products found this statement unexpected because it had never before been mentioned. Air Products stated that if construction traffic would complicate the progress of the helium project then it considered the delay reasonable. Also, discussion of a potential problem for movements of liquid helium containers to and from the helium plant and markets was a Sonatrach concern. Air Products recommended adequate on-site liquid helium storage of over three days of production to handle the potential problem. Sonatrach was responsible for: (1) project definition; (2) basic cost data; (3) defining capacity; (4) interface with existing facilities; and (5) site selection. Sonatrach was now seriously considering a 300 MMcf per year plant. The remainder of the meeting involved a review of the draft helium supply contract. Sonatrach suggested that helium price changes be related to price changes in the European retail market, but was concerned that the new Polish helium would lower prices. In closing, Air Products requested that Sonatrach select a partner because engineering and the final investment work was costly.

Air Products' estimating and project personnel met with Sonatrach's team in June 1984 to help Sonatrach prepare a full cost estimate for the helium project. In July, Reinoehl, as requested, mailed Sonatrach payback calculations for the helium project with and without liquid nitrogen for helium container shields and the local nitrogen market. The final helium project investment and operating costs for a joint venture were developed by Air Products, and the final actual cost for the helium project was agreeable to the parties.

In August 1984, Air Products' personnel from Europe visited Algiers to discuss the helium project with Sonatrach. Sonatrach was pleased with project work prepared by Air Products and stated that the project was approved by Sonatrach and was to be presented to the Algerian Minister in September for inclusion in the 1985 budget. Commercial discussions included the larger plant size, project cost details, liquid helium price, and a new financing plan.

By August 1985 the parties decided to prepare and sign a Protocol covering the major points of four future agreements: finance, construction, operating, and purchase of liquid helium. The Protocol allowed the parties to have a legal document for presentation to the Boards of Sonatrach and Air Products for approval, and thereby the ability to spend money to define the investment and profitability of the helium project. The liquid helium plant size changed to two cases (approximately 200 MMcf per year and 300 MMcf per year) and 50 tonnes per day of liquid nitrogen was added to the plant for the local market. The tail gases of LNG-2 (GL2Z), when all six trains were operating at capacity, were estimated to contain about 900 MMcf per year of helium and the remainder was nitrogen with some hydrocarbons. Air Products informed Sonatrach that the worldwide helium market had grown at a CAGR (compound annual growth rate) of over 7 percent since 1981 and Air Products was considering a 300 MMcf per year helium plant at Liberal, Kansas utilizing a crude helium supply from the Panhandle eastern conservation program plant. The US Congress had cancelled purchases of crude helium production available from all five of the US Helium Conservation Program plants.

Figure 4.2 Pipe Bridge module with air coolers on top in route to the Helios plant site.

Due to worldwide helium demand Air Products expanded its Texas plant in 1986 and also committed to install a 300 MMcf per year helium plant at Liberal, Kansas. The Liberal plant came on stream in 1991. Thinking ahead because of the Sonatrach negotiations, Air Products had decided to design the Liberal plant to the Seismic Zone #5 requirements that would be needed for a plant to be constructed at Arzew, Algeria.

At its regular meeting in October 1988, the Air Products Board of Directors voted to authorize the proper officers of the company to execute the certain protocols with Sonatrach.

Name, organization and financing of the joint venture project

Not much more happened with the helium project until, in early 1989, Sonatrach asked Air Products to meet with L'Air Liquide because it wanted them to be partners in the helium joint venture. Air Products and L'Air Liquide held meetings in August 1989. Sonatrach was responsible for the helium project, the three-party joint venture officially named Helios. Helios was later incorporated in April 1991 under the laws of the Democratic and Popular Republic of Algeria to engage in the construction and operation of a liquid helium and nitrogen production facility to be located at Arzew, Algeria. With the inclusion of L'Air Liquide, Air Products proposed to Sonatrach two 300 MMcf per year liquefiers for the helium project at Arzew that would be identical to Air Products' new and already operating liquefier at Liberal, Kansas. The Arzew purifiers would be larger since the tail gases supply contained less helium (8–10 percent) than Air Products' Liberal plant's crude helium supply (50–60 percent).

Air Products and L'Air Liquide formed a joint venture named Helap SA in late 1989 to support Helios and to: (1) handle all liquid helium product purchases; (2) provide technical support to the helium plants; and (3) provide training for the Algerian operating personnel for the total helium facility.

Air Products' Corporate Treasury Department was responsible for structuring the helium project financing, leading the negotiations and securing lines of credit, providing for maximum borrowings with various commercial banks, the International Finance Corporation (IFC), the Exim Bank and Banque Exterior D'Algerie (BEA). Air Products attended two meetings in October 1989, the first with L'Air Liquide, and the second a joint meeting with Sonatrach's Financial Director to discuss all financial terms and conditions.

A two-day meeting in January 1990 was then held in Algeria by representatives of Air Products, L'Air Liquide and Sonatrach to agree on a global estimate for the helium project. The global estimate included purchased equipment, construction, site services and support, project reserves and Algerian duties and taxes. In May 1990, a preliminary financial plan was prepared by Air Products and presented in Algiers to Sonatrach, and also used for meetings with banks in Paris and London.

In September 1990, Air Products developed a helium demand history and forecast for worldwide helium at the request of Sonatrach management. Since

1972 worldwide helium demand growth was nearly 10 percent GAR and Air Products forecasted that CAGR for the next 20 years would be in the range of 3 to 4 percent (the actual CAGR to 2009 was 3.5 percent).

In December 1990, a financial meeting was held in Oran, Algeria to allow the helium project to proceed. Air Products presented project financing. Helios, SPA was incorporated on April 29, 1991 as a company limited by shares under the laws of the Democratic and Popular Republic of Algeria to engage in the construction and operation of a liquid helium and nitrogen production plant at Arzew, Algeria.

In late January 1991 the Algerian political situation was of concern to the partners because the Front Islamic Salvation (FIS) party held an overwhelming lead in Algeria's first free legislative elections and were poised to sweep second-round voting in January 1992 and therefore gain control of the Algerian National Assembly. President Chadli Bendjedid's ruling party, the National Liberation Front (FNL), which had governed since Algerian independence in 1962, took only 16 seats in the first-round voting. President Bendjedid resigned prior to the January's scheduled second round of voting and military and civilian leaders formed a security and presidential council and ended the elections. Mohamed Boudiaf, an ex-FNL dissident, returned to Algeria after 27 years of exile in France to be sworn in as head of a five-man presidential council. Most FIS leaders were arrested and accused of insurrection. Boudiaf declared a 12-month state of emergency and the courts ordered the dissolution of the FIS. Boudiaf was assassinated on June 29, 1992. There was extremist violence and sabotage in Algeria in 1992 and during the following few years, which had the effect of slowing down the helium project because of concerns about the security of foreign personnel traveling in and out of Arzew.

At a Helios partners' meeting in early March 1991, Air Products presented the final finance plan for the helium project with a forecasted commencement date of March 31, 1991. Sonatrach and Banque Exterior D'Algerie (BEA) held additional financial meetings in Algeria during April 1991. Air Products and L'Air Liquide personnel met with the IFC in Paris to negotiate a loan. Sonatrach, together with L'Air Liquide and Air Products, met with the Exim Bank to complete the financing of the helium project by the end of May 1991.

The Helios helium project at Arzew becomes a reality

With an agreed helium project investment budget and finance plan in place, the Helios partners, Sonatrach, Air Products and L'Air Liquide, were prepared to move forward with the world's first large helium project with crude helium supplied from the tail gases of the LNG-2 (GL2Z) facility at Arzew, Algeria.

In mid-1991, Helap SA presented to Helios a plan to modularize the two-train helium facility for which approval was slow in coming, due mainly to the political unrest in Algeria. Verbal agreement of the modularization plan's commercial terms was reached in meetings at Air Products in May and June. In September, Air Products' Board of Directors authorized the sale of the helium equipment for the facility and construction services to Helios. The modularization plan was further discussed by the Helios parties in a meeting at Air Products in December, with agreement to be finalized at a meeting in January 1992.

The modularization plan was two steps ahead of: (1) "stick-building a facility on-site" by shipping all the individual parts to Arzew and building the total facility piece by piece; and (2) "skid-mounting" many of the individual units, with auxiliary equipment arriving at the facility site mounted on one skid. Air Products designed and provided each of the two-train helium plants in large ocean-shipping modular units. The benefits of modularization were lower overall cost, schedule improvement, US factory assembly, reduced site construction risks, project organization simplification, and reduced on-site time and labor. All modules were fabricated for the two-train facility by Spartan Constructors, Savannah, Georgia, under Air Products' supervision. During the summer of 1993, the modules for both plants were shipped to Arzew on the oceanworthy ship *Submerger II* out of Hamburg, Germany. Air Products was responsible for the modularization fabrication and ocean shipments portion of the helium project. There were two shipments from Savannah to Arzew, including complete cold boxes, liquefiers, a control and analyzer building, compressor and switchgear buildings, pipe racks and air-coolers bridges, pressure-swing absorbers, two Gardner Cryogenics-manufactured 30,000-gallon liquid helium storage tanks, a two-bay container-loading building, plus a minor amount of miscellaneous equipment. Each of the shipments of the modules from Savannah to Algeria took about three weeks. Set-up of all the modules and installation of other loose equipment in each shipment took place on prepared foundations previously contracted for Air Products under company supervision. Air Products was responsible for the set-up of the two-train facility, with each train requiring approximately two weeks to install after each shipment arrived at Arzew. Modularization proved to be a very successful approach for constructing the Helios facility in Algeria. The modularization option, when compared with the "stick-building" or "skid-mounted" options for constructing the total facility, shifted work from the field to the controlled environment of a fabrication shop, reduced risk and made it much easier to attain the quality needed for a reliable liquid helium production facility. On top of all the manufacturing benefits, the modularization plan was estimated to have saved the overall project about US$10 million.

Operator training, facility commissioning and performance test

Since Air Products was operating a very similar 300 MMcf/year helium facility at Liberal, Kansas, with an identical liquefier and container-loading and storage equipment, Helios accepted Air Products' offer to train its operators for the Arzew helium plants. They sent 18 Sonatrach personnel to the United States for helium plant operations training between May and August 1993. The visiting Sonatrach personnel, all native Algerians, spent their first four weeks at Air Products in Allentown, Pennsylvania for preliminary training in the Operation Department and one week in Philadelphia, Pennsylvania for Distributed Control Systems (DCS) training. Most of the equipment and plant operating controls for the two Arzew plants were nearly identical to those at the Air Products' Liberal plant. Final training was at the Liberal plant with hands-on training on a round-the-clock basis with the Air Products operators, including planned plant start-ups and shutdowns, preventive maintenance and actual maintenance training. Air Products Global Helium Operations Manager was responsible for the set-up, administration and supervision of the training program for all future plant operating and supervisory personnel for Algeria. He recommended to Sonatrach and to Helios the best of the trained personnel from the program for lead operator and supervisory positions. The majority of the plant operating candidates qualified, but a few did not meet the standards set for the program by the management of both Sonatrach and Air Products.

The Arzew facility commissioning began in early 1994 and the liquid helium production met product specification. The total cost of the helium project was very close to the global estimate set by the Helios partners at a meeting in Algiers in January 1990. The 50 tonnes per day of liquid nitrogen production, some for liquid helium container shields, but mainly for sale to the local merchant market, required modifications to the cold boxes in order to meet the contracted merchant product specification. These modifications had to be designed, reviewed and installed before Helios would allow the final performance test for acceptance of the facility. The Arzew facility was completed in 1995 and first produced liquid helium for export in late 1995. Liquid nitrogen modifications were completed in 1996. The final facility performance test was completed in March 1998 and the results were acceptable to Helios. Thereafter, Helios took over full responsibility for operation of the Arzew helium facility. Helap SA entered into ocean and inland transportation agreements for the transport of primarily 5,000-gallon and 11,000-gallon liquid helium cryogenic ISO containers, manufactured by Air Products' Gardner Cryogenics Division, between Marseilles, France and Algeria. The ISO containers were shipped to Marseilles and transported to customers and transfills in Europe, by road on flatbeds with tractors or transferred to the major container port at Algeciras, Spain for shipment to South Africa, South America and North European ports in the United Kingdom, France and the Netherlands.

The Arzew two-train helium facility has been able to operate at a capacity of at least 600 MMcf per year provided appropriate tail gases were available from LNG-2 (GL2Z). The six LNG trains require downtime for maintenance, which usually is scheduled during low LNG demand. During LNG downtime the quantity of tail gases to the helium facility is reduced, and therefore less liquid helium is produced. Since 1995 through 2009 the Helios helium facility at Arzew has shipped over 5 Bcf of liquid helium, mostly to the European Union market or the "Common Market" as it was known.

The present and future for helium from Algeria

In May of 2002, Sonatrach and Linde AG formed two joint ventures (Helison Production SPA and Helison Marketing Ltd) to extract helium from the tail gases of two LNG plants at Skikda, Algeria. Each LNG plant had three trains. GL1K came

Figure 4.3 Artist's conception prepared prior to the construction of the Helios helium plant facility at Arzew, Algeria.

on stream in 1972, was installed by Technip using the Teal process, and had a rated LNG capacity of 2.8 million tonnes per year. GL2K came on-stream in 1981, was installed by Pritchard using the Prico process, and had a rated LNG capacity of 3.0 million tonnes per year. The Skikda helium plant had a capacity of 550 MMcf per year and came on stream in 2007. Unfortunately, Sonatrach suffered a major explosion at Skikda, which destroyed parts of both LNG plants so that only three of the six trains at the site were operable, but with major safety concerns. In 2009, Sonatrach ordered from Air Products a single new LNG train (C3MR/SplitMR™) with a capacity of 4.5 million tonnes per year. The new LNG train is scheduled to operate in 2012–13 to provide enough additional tail gases to the Helison helium plant for it to run at capacity.

Since 2008, Sonatrach has been negotiating with several industrial gas companies, including Air Products, for a second helium plant project to be located at Arzew. The new helium plant, if approved by Sonatrach management, would use tail gases from LNG-1 (GL1Z) and/or from new LNG trains, most likely similar to the recent train purchased for Skikda in 2009.

Algeria had estimated natural gas reserves of 159,000 Bcf as of January 1, 2010. The major natural gas field is Hassi R'Mel with 0.19 percent helium, but additional fields to the east and south have been discovered that also contain helium. Sonatrach currently has two major natural gas pipelines operating to supply Europe and three more are in various stages of development. Sonatrach

Figure 4.4 View of the Helios completed helium plant, looking west towards the LNG-1 facility at Arzew, Algeria.

can consider recovering crude helium to enrich the feed gas to its LNG facilities and/or store crude helium in a nearby depleted gas field for the future. Sonatrach has enough helium-bearing natural gas reserves to supply the major portion of European helium demand for the next 50 years. The potential helium competition to Sonatrach is from Qatar and Russia.

Reference

Hammoutene, R. (1983) "Helium", *Algerie Energie* (No. 1, June 1983) (Magazine of the Algerian Ministry of Energy and Mining)

Acknowledgment

The author is most grateful to Air Products and Chemicals including their permission to use the company images reproduced in this chapter.

5 LNG

The global liquefied natural gas market

Andrew Flower

Introduction

Natural gas was first liquefied over a century ago to produce helium from the trace elements found in most natural gas fields. There were no uses, at that time, for the LNG, which was effectively a by-product of the helium process. By the 1930s, as natural gas became a significant fuel, liquefaction began to be used as a way of storing natural gas, taking advantage of the approximately one-six-hundredth reduction in volume that occurs when methane is turned into a liquid by cooling it to around –161 °C.

Companies marketing natural gas to downstream consumers need to be able to meet the surges in demand that inevitably occur when outside temperatures are very low. Storing gas in its gaseous form requires large volumes of storage capacity. One alternative is to build a peak-shaving LNG plant, which liquefies gas during the summer months when demand is low and stores it in insulated tanks until the winter, when it is regasified to meet the peaks in demand on cold days. The first such facility, in West Virginia, USA, came into operation in 1938. A similar facility in Cleveland, Ohio came on stream in the early 1940s and was the site of the first major accident involving LNG, when a storage tank ruptured spilling its contents. The LNG vaporised and ignited, killing workers in the plant. Some of the liquid also flowed into the sewers, evaporated and entered homes where it exploded. The total death toll was 128.

This accident, which was the result of poor design and the use of materials not suitable for storing a liquid at very low temperatures, set the development of LNG back by several years. However, in the 1950s LNG began to be seen as a way of transporting natural gas over long distances by sea or by inland waterway. A pilot project, involving the construction of a small liquefaction unit in Louisiana and the installation of LNG tanks in a bulk carrier that transported the LNG to a temporary receiving facility at Canvey Island in the UK, showed that LNG was a viable technology to move natural gas by sea. It led to the development of the world's first commercial LNG export plant at Arzew in Algeria, a plant that continued in operation until 2010. The first LNG cargo was delivered from Arzew to Canvey Island by the *Methane Princess*, the world's first purpose-built LNG ship, where it arrived on 12 October 1964.

In the 47 years since those humble beginnings, LNG has expanded into a global business. In mid-2010, Peru became the world's eighteenth LNG-producing country. At the end of 2011 26 countries were importing LNG. The world's LNG fleet had grown to over 360 vessels, with the largest in operation having a capacity some ten times that of the *Methane Princess*. The total volume of LNG traded in 2011 was 242 million tonnes (327 billion cubic metres), which represented just under 10 per cent of commercial natural gas production. More than 100 years after the liquefaction of natural gas was first used to produce helium, the LNG business continues to contribute to the supply of the gas, with liquefaction plants in Qatar, Algeria and Australia producing helium.

As a rule of thumb, LNG economics are better than those of pipeline natural gas delivery where the distance between liquefaction plant and receiving terminal is greater than around 4,000 km if the alternative is an onshore pipeline, while compared with an offshore pipeline the break-even distance is around 1,500 km.

LNG activity is expected to continue increasing as the demand for natural gas grows and importing countries have to look to more distant sources of supply to meet their needs. However, the rate of expansion is uncertain as the business faces the challenge of high investment costs, policies aimed at reducing the use of hydrocarbons for environmental reasons and the development of new long-distance pipelines to deliver gas to market. The so-called shale gas revolution, which has seen US natural gas production growing again after many years of stagnation and decline has resulted in the United States being seen as a potential exporter of LNG rather than an importer. It has also raised the question of whether transferring the technology will have a similar impact in the other LNG-importing countries in Europe, South America and Asia.

The LNG chain

Bringing LNG to market involves the development of a chain of activities, each of which requires the investment of large sums of money and the skills and expertise of people from many disciplines. Figure 5.1 shows the links that make up the chain:

- ***Upstream***: the production of natural gas and its initial treatment – for example, the separation of condensate and the extraction of water – and the pipeline to the liquefaction plant.
- ***Liquefaction***: facilities to receive the gas and to treat it (by removing any remaining condensate and water, plus the removal of any contaminants such as carbon dioxide, hydrogen sulphide and mercury), the heat exchangers and turbines and compressors that drive the liquefaction process, the storage tanks for LNG and LPGs (propane and butane), if they are separated from the LNG, and the jetty and berths for loading the ships.

  Most LNG plants have two or more liquefaction trains, each of which can take the gas from the upstream production facilities, treat it and liquefy

	Upstream	Liquefaction	Shipping	Regasification	Total
Gas use		10–14%	1.5–3.5%	1–2%	12.5–19.5%
Capex	\$2–6bn	\$6–10bn	\$1–2.5bn	\$1–1.5bn	\$10–20bn
Unit cost in US\$/MMBtu	\$1–3	\$3–4.5	\$0.8–1.5	\$0.4–0.8	\$5.2–9.8

Capex estimates based on a 8 mtpa project

Figure 5.1 LNG chain gas use, capital investment and unit costs.

it independently of the other trains. This means that individual trains can be shut down for planned or unplanned maintenance while the other trains in the same plant continue to operate. The size of trains has increased as the technology has developed. The earliest LNG trains had a capacity of 0.3 million tonnes per annum (mtpa) or 0.4 billion cubic metres per annum (bcmpa). In 2012 the largest trains in operation have a capacity of 7.8 mtpa (10.5 bcmpa).

- ***Shipping***: the tankers transporting LNG from the liquefaction plant to the LNG receiving terminal. The ships have specially designed tanks to keep the LNG cold during the voyage. However, there is some heat ingress, which causes a small percentage (typically around 0.1 per cent to 0.15 per cent of the volume) to boil off each day.

  The first purpose-built LNG ship, the *Methane Princess*, had a capacity of 27,400 m^3 of LNG (approximately 12,500 tonnes per voyage). The largest ships in operation in 2012, the so-called Q-max ships, can carry nearly ten times as much LNG on each voyage.
- ***Regasification***: An LNG receiving terminal has facilities to berth and unload the ships, to store the LNG and to regasify it as required by the market. The regasification process involves applying heat to the LNG using seawater, burning some gas to heat water or, in some newer terminals, using the ambient air temperature. Regasification terminals vary in size from those that receive less than 1 mtpa (1.35 bcmpa) to those that can receive as much as 30 mtpa (40 bcmpa). The majority of LNG-receiving terminals are on shore, but in recent years a number of facilities have been developed using LNG ships as floating storage and regasification units (FSRUs). In addition, there is one offshore terminal on a gravity-based structure (i.e. a support structure held in place by gravity) off the coast of north-east Italy.

Figure 5.1 also shows an estimate of the gas used in each stage of the process, together with the capital cost and the revenues required to remunerate the investment and cover the operating costs. These costs are indicative and the range is necessarily wide since there are considerable differences amongst projects depending on such factors as the composition and location of the gas reserves, the design of the liquefaction plant and the distance from the liquefaction plant to the market.

The total capital cost of the LNG chain is estimated to range from US$10 billion to US$20 billion for a chain producing 8 mtpa (10.8 bcmpa). The price needed to remunerate the capital investment and cover the operating cost ranges from US$5.2/million British thermal unit (MMBtu) to US$9.8/MMBtu. An LNG chain with costs at the bottom end of the range would be based on low-cost gas reserves, possibly with significant condensate content in the gas that would generate additional revenues to remunerate the upstream investment, a low-cost plant and a location close to the market. There are very few projects which enjoy these advantages, and costs are generally in the middle or upper part of the range.

The largest element of cost is the liquefaction plant, which typically represents around 50 per cent of the total for the entire chain. The facilities in the producing country (the liquefaction plant and upstream gas production) account for around 75 per cent of the total chain costs, with shipping constituting 10 per cent to 15 per cent and the receiving plant in the importing country 5 per cent to 10 per cent. The total use of gas in the chain from the inlet to the liquefaction plant to the outlet from the receiving terminal is between 12.5 per cent and 19.5 per cent. The main use of gas is in the liquefaction plant, where it is consumed in the turbines that drive the refrigeration process and for the production of the power required by the plant. In the receiving terminal, 1 per cent to 2 per cent of the gas is used in the vaporisers and to produce power. On the ships, the gas which boils off during transit is used by the ship's engines, although around 15 per cent of the world fleet (mainly the vessels of over 200,000 m^3 capacity used by Qatar's two liquefaction plants) has on-board reliquefiers.

Constructing the liquefaction plant currently takes between 45 and 54 months from a final investment decision (FID) to the loading of the first cargo, and generally determines the time taken to implement an LNG chain. In most projects, the upstream facilities are developed in parallel with the construction of the liquefaction plant. LNG ships can be built in around 27 months using an existing design, with up to nine months extra if a non-standard capacity or design is required. The timeframe for the construction of an LNG receiving terminal is 36 to 42 months from FID. However, the planning for a new LNG project can add many years to the timeframe.

The development of the LNG business

A total of 242.4 million tonnes (327.2 billion cubic metres) of LNG was traded internationally in 2011. The build-up of activity by importer (Figure 5.2) shows

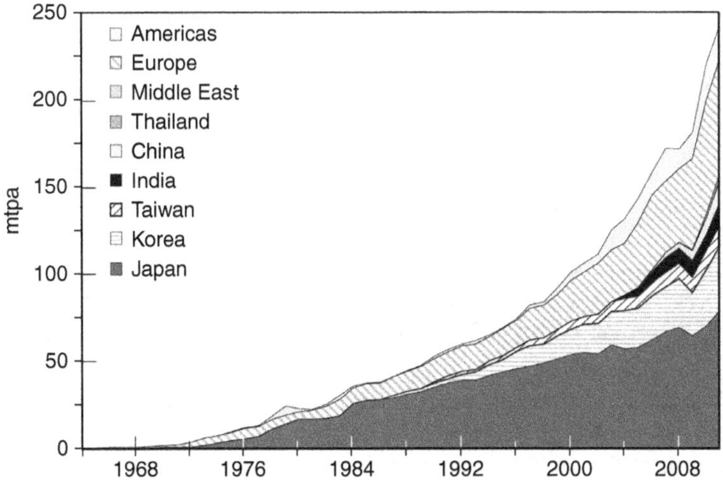

Figure 5.2 LNG trade by importer, 1964 to 2011.

that Europe, which became the first region to receive LNG in 1964, has seen a steady increase in activity, with imports reaching 65.8 million tonnes (88.8 billion cubic metres) in 2011, some 27.1 per cent of the global total. France joined the UK as an LNG importer in 1965, and subsequently Spain, Italy, Belgium, Turkey, Greece and Portugal started to import LNG. The Netherlands became Europe's ninth LNG importer when Rotterdam's Gate terminal received its first cargo in June 2011.

Europe's LNG imports increased by 13.5 million tonnes (18.2 billion cubic metres) between 2009 and 2011, largely because Qatar increased its deliveries to the UK, Belgium and Italy as it expanded its production and brought into operation new receiving terminals in the UK and Italy. LNG accounted for over 16 per cent of European gas supply in 2011.

Imports of LNG into Asia commenced in 1969 with deliveries into Japan from a small plant at Kenai in southern Alaska. Japan's need for LNG was driven by a policy of reducing its dependence on oil for strategic and environmental reasons. Japan has very limited reserves of natural gas and is too far from major gas reserves in eastern Russia for pipeline imports to be a feasible option. In the 1960s LNG was the only way for it to access natural gas supplies, and this remained the case in 2012, with no active plans for pipeline imports. The country quickly became the world's largest importer of LNG, accounting for over 70 per cent of world imports by the mid-1980s. It has remained the largest importer but its share of global LNG trade had fallen to 31.7 per cent by 2010. The earthquake and tsunami that caused death and devastation along the northeast coast of the main island, Honshu, triggered the world's worst nuclear crisis since Chernobyl in 1986. By the end of 2011 all but six of Japan's nuclear units were shut down either directly as a result of the earthquake and tsunami or in the aftermath as the government imposed more stringent safety requirements before units taken off-line for maintenance were allowed to restart. Japan's power generators increased their use of LNG to help replace lost nuclear output and LNG imports increased by 12 per cent in 2011.

South Korea became the second LNG importer in Asia in 1986, and is now the world's second largest after Japan. It was followed as an importer by Taiwan (1990), India (2004), China (2006) and Thailand (2011). Malaysia and Singapore were both constructing LNG receiving terminals at the end of 2011 and were due to start importing in 2012 and 2013, respectively. Indonesia was also developing its first LNG receiving terminal to take delivery of domestically produced LNG. The amount of LNG imported by Asian buyers in 2011, 153.7 million tonnes (207.5 billion cubic metres), represented 63.5 per cent of the global total. A surprising addition to the regions importing LNG is the Middle East, given its abundant gas reserves. Kuwait started to receive LNG in 2009 and Dubai followed in 2010. Both countries require LNG mainly in the summer months, to meet the rapidly increasing demand for power for air conditioning. The lack of a pipeline network in the Gulf region means they have had to turn to LNG and both have opted to use a ship as an FSRU, since this approach enables imports to start much more quickly than would have been the case with a conventional onshore facility. In 2011, Kuwait imported 2.4 million tonnes (3.2 billion cubic metres) and Dubai imported 1.2 million tonnes (1.6 billion cubic metres) and the region's share of the global total was 1.5 per cent.

The Americas accounted for 8.0 per cent of global LNG imports in 2011. Until 2000, the USA was the only importer of LNG in the region but it was joined first by Puerto Rico and then by the Dominican Republic, Mexico, Argentina, Brazil, Canada and Chile. The new importers accounted for 68 per cent of the total LNG received by the Americas in 2011.

The USA has experienced something of a rollercoaster ride in terms of the role of LNG in its overall natural gas supply. In the 1970s, LNG was seen as necessary for the country to meet growing natural gas demand as domestic production levelled off under the impact of wellhead price controls. Four LNG receiving terminals were brought on stream in the 1970s and early 1980s and long-term contracts were signed with Algeria. However, increasing prices and the deregulation of the natural gas business boosted supply, and LNG imports were no longer needed. Three of the terminals were mothballed for extended periods and just a few cargoes were imported each year over the next 20 years, generally in the winter months when natural gas demand and prices spiked. Around 2000, there was a second wave of interest in LNG to meet the forecast of a growing gap between increasing natural gas demand and declining domestic production. The existing terminals were reopened and expanded, and eight new terminals were brought into operation between 2005 and 2011.

The unexpected surge in the production of natural gas from shale deposits took most observers by surprise and has led to a situation in which the USA has the reserves to meet projected demand for several decades and no longer needs LNG imports. Prices of natural gas have been the lowest of all the LNG importing countries since 2010 and the owners of the LNG terminals have been left with assets that are no longer needed for the purpose for which they were built. The owners are now looking at converting them into facilities to export US natural gas production as LNG using the existing infrastructure including the storage tanks, jetties and berths and adding liquefaction trains. Plans have been announced for five of the import terminals to be turned into export plants and there have been proposals for the development of green field facilities on three sites originally earmarked for import terminals. In total, and in the unlikely event that all the proposed projects were to be developed, the USA would be able to export around 130 million tonnes per annum (175 bcmpa) of LNG.

Figure 5.3 shows the sources of LNG supply by country from 1964 to 2011. Qatar, which started producing LNG at the end of 1996, is now the world's largest producer by a wide margin, with exports of 76.2 million tonnes (102.9 billion cubic metres) in 2011. It has 14 liquefaction trains in operation at its Qatargas and RasGas sites with a total capacity of 77.5 mtpa (105 bcmpa).

Malaysia was the world's second-largest producer in 2011, with output of 24.7 mt (33.3 bcm). Indonesia was for many years the world's largest producer but its output has been in decline since 1999 as the gas reserves in fields supplying its two original liquefaction plants at Arun in the north of Sumatra Island and at Bontang on Borneo Island, which came on stream in the 1970s are depleting. The decline was halted temporarily in 2010 following the start-up of the country's third plant at Tangguh in West Papua but it resumed in 2011 when production was 21.7 mt (29.3 bcm). In addition to the shortage of reserves around the Arun and Bontang plants, the government is giving priority to domestic gas use over LNG exports. This is becoming an increasing trend around the world as

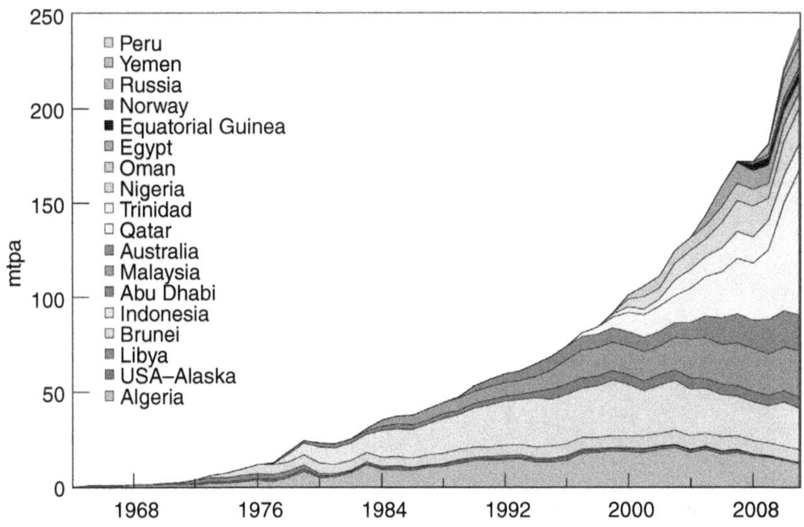

Figure 5.3 LNG trade by exporter, 1964 to 2011.

governments question whether exporting their natural gas as LNG is the best option for limited reserves when they face challenges in generating economic activity and creating jobs for their growing workforces. In 2011, LNG production in Oman, Egypt and Algeria was below the available capacity because governments opted to supply gas to the domestic market rather than export it as LNG.

A feature of the last decade has been the increasing role played by producers in the Middle East in global LNG supply (Figure 5.4). In 2000, the Pacific Basin was the dominant producer, accounting for 57 per cent of total output. By 2011 its share had fallen to 35.8 per cent. The share of the Middle East had increased over the same period from 17.3 per cent to 40 per cent, largely a result of the growing output from Qatar. The Atlantic Basin's contribution declined from 25.7 per cent in 2000 to 24.2 per cent despite production building up from Trinidad and Tobago (1999) and Nigeria (1999), and plants in Egypt (2005), Equatorial Guinea (2007) and Norway (2007) coming into operation. The increasing role of the Middle East has added new flexibility to global LNG supply since producers in the region are approximately equidistant from markets in Europe and North-East Asia, and are able to switch cargoes between destinations without a major disruption to shipping programmes.

In 2010 (the most recent year for which data on pipeline gas supply was available), 9.4 per cent of world natural gas production was delivered to market as LNG. A further 21.4 per cent was traded internationally by pipeline, with the remaining 69.2 per cent being consumed in the country in which it was produced (Figure 5.5). The natural gas business is very different from the oil industry, where around two-thirds of production is traded internationally, largely by ship.

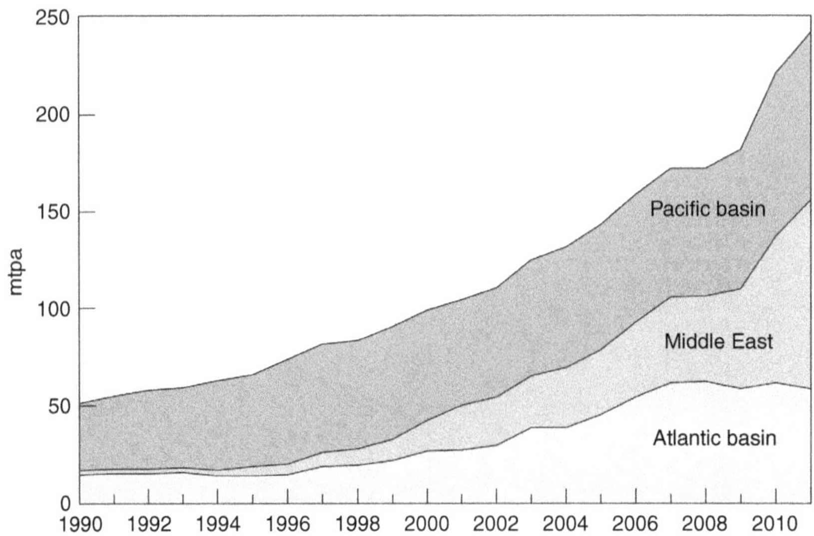

Figure 5.4 Regional LNG production, 1990 to 2011.

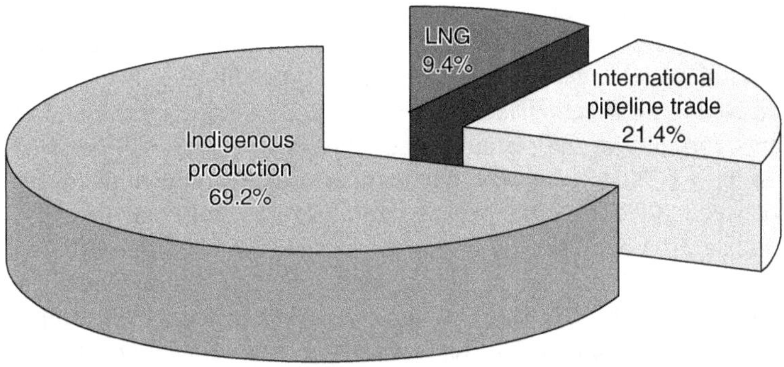

Figure 5.5 World natural gas supply in 2010.

As a result, oil has become a global commodity whereas the more limited international trading of natural gas has meant it is a regional and often national business, with prices varying widely between markets.

Liquefaction

As Table 5.1 shows, in January 2012, the installed liquefaction capacity globally was an estimated 273 mtpa (368 bcmpa). However, actual output was significantly below that level. In 2011, LNG production was about 89 per cent of capacity. The difference between available capacity and actual production is partly explained by the last of Qatar's trains being in build-up mode in the early part of 2011, but it is also due to the priority given to domestic demand in some countries, while technical and gas supply problems caused unscheduled shutdowns in production at other plants.

A further 84 mtpa (113 bcmpa) of capacity was under construction in January 2012. The new capacity, the majority of which is in Australia but trains are also under construction in Algeria, Angola, Papua New Guinea, and Indonesia, will come into operation by 2017. These additions will take the world's installed liquefaction capacity to just over 350 mtpa (472 bcmpa), after taking into account

Table 5.1 Global liquefaction capacity, January 2012

Region	*In operation*		*Under construction*		*Planned*		*Total*	
	in mtpa	*in bcm/a*	*in mtpa*	*in bcm/a*	*in mtpa*	*in bcm/a*	*in mtpa*	*in bcm/a*
Pacific Basin	94	156	70	94	118	159	281	379
Middle East	101	136	0	0	22	30	123	166
Atlantic Basin	78	105	14	19	143	192	235	317
Total	273	368	84	113	282	381	639	862

facilities in Alaska, Algeria and Indonesia that will probably have been shut down permanently by that time.

Production over the period to 2016 will largely be determined by the plants currently in operation and under construction, since it typically takes at least four years to build a new liquefaction facility. Even if FID is made by the end of 2012 for plants currently at the planning stage, the earliest they will begin to contribute to supply is 2016. Global production increased by a record 60.5 mtpa (54 bcmpa) between 2009 and 2011 but the rate of increase will slow over the period 2012 to late 2104 because only four new trains are due on stream between 2012 and 2014 and two of those are in Algeria where they may replace old trains rather than add to capacity. By 2015 output is expected to reach 290 million tonnes (391 billion cubic metres). This represents an average annual growth rate of 5.5 per cent over the period from 2010, slower than the 7.7 per cent per annum recorded between 1980 and 2010.

Beyond 2015 the expansion of global LNG supply will depend on the rate at which new liquefaction capacity is commissioned. As Table 5.1 shows, nearly as much new capacity was being planned in January 2012 as was in operation and under construction at that time. The total planned capacity shown in the table does not take into account some of the more speculative projects in the USA and Canada and some which could be eventually developed based on more recent major gas discoveries offshore East Africa and in the eastern Mediterranean.

Table 5.2 lists the main countries in which new liquefaction plants were being planned in January 2012. Australia led the way, with 75 mtpa (102 bcmpa) of capacity planned to be added to the 20 mtpa (37 bcmpa) in operation and the nearly 70 mtpa (94 bcmpa) under construction. The USA, Nigeria and Russia followed Australia, with between 41 mtpa and 59 mtpa of capacity planned in each country. In total, the top four countries in Table 5.2 accounted for over three-quarters of the planned capacity.

In April 2005, Qatar, the world's largest LNG producer, instituted a moratorium on the development of further production from its giant North field, which has estimated reserves of 2.5 trillion cubic metres (15 per cent of the world's

Table 5.2 Countries planning new liquefaction capacity

Country	*Planned New Capacity*	
	in mtpa	*in bcmpa*
Australia	75	102
USA	59	80
Russia	47	64
Nigeria	41	55
Canada	15	20
Qatar	12	16
Others	33	45
Total	282	382

total). The Qatari government's objective was to evaluate the performance of the North field to ensure that it was not being over-exploited to the detriment of its long-term performance. The studies have taken much longer than expected, and at the beginning of 2012 the expectation was that the moratorium would not be lifted until 2015 at the earliest. The only indication of the possible expansion of liquefaction capacity after the moratorium is lifted has been a plan to "de-bottleneck" (i.e. remove constraints in the production that have been identified during operations) each of the six 7.8 mtpa megatrains that are in operation. It is estimated that the debottlenecking will add 2 mtpa (2.8 bcmpa) to the capacity of each train. Qatar has the reserves and the infrastructure to support the addition of more liquefaction trains if the decision is made for its LNG capacity to be increased further.

There is considerable uncertainty over just how many of the world's planned projects will be developed and when they might come on stream. Furthermore, projects will almost certainly be added to the list as new reserves are discovered. However, there are major challenges in taking a new project from the planning stage to the FID.

The cost of building new liquefaction capacity in 2012 is between three and five times the level in 2005 as a result of higher raw materials and equipment costs, higher labour costs because of the shortage of people with key skills and the more challenging locations for new projects, where environmental issues are frequently a key concern for developers. The impact of the economic recession on costs has been, at best, marginal, and costs in 2012 remain close to the record levels reached in 2008.

The escalation in costs, which in many markets has not been matched by increases in prices, is one reason that FIDs for new liquefaction capacity slowed between 2006 and 2009. The priority given by some governments to domestic gas consumption over the export of gas as LNG was, as already discussed, another reason for the slowdown in the commitment to new liquefaction capacity.

There has been a recovery in FIDs since 2009, with seven Australian projects reaching that stage. It has also seen the first commitment to a floating liquefaction unit and to projects supplied with coalbed methane (CBM). The large number of projects at the planning stage suggests that the second half of the 2010s should see a continued growth in liquefaction capacity but many of the potential developments are challenging and the growth in output is unlikely to match the rates seen in 2010 and 2011.

Figure 5.6 shows a forecast of LNG production over the period from 2010 to 2020. The production from plants in operation declines marginally as older plants in Alaska (Kenai), Algeria and Indonesia (Arun) are closed. The build-up in production from plants under construction has been developed taking into account the expected start-up date and making assumptions about the rate at which output will build. Finally, the production from plants at the planning stage has required assumptions to be made as to which plants will eventually go ahead and when they will start up.

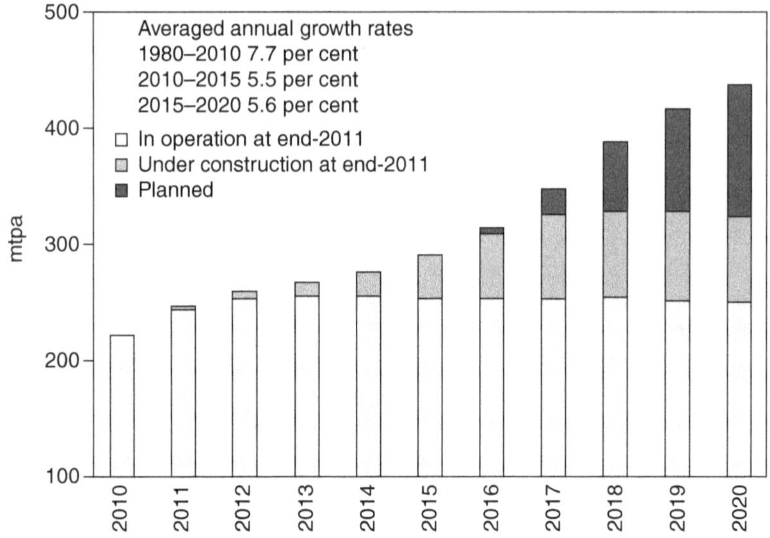

Figure 5.6 LNG supply forecast 2010 to 2020.

Overall, LNG production is forecast to increase at a rate of 5.6 per cent per annum from 2015 to 2020, taking total production in 2020 to around 430 mtpa (580 bcmpa).

LNG shipping

The LNG shipping industry responded rapidly to the increase in LNG activity after 2000, nearly trebling the number of ships in operation between 2002 and 2011 (Figure 5.7). Numbers were boosted by speculative orders (i.e. those unrelated to a long-term commitment for use of the vessel). This was the first time speculative orders had been seen since the early 1980s, when some ship owners incurred major losses as speculative ships failed to find employment and several were laid up for many years. Oil and gas majors, such as Shell, BP and BG, also ordered ships to support their LNG trading activities. These orders were in addition to those from projects and LNG buyers to support sales and purchases under long-term contracts, which were the mainstay of ship orders before the year 2000.

The number of tankers entering service peaked at 52 ships in 2008 and started on a downward trend in 2009. Over 60 orders for new ships were placed in 2004 but, as a surplus developed, companies refrained from placing new orders. No orders were placed between May 2008 and January 2010, and the order book for new ships had fallen from a peak of over 140 ships in early 2007 to just 22 ships in early 2011. However, the growth in production in

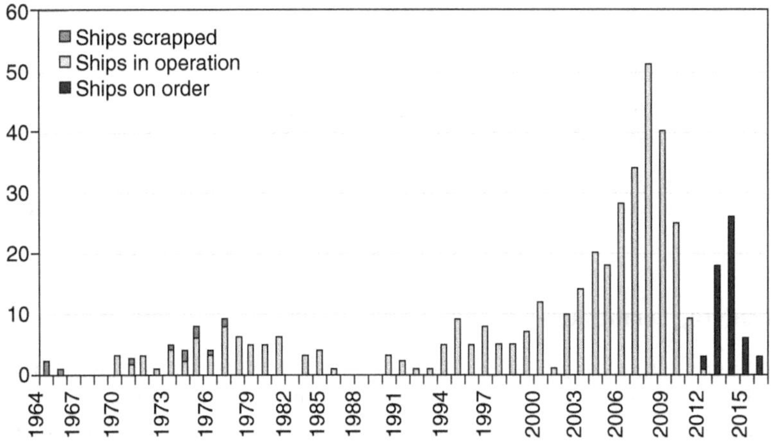

Figure 5.7 Number of new ships entering service, 1964 to 2016.

2010 and 2011 outpaced the growth in shipping capacity and the Fukushima crisis created the requirement for more shipping capacity to transport LNG cargoes from the Atlantic Basin to Japan. As a result, LNG ship charter rates more than trebled between mid-2010 and mid-2011 and ship-owners responded by ordering around 50 new ships in 2011, about half of which were on a speculative basis.

The average capacity of vessels has also increased, from 112,000 m^3 for those entering service before 2001 to 161,700 m^3 for ships entering service between 2002 and 2011. The average was boosted by Qatar's 31 Q-flex vessels (capacity 210,000 m^3 to 217,000 m^3) and 14 Q-max (capacity 263,000 m^3 to 266,000 m^3) vessels. As a result, the total capacity of the fleet had increased from 14.2 million cubic metres at the end of 2000 to 54.1 million cubic metres at the end of 2011. Not all of the ships are available for trading since ten were operating as FSRUs or were in a ship-yards being converted.

Available shipping capacity increased at a much faster rate than global liquefaction capacity over the first decade of the twenty-first century. In part, the increase in shipping capacity was needed to support the longer shipping distances resulting from Middle Eastern producers increasing their share of global production. In addition, more ships are needed to support the growth in short-term trading of LNG, which inevitably makes less efficient use of ships than deliveries under long-term contract, where regular voyages between the liquefaction plant and a receiving terminal can be planned well in advance. Some of the demand for short-term contracts or individual spot cargoes comes from competitive liquid gas markets such as those in the US and the UK. The increase in short-term activity – from 5.3 per cent of total trade in 2000 to 18.8 per cent in 2010 – has provided producers with opportunities to market

LNG cargoes not needed by their long-term buyers. This can support the development of new projects, since the opportunity to market the planned output on a short-term basis can allow projects to reach FID with a smaller proportion of their output committed on a long-term basis.

The availability of shipping capacity has not been a constraint on the development of new liquefaction projects. Shipyards in Japan, South Korea and China have the capacity to build, in total, over 50 ships per year, with the construction of a new vessel taking a maximum of 36 months from the placing of an order.

LNG receiving terminals

Receiving terminal capacity also increased more rapidly than liquefaction capacity between 2000 and 2011, as new terminals were built and those already in operation were expanded. Demand for LNG increased for a number of reasons:

- countries which lack access to pipeline gas expanded their capacity to import LNG to meet growing natural gas demand (e.g. Japan, Korea, Taiwan);
- countries wanted access to LNG to increase the security of natural gas supply (e.g. Singapore, Brazil, Poland);
- new importers needed LNG to meet growing demand that could not be satisfied by domestic production or pipeline imports (e.g. China, India);
- countries where domestic production is in decline (e.g. UK, Malaysia) looked to LNG to boost natural gas supply;
- some companies sought access to flexible markets (e.g. UK, USA, the Netherlands) to support their LNG trading activities.

The capacity of a receiving terminal depends on a combination of the size of the regasifiers, the number of berths, the size and number of storage tanks and the characteristics of the market into which the LNG is delivered (seasonality, daily swings in demand, the role of LNG and the availability of pipeline gas). Most terminals are built with spare capacity to ensure that the peaks in demand for natural gas can be met. As a result, it is difficult to establish a reliable estimate for the global re-gasification capacity but most estimates show that it is well in excess of liquefaction capacity which means that it has not been a constraint in the level of LNG activity.

At the end of 2011 there were 26 countries importing LNG and the number is set to increase, with many countries either constructing import terminals or actively planning to do so. The use of ships as FSRUs has reduced the investment cost compared with building a conventional onshore terminal and it can also shorten the time needed to implement a project from five or more years to as little as 12 months.

LNG prices

There are two basic ways in which most LNG (and natural gas) is priced:

- ***Formula-based pricing***: The LNG is linked to the price of another commodity, typically crude oil or refined oil products. This approach is used in markets where there is no gas-on-gas trading to set a price based on the supply and demand for natural gas. It is used in Asia and in Europe outside the UK, Belgium and the Netherlands.

  In Asia the linkage is generally to JCC – the Japanese Customs-cleared crude oil price, often referred to as the Japanese Crude Cocktail. This is the average price of crude oil imported into Japan each month, as published by the Japanese Ministry of Finance. It was first adopted in contracts with Japan but is now also used by buyers in Korea, Taiwan, China and India. It is based on the large volume of crude oil imported into Japan each month and is therefore seen as being representative of crude oil prices in Asia. The main exceptions to the use of JCC are Indonesian contracts, which have used the average Indonesian crude price (ICP). The Brent crude oil price rather than JCC has also been used in at least two contracts.

  In Europe, pipeline natural gas prices have traditionally been linked to a mix of gas oil and fuel oil prices, the main products competing with natural gas in the domestic and commercial and industrial sectors, respectively. LNG had to adopt the pricing approach used by pipeline gas. However, more recently, the linkage in LNG contracts has increasingly been to Brent crude oil, which is a widely traded commodity and allows buyers and sellers to manage the price risk using forward prices for crude oil on the trading exchanges.

  It is estimated that in 2011, approximately 75 per cent of LNG sales used an oil-linked approach.
- ***Market-based pricing***: In markets where natural gas is actively traded, the price of LNG is linked to market (or hub-based price) to ensure it is competitive with other sources of natural gas supply. In the USA, prices are linked to the Henry Hub price, and buyers in the rest of the Americas have generally adopted some form of linkage to Henry Hub prices in their contracts. Natural gas is actively traded at Henry Hub, a physical trading hub in Louisiana that is used as the reference point for natural gas futures contracts bought and sold on the New York Mercantile Exchange (NYMEX).

  In the UK the linkage is to the National Balancing Point (NBP) price. NBP is a virtual point in the UK gas network, which is used as the basis for the trading of natural gas on the Intercontinental Exchange (ICE). Any gas entering the UK gas transmission network is deemed to be at the NBP and is traded on that basis. The Zeebrugge Hub, a physical hub in Belgium, is used for LNG delivered to the Zeebrugge terminal, while the Dutch Title Transfer Facility (TTF), which operates in the same way as the NBP, is the basis for pricing LNG delivered to the Gate terminal in Rotterdam.

  Approximately 25 per cent of LNG production is currently sold linked to prices at a trading hub.

Figure 5.8 compares the monthly prices in the UK, USA, Germany and Japan from January 2009 to November 2011. The UK prices are the average NBP prices and for the USA they are the average Henry Hub prices; since LNG prices are linked to these prices, they are an indicator of the price of LNG delivered to the UK and the USA. The Japanese prices are the monthly average LNG import price as reported by the Ministry of Finance, and the German prices are the border prices for natural gas, which are representative of natural gas price in continental Europe, including LNG prices. The price of Brent crude oil in $/MMBtu is also included for comparison.

The figure shows that since the middle of 2009, natural gas and LNG prices in the four markets have been at a discount to oil prices on a Btu basis. Prices have been highest in Japan and Germany, where they are linked to oil prices. From mid-2009 to mid-2010, UK and US prices moved closely together but since mid-2010, the abundant gas supply as a result of the shale gas revolution has held US prices below $5/MMBtu while UK prices have increased and by late 2011 they were more than double US prices. UK prices have been at a discount to European prices throughout the period shown in Figure 5.8. This difference is putting pressure on oil-linked prices in Europe, where buyers with long-term contracts to purchase pipeline gas or LNG at oil-linked prices are facing competition from companies that purchase supplies of natural gas at the NBP price. In Asia, the lack of pipeline connections between the markets and the lack of any trading hubs that could provide an alternative to oil-linked prices have meant that there is less pressure for a change in the pricing methodology. Long-term deals for LNG supplies signed between 2009 and 2011 by buyers in Japan, Korea, China, India and Taiwan are all understood to have retained the linkage to oil prices.

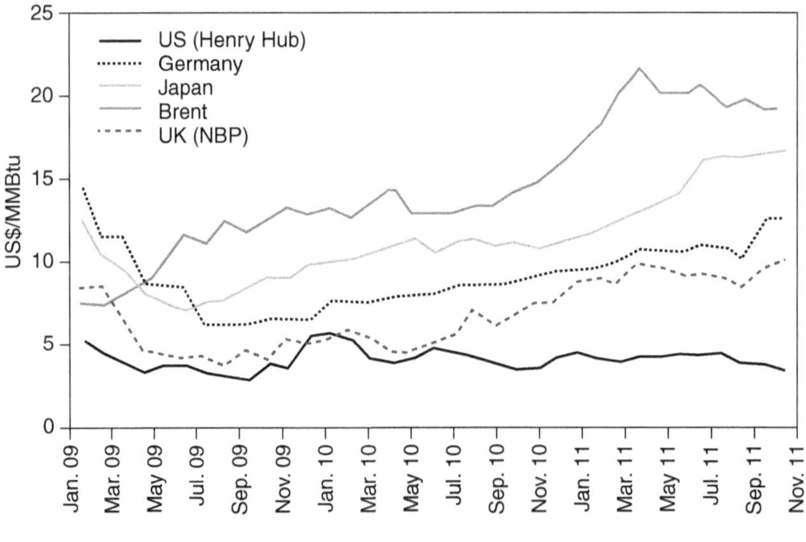

Figure 5.8 Natural gas prices and LNG in the UK, the USA, Japan and Germany: January 2009 to November 2011.

Future price expectations will be a major factor in the decision on committing to the development of new LNG projects. As Figure 5.1 shows, new projects need prices in the range of US$5–10/MMBtu to remunerate the capital investment and cover the operating costs. Prices in the USA have been too low to support a new project while those in the UK have moved to a level that would probably support a new project but the volatility of the NBP price would be an issue. This has made it difficult for investors in new projects in the Atlantic basin to commit the billions of dollars needed to develop the full LNG chain, unless they are prepared to assume that prices will improve and remain at high levels for 20 years from 2016, the period over which a new project will need to earn the revenues to remunerate the investment.

However, prices in Asia were at a level in 2010 and 2011 that would cover the fully built-up costs of even the highest-cost projects. This has supported FIDs on projects in the Pacific basin since 2009, a period when there have been no FIDs on new capacity in the Atlantic basin.

What are the threats to LNG demand growth?

The demand for LNG will depend on the demand for natural gas and the extent to which it can compete with pipeline gas to meet that demand. LNG is generally most competitive over long distances and where pipelines are not technically feasible because of deep water or the need to transit other countries with the potential risk of the interruption of supply, as European buyers have experienced with Russian pipelines crossing the Ukraine and Belarus.

There are risks to the growth of natural gas demand as governments turn to conservation and renewable energy in response to global warming. However, natural gas has lower carbon dioxide emissions than the other hydrocarbon fuels, oil and coal. Furthermore, its flexibility gives it a potential role as a transition fuel in power generation, backing up the intermittent nature of wind and solar power.

As discussed above, many countries are turning to LNG because domestic gas production is not available or is in decline. Having LNG import facilities increases a country's access to sources of natural gas supply and, hence, its security of supply. However, this does not mean that LNG will have its own way in competition with pipeline gas for markets. In Europe, new pipelines are being planned connecting natural gas-importing countries with producers in Central Asia and possibly the Middle East. A pipeline from Russia across the Baltic Sea to Germany came into operation in 2011 and a pipeline across the Black Sea is being planned to avoid the transit issues that have been experienced in recent years. The pipeline from Central Asia to China which started-up in 2009 may impact Chinese LNG demand, as will any future development of supply from the large but untapped gas reserves in Eastern Siberia.

However, the main threat to LNG will probably come from unconventional gas and, in particular, shale gas. The so-called "shale gas revolution", which started largely unnoticed in the mid-2000s in the USA, has transformed natural gas markets in North America and has turned the US from a potentially large LNG importer into a potential exporter.

The increase in the production of shale gas was the result of technical developments in horizontal drilling and in the fracturing of the shale to allow the natural gas to flow. The result had been that US domestic natural gas production, which had been flat or declining since 1990 (Figure 5.9), started to increase in 2005. In 2011, it reached an estimated 650 billion cubic metres, some 135 billion cubic metres above the 2005 level, an average rate of increase for the six-year period of 4 per cent per annum. Shale gas production grew from 22 billion cubic metres in 2005 to 153 billion cubic metres in 2011.

A key question for the future of LNG demand is whether shale gas could transform natural gas supply elsewhere in the world. There are large shale deposits in Europe, China, Argentina and India, all key markets for LNG. Many of the major oil and gas companies, having bought into the independent companies that have been at the forefront of the development of shale gas in the USA, are now looking at developing shale gas in these regions. However, the evaluation of the prospectivity is at an early stage, with few wells having been drilled as yet. It is unlikely that progress will be as rapid as it has been in the USA, where there was already an infrastructure of exploration companies and active drilling rigs and an abundance of water, which is needed in large quantities for the fracturing process. Furthermore, it is generally more straightforward to obtain permits for exploration and production in the USA than elsewhere in the world. It is too soon to estimate the cost of shale gas production outside the USA, and therefore how competitive with LNG it will eventually be.

The expectation is that shale gas will eventually contribute to natural gas supply in markets outside the USA, but it will take time for the supply to be developed and it is unlikely to have the same impact as it is having in the USA. Competition from shale gas will not be a factor for LNG projects targeting FID

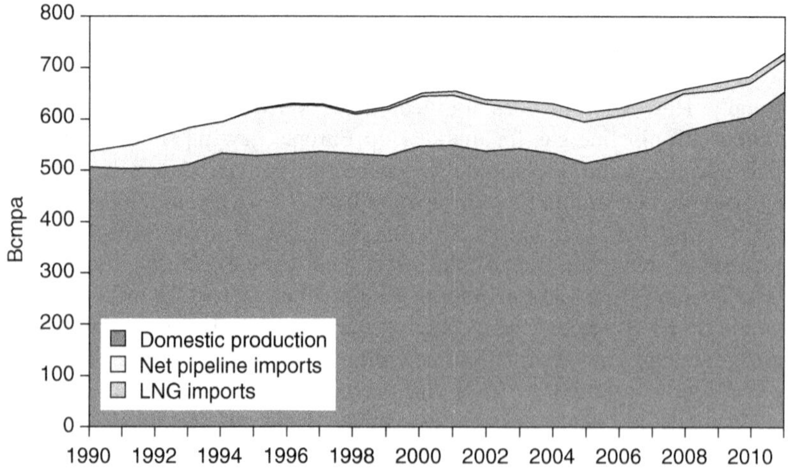

Figure 5.9 US natural gas supply, 1990 to 2011.

in the next few years, but for those projects with a longer time horizon (say around 2020 or later) it could provide additional competition for market share.

On the other hand, unconventional gas can also contribute to the expansion of LNG production, as is being seen in Queensland, Australia, where three major projects are now under construction and two more are being planned near the port of Gladstone with coalbed methane as the source of gas supply. It is also possible that production at the Bontang plant in Indonesia could be boosted by developing coalbed methane reserves located close to the plant.

Conclusion

The LNG business has grown rapidly since the first commercial cargo was exported from Algeria to the UK in 1964. By 2011, nearly 10 per cent of the world's gas consumption was delivered to market as LNG and the share is set to increase as new liquefaction capacity comes on stream, with Australia the leading contributor in the 2014 to 2017 timeframe.

In early-2012, liquefaction projects at the planning stage had the potential to more than double available capacity. However, many faced significant challenges, especially with capital costs remaining at or close to the record levels reached in 2008, despite the economic recession. The policy of some governments of prioritising domestic gas consumption over LNG exports is another challenge faced by some projects, many of which are in difficult locations where environmental issues have to be addressed.

There appears to be a growing demand for LNG imports in markets across the world, as buyers have to turn to more distant sources of supply to meet demand in their downstream markets. Furthermore, LNG can widen the access to sources of natural gas supply and thereby increase security of supply, an important consideration for countries heavily dependent on natural gas imports.

The outlook for LNG demand could change if shale gas deposits located in many natural gas-consuming countries can be exploited at a competitive cost. Between 2005 and 2011, the development of shale gas in the USA has resulted in it becoming a potential LNG exporter rather than an importer. Outside the USA progress has been slower, and the drilling and appraisal of shale gas deposits has only just started. The expectation is that shale gas production outside the USA will increase as the technology is transferred from the USA, but it will take time. Therefore, shale gas could be a strong competitor for market share with LNG in some markets after around 2020, but it is unlikely to have a significant impact for projects planning to reach FID over the next few years.

Reference

US-EIA (2011) "Annual Energy Outlook 2011", US Energy Information Administration, www.eia.doe.gov/oiaf/aeo/demand.html

6 Helium in Russia

Benjamin Hooker

Russia is a riddle wrapped in a mystery inside an enigma.
Winston Churchill

Introduction

For over 50 years Western Siberia has been the cornerstone of Russian oil and gas production meeting the domestic and export needs of the country. Even today, with its major fields entering a stage of declining production, the area accounts for some 90 per cent of gas output. To maintain its reserve base Russia is looking ever northward and eastward, but both these areas provide formidable technological challenges and will require enormous investment (Poussenkova, 2007: 4). This paper's aim is to quantify Russia's significant helium reserves, identify issues associated with the development of those resources and speculate as to Russia's significance in future global helium supply.

Russia's reserve base

Russian helium reserves account for some 13 per cent of the world's total helium resources (Table 6.1), but these figures should be treated with caution as a direct comparison between countries may not be possible. The United States figure of 744 billion cubic feet, for example, incorporates 153.2 Bcf of measured reserves,

Table 6.1 Global helium reserve base in billion cubic feet (Bcf), as at 31 December 2006

United States (40 per cent)	744
Qatar (19 per cent)	364
Algeria (16 per cent)	296
Russia (13 per cent)	245
Canada (4 per cent)	72
Poland, China and others (8 per cent)	151
Total	1,872 Bcf

Source: Pacheco (2009).

192.2 Bcf of probable resources, 213.8 Bcf of possible resources and 184.4 Bcf of speculative resources that may be uneconomic or even impossible to develop. In addition to this, countries with significant known helium reserves (Iran, Libya and Australia) are not mentioned (Pacheco, 2008: 78–9, 2009: 74–5, 2010, 72–3; US Department of Interior, 2008: 7; Alger, 2009: slide 20).

Where are Russia's helium reserves located?

When helium was deemed a strategic commodity by the United States government and responsibility for establishing a domestic source given to the US Bureau of Mines (BOM) during World War I (National Research Council, 2000: 16), Russia responded by placing the responsibility for helium field identification with the Petroleum Scientific Research Institute of Geological Exploration (VNIGRI), mirroring the work of the BOM (which was renamed in 1946 as the Bureau of Land Management (BLM)). Headquartered in St Petersburg, VNIGRI has conducted over 35,000 helium gas analyses from 15,000 separate locations and identified over 250 areas with some helium development potential (Ben'yaminovich, 1995b: 64–6; Yakutseni, 1995: 61–3).

Russia's main helium reserves are concentrated in three regions (Figure 6.1) – Northern Caspian, Yakutia and Eastern Siberia, with typical helium concentrations in gas and condensate fields of 0.03–0.06 per cent, 0.3–0.6 per cent and 0.2–0.3 per cent helium, respectively (Yakutseni, 2001: 10–22; Udut, 1995: 96–100).

Within each of these regions, the most significant and active helium projects are centred on the city of Orenburg in the Northern Caspian, the Kovykta field in Eastern Siberia and the Chayanda field in Yakutia (Figure 6.2). The distance between these areas is vast; one possible export route to market via Sakhalin is 1,650 miles as the crow flies (Kanekiyo, 2005: slides 1–9).

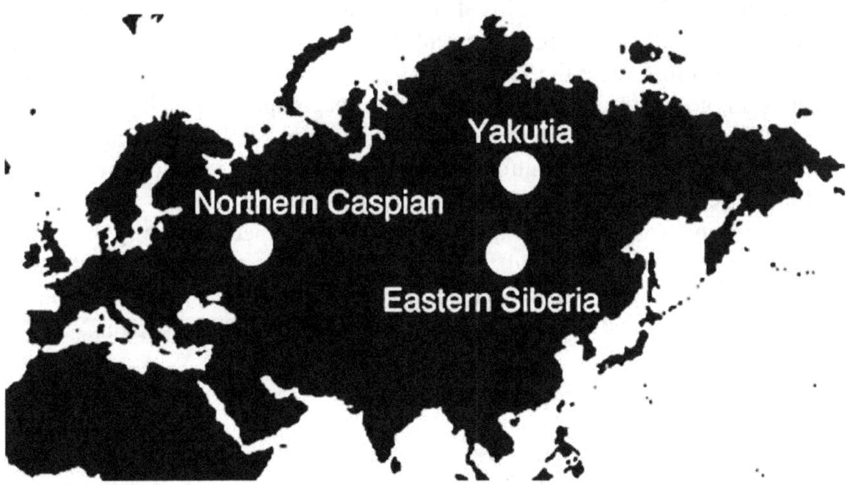

Figure 6.1 Russia's three main helium regions (source: *Eleuthera Capital*).

Figure 6.2 Active Russian helium projects (source: *Eleuthera Capital*).

Orenburg – Northern Caspian

Eleuthera Capital (a Guernsey-registered helium exploration and production company) estimates the total recoverable helium reserves of the Orenburg area at 3 Bcf. Commissioned in 1977 to process natural gas, the Orenburg gas processing complex is currently Russia's largest source of helium. The helium plant was designed to extract and purify 300 million standard cubic feet of helium per year (MMscf/y) from an initial concentration in the raw gas of between 0.05–0.06 per cent. Closed in 2004 after a fatal explosion and reopened again in 2006, the plant currently produces 234 MMscf/y. Although there are plans to double the feed gas supply from the Karachaganak field in Kazakhstan, no helium has been identified in its composition. With a production decline of approximately 5 per cent per year in the helium-rich gas, the entire gas processing plant requires significant refurbishment and replacement with an anticipated start-up of 2015 and an estimated total cost approaching US$735 million. Consequently, although we would expect the new plant to be built in parallel, there is bound to be some disruption to helium supply and production within the next five years, during the refurbishment programme (Andreev, 1995: 73–85; Ben'yaminovich, 1995a: 86–8; Stepanyuk, 1995: 67–9).

Chayanda – Yakutia

Chayanda has been identified as a potential future helium reserve with an estimated total of 31 trillion cubic feet of gas in place. At 0.58 per cent, Chayanda has some of the highest concentrations of helium to be found in Russia. Whilst reserves cannot be accurately estimated at this time due to a lack of well control and the processing of further seismic data, based upon a 50 per cent recovery rate, Eleuthera would estimate the field to have a total recoverable helium reserve of approximately 90 billion cubic feet. In 2007 the field had its status raised to 'strategic'; meaning that the rights to develop the field can be awarded without a tender. In April 2008 Gazprom was awarded the rights to develop Chayanda. The project envisions not only the development of the field but also the creation of a new natural gas processing plant and transportation system. The plant will be built either in Lensk or Aldan along the possible eastward route for a future natural gas pipeline. Currently in planning and with an associated total budget cost in excess of US$4 billion, given the lack of infrastructure and production-related difficulties, Eleuthera's estimated start-up for this project would be 2020, although the Russian government is currently targeting 2016–17. Of the total budget, Eleuthera would expect initial drilling to cost US$600 million, field facilities US$2,811 million, a pipeline to Talakan US$576 million and a further US$110 million for the gas processing plant. The field is part of the Eastern Gas Program, which involves creating an integrated system for producing and transporting gas in Eastern Siberia and Russia's Far East, with the possibility of exporting gas to China and other Asia-Pacific countries. The entire programme is expected to cost an estimated US$93 billion and has yet to be approved (Vasenev, 2006: slides 1–17; Timoshilov, 2008: 1–10).

Kovykta – Eastern Siberia

Along with Chayanda, Kovykta is one of the largest undeveloped natural gas fields in the Irkutsk region of Eastern Siberia and has been identified as another potential future helium reserve. Geologists believe the reservoir could be as large as 125 miles in diameter, or approximately 12,000 square miles, and could require 50,000 wells and 30–50 years to develop fully, with an initial budget of US$12 billion required. The project again relies upon exporting natural gas to Asia-Pacific countries and involves not only the development of the Kovykta field but also the laying of a new 2,500-mile circuitous pipeline in order to take the natural gas to market. Kovykta's potential total recoverable helium reserves are estimated at 94 Bcf (total gas 55 Tcf). One possible route for the pipeline would take it overland from Irkutsk in Russia through China and onward to South Korea, via a 310-mile subsea pipeline across the Yellow Sea (Figure 6.3) (Griffith, 2005: slide 7). An alternative route would take the natural gas to market further north, to Sakhalin (Figure 6.4) (Sagers, 2007: slide 4).

The history of the Kovykta field has been dominated by technical difficulties, political changes and other uncertainties.

Development of the Kovykta Field

1987–2004: The field was originally discovered in 1987. Located in the northern part of the Irkutsk Oblast, in the Zhigalovo and Kazachinsko-Lensk districts, it was initially assessed to have a potential of 7 Tcf recoverable reserves, classifying it as a modest 'regional project'. In 1997 BP became interested when the reserve was estimated to hold potentially 35 Tcf of recoverable natural gas. It was also found to contain 0.28 per cent helium and potentially 40 per cent of Russia's helium reserves. In 2003 TNK-BP held the majority interest (62.89 per cent) in the Kovykta development company RUSIA Petroleum, and in 2004 a trilateral study between RUSIA, CNPC (China) and Kogas (Korea) suggested that the final market place for Kovykta gas be 10 per cent in Irkutsk (retained for local market), with 60 per cent sold to China and 30 per cent to South Korea.

2005–2006: In 2005, with its export monopoly under threat, Gazprom claimed that the production sharing agreement for Kovykta ran counter to state interests, and in 2006 Russian prosecutors threatened to withdraw RUSIA's licence over alleged violations of environmental law. In December of that same year the Russian Federal Nature Utilization Control Service (Rosprirodnadzor) sent documents to the Prosecutor General's Office alleging violations committed during the construction of a gas pipeline running from Kovykta to the Zhigalovo district. At the same time an official stated that TNK-BP must accommodate Gazprom into the Kovykta gas project or face further sanctions for licensing non-compliance.

2007–2009: During early 2007, Rosprirodnadzor claimed that RUSIA had only extracted 33.8 million cubic metres of natural gas against its yearly anticipated commitment of 9 billion cubic metres (targeting the expected requirements of the Irkutsk region) as stated under the terms of the 2006 contract and, as a result, Gazprom should receive a maximum stake in RUSIA. RUSIA was given three months to remedy the situation or have its licence revoked (an impossible task as there was no demand for the gas locally). Applying more pressure, Gazprom stated in May 2007 that it was not interested in taking part in Kovykta as the company did not estimate the possibility of successfully marketing the gas to be very high. By May TNK-BP had offered to sell its Kovykta stake and 50 per cent of Eastern Siberian Gas Company to Gazprom for US$700–900 million (valued at US$600–700 million by Gazprom). During February 2008 TNK-BP raised the stakes, claiming its interest in the field was now worth US$1 billion. By May of the same year, Gazprom said that it was backing out of the deal, and by November it was struggling to raise the necessary financing for its projects amidst the financial crisis. By February 2009 a deal was reported as 'unlikely' to be completed (Poussenkova, 2007: 48–52; Kashirtev, 2007: 1–14; Kontorovich, 2007: 74–81; Paik, 2008: 5–8; Griffith, 2005: slides 4–9). Eleuthera's estimate is that the start-up for the project will not be before 2017.

Despite the enormous potential, development of Eastern Siberia and Yakutia is fraught with formidable problems (Poussenkova, 2007: 14–15).

Geographical problems

Eastern Siberia and Yakutia are characterized by large areas of permafrost with associated harsh climatic conditions (annual air temperature spread +70 °C to –70 °C), with resulting dramatic increases above normal construction and raw

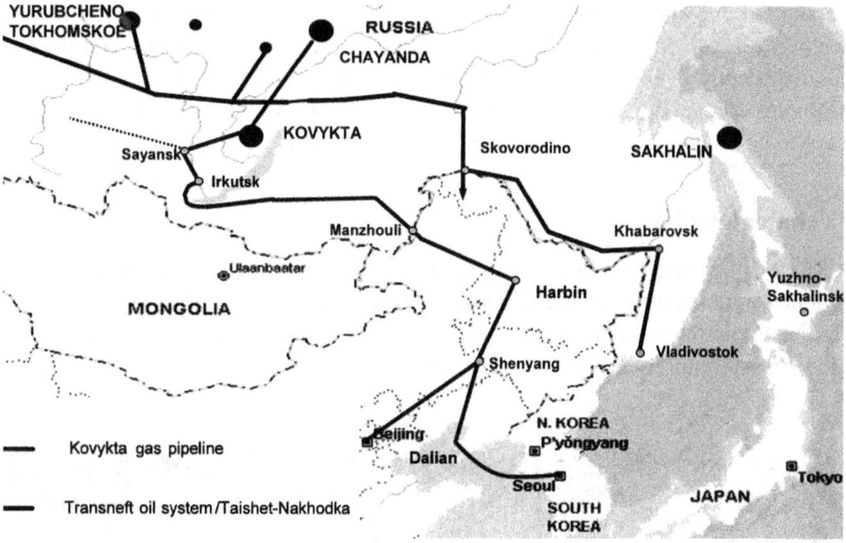

Figure 6.3 Possible pipeline routes to Korea (source: TNK-BP).

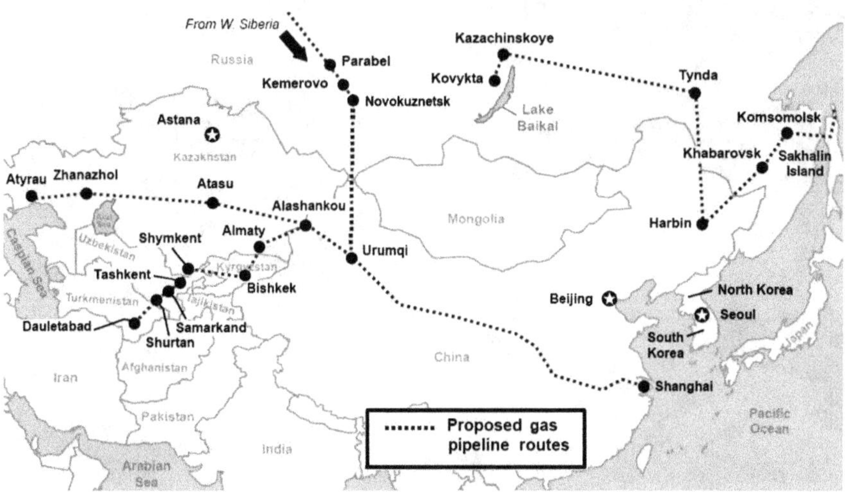

Figure 6.4 Possible pipeline route to Sakhalin (source: CERA).

material costs (Figure 6.5). Chayanda and Kovykta are located at potentially prohibitive distances from existing pipelines. The additional problem of a limited transportation infrastructure – in particular, inadequate roads, bridges and electricity supply – means that initial significant infrastructure investment would be essential to develop either field. With its fragile ecosystems, 'no-spill' and environmentally friendly standards and procedures must be adopted in all phases of project development, to protect these natural areas of global significance. This again will increase costs.

As areas, Eastern Siberia and Yakutia are characterized by one of the lowest population densities in the world, with a trend towards depopulation resulting in a chronic shortage of qualified personnel in the region. Oil service companies created locally over the last two decades have ceased to exist. Additional social problems created by drugs, alcohol, organized crime and the 'oil' dollar will also have to be considered and must not be underestimated.

Technical problems

Reservoir pressure is generally below hydrostatic pressure, and associated well production tends to be relatively small. Development of the fields would require significant additional expenditure for the purchase of compressors, in addition to the stations required along the lengthy pipelines.

Figure 6.5 Eastern Siberia (source: UKAEA).

As the fields contain multiple valuable products (oil, methane, ethane, propane, butane and helium), which must be produced simultaneously, the construction of oil pipelines, gas pipelines and condensate pipelines will be required (again, significantly raising costs). These would be required in addition to any pretreatment, dehydration, sour gas removal, helium purification, liquefaction, storage and transportation equipment. These same large-scale extraction and processing facilities, coupled with transportation and pipeline costs, may make the projects economically unviable (Sagers, 2007: slides 3–10). The author is unaware of the existence of any railcars capable of transporting liquid helium, but is aware of some early stage prototypes.

Historical exploration of the territory has been minimal with only very limited well control, creating a huge gap between explored reserves and those forecasted. There also appear to be few incentives, if any, for international subsurface users to invest in exploration.

Legal and financial problems

Indigenous rights must be taken into account as well as the associated political risk and legal obstacles created with the development of 'strategic fields'. Final price negotiations are also difficult as Gazprom holds the export monopoly.

Enormous expenditures are required not only for the development of hydrocarbon reserves but also for essential transportation and social infrastructure (Kashirtsev, 2007: 2–14). Exploration costs are likely to be at least treble that of other areas, political risk is high and is not to be underestimated, and available project finance and access to capital markets is at an all time low. Project economics are also heavily dependent upon the economies of China and Korea and their associated demand for energy.

Likely outcome

The development of Kovykta and Chayanda will only become attractive to investors if the gas is exported to China and Korea (Paik, 2008: 40–2), as this is perceived as its only significant market. Gazprom speculates that China will require its eastern gas only after 2015 and South Korea no earlier than 2020, making development financially attractive only at that point.

It seems, therefore, that the implementation of Kovykta and Chayanda will hinge primarily on the results of negotiations with China. To date no agreement has been reached, and negotiations deadlocked and hugely influenced by the agreement and deliveries from Sakhalin 1 between Exxon Mobil and China National Petroleum Corporation (CNPC), reducing the short-term market potential that would have justified developing the fields. Gazprom's focus, therefore, appears to be westward (Yamal and Shtokman) at the current time rather than eastward (Kovykta and Chayanda), with the result that the development of Kovykta and Chayanda will be delayed until 2017–30, hindering the region's development.

The impact of Russian helium on future global helium supply

In order to predict and illustrate the impact of Russian helium on future global helium supply (Figure 6.6), I have made the following assumptions.

- Helium plants worldwide, both current and projected, achieve 95 per cent of nameplate capacity.
- New plants scheduled to start up within the next five years in Russia, USA, Algeria, Qatar and Australia come on stream on time.
- Projected maximum withdrawal rates from Cliffside of 2.1 Bcf per annum (straight line) are unachievable in the longer term, and a more realistic formula of 20 per cent of the remaining/declining total reserve once it has been depleted to 10.5 Bcf has been used.
- Hugoton depletion rates (excluding coalbed methane) have been estimated at 15 per cent per annum.
- Orenburg depletion rates have been estimated at 5 per cent per annum.
- Global demand for helium remains flat during 2009 and 2010, increasing by 4 per cent per annum thereafter.
- Russian production remains reliant upon Orenburg between 2010 and 2016, with one other source coming on stream at 270 MMscf per year in 2017.
- The reported reserves for Cliffside are not overstated and the 2.5 per cent margin of error based upon historical withdrawal and reinjection rates is valid.

Raw data for Figure 6.6 was compiled from the following sources: Pacheco, 2007: 76–7, 2008: 78–9, 2009: 74–5, 2010: 72–3; Garvey, 2007: 1–5, 2003: 32–3, 2004: 34–5, 2005: 31–3, 2007: 22–3, 2008: 30–4, 2009: 24–8.

Although it seems clear that Russia is blessed with significant reserves of helium, there are formidable technological challenges associated with developing its resources and these can only be overcome with enormous investment. As can be seen in Figure 6.6, Russian production will have a negligible effect on global helium supply in the short term, but in the long term its reserves are expected to become significantly more important.

If plants in Russia, USA, Algeria, Qatar and Australia can come on stream on time, pressure in the helium supply chain will continue to be relieved between 2010 and 2015. Should start-up dates for those plants fall behind by a year or demand rise by more than 4 per cent, the developing shortfall will be brought forward by one to two years. It has been said to me a number of times by employees of industrial gas companies that after 2007, the recession was the best thing that could have happened to the helium supply industry in 2008 and 2009.

Any shutdown of the larger worldwide helium plants for annual maintenance or equipment failure after 2014 will impact supply significantly. The US federal reserve of Cliffside should work very effectively as a flywheel (National Research Council, 2000: 22–4) but new sources must be found and developed at a faster rate than it is being depleted.

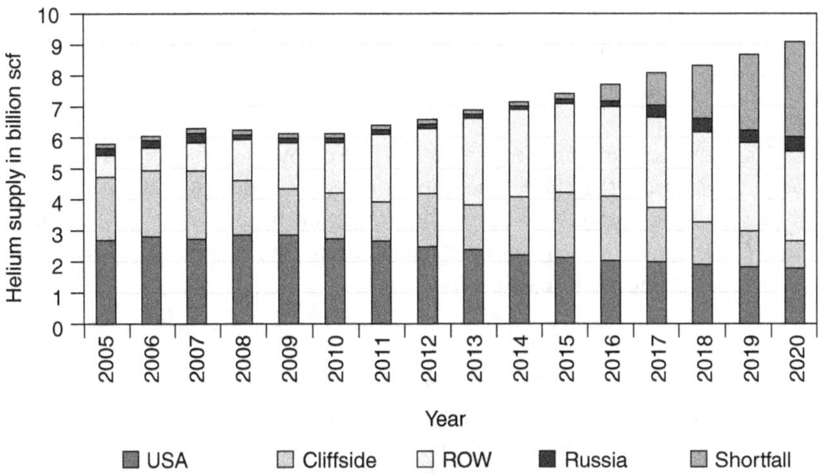

Figure 6.6 The impact of Russian helium (source: *Eleuthera Capital*).

As the Committee on Understanding the Impact of Selling the Helium Reserve has stated,

> in response to its charge, the committee finds that selling off the helium reserve, as required by the 1996 Act, has adversely affected critical users of helium and is not in the best interest of the U.S. taxpayers or the country.
>
> (National Academy of Sciences, 2010, sect. 0–1)

> Helium from BLM's Federal Helium Reserve is not 'new' helium; it is a supply that was built up from decades of federal helium conservation and storage programs. Obviously, this imbalance of supply and increased demand is not sustainable in the long term; additional sources of helium are needed across the globe – and the sooner the better.
>
> (Theiss, 2007: 32)

Significant investment in Cliffside will be required in the longer term to maintain helium production (some edge wells have watered out), increasing US Treasury debt and raising the baseline price of helium. It takes at least three years to develop new sources, and without further exploration activities, a peak helium production scenario is looming within five years. Consequently, with the continued depletion of Cliffside, the world's reliance for helium will become more weighted in favour of production from Algeria and Qatar. The development of Iran's South Pars field would probably produce the biggest single impact on global helium supply within the next decade, although it is unclear whether anyone would be able to sell Iranian helium for fear of falling foul of United Nations sanctions.

The future

Leslie Theiss is Manager of the Helium Operations Office for the US Government Bureau of Land Management in the Amarillo Field Office. She has offered some portentious observations on the future supply of helium.

> Unless new helium supplies are found and/or overseas helium production output increases, the status quo cannot satisfy continuing helium demand throughout the world.... We have no projection for an easing of the shortage. Clearly new helium sources, substitution, and conservation measures are needed to meet future demands.
>
> (Theiss, 2007: 34)

In a subsequent interview article in the *Chicago Tribune*, she went on to say 'We're pedaling as fast as we can here, but we just can't produce enough. One-third of the world's helium comes from our little place here. That's kind of frightening' (Theiss, 2007). The world should be listening.

Qualifications

The factual details included have all been taken from sources that are believed to be reliable but the author makes no representation as to their accuracy. The assessments and conclusions reached are those made by the author after considerable research but no representation is made that they are endorsed by any other party or that any other party would necessarily reach the same assessments or conclusions from the same information. The author accepts no responsibility or liability for any action that the reader may take as a result of this chapter.

References

Alger, M. (2009) 'Natural Gas Liquefaction', Presentation in Qatar, 10–12 January, slide 20.

Andreev, I. (1995) 'Helium Industry in Russia and Experience with the Development and Operation of Helium Equipment', *Chemical and Petroleum Engineering*, Vol. 31(1–2), pp. 73–85.

Ben'yaminovich, O. (1995a) 'Industrial Processing of Helium Containing Natural Gas and Petroleum Associated Gas in Russia', *Chemical and Petroleum Engineering*, Vol. 31(1–2), pp. 86–8.

Ben'yaminovich, O. (1995b) 'Helium in Russia', *Chemical and Petroleum Engineering*, Vol. 31(1–2), pp. 64–6.

Garvey, M. (2002) 'Helium Market Update: Demand Growth Outpacing Supply', *Cryogas International*, February, pp. N1–N5.

Garvey, M. (2003) 'Respite in Tight Helium Supply but Demand Expected to Pick Up Mid'03', *Cryogas International*, February, pp. 32–3.

Garvey, M. (2004) 'Helium Update: New Crude Enrichment Dedicated', *Cryogas International*, June, pp. 34–5.

Garvey, M. (2005) 'Helium Market Update: Rising Costs for Crude Driving Prices Higher', *Cryogas International*, July, pp. 31–3.

Garvey, M. (2006) 'The Helium Market: Demand Floats Higher as Supplies Sink', *Cryogas International*, October, pp. 18–20.

Garvey (2007) Garvey, M. 'Up, Up, and Away – Where has all the helium gone?', *Cryogas International*, October, pp. 22–3.

Garvey, M. (2009) 'World Helium Market Update', *Cryogas International*, October, pp. 24–8.

Griffith, H. (2005) 'OJSC RUSIA Petroleum, Kovykta Project Development', Presentation to James A. Baker III Institute and the Carnegie Endowment for International Peace, slides 1–11.

Kanekiyo, K. (2005) 'Energy Outlook of Northeast Asia and Regional Energy Partnership', The Institute of Energy Economics Japan Presentation, 17–18 November, slides 1–9.

Kashirtsev, V. (2007) 'Independent Study of Optimum Transportation Corridors for Kovykta Gas Exports, Key Conclusions and Recommendations', pp. 3–14.

Kontorovich, A., Kashirtsev, V., Korzhubaev, A. and Safronov, A. (2007) 'The General Plan for a Gas and Oil Complex in East Siberia and the Republic of Sakha (Yakutia) and its Significance for Russia's Socioeconomic Development and Energy Security', *Far Eastern Affairs*, 1, pp. 74–81.

National Academy of Sciences (NAS) (2010) *Selling the Nation's Helium Reserve*, The National Academies Press, Washington, DC, http://www.nap.edu/catalog/12844.html (accessed 22 December 2010).

Pacheco, N. (2007) 'Minerals Commodity Summaries, 2007', US Geological Survey, pp. 76–77.

Pacheco, N. (2008) 'Minerals Commodity Summaries, 2008', US Geological Survey, pp. 78–79.

Pacheco, N. (2009) 'Minerals Commodity Summaries, 2009', US Geological Survey, pp. 74–75.

Pacheco, N. (2010) 'Minerals Commodity Summaries, 2010', US Geological Survey, pp. 72–73.

Paik, K. (2008) 'Northeast Asian Countries' Oil and Gas Relations with Russia', Presentation in The Hague, Oxford Institute for Energy Studies, 22–3 January, pp. 1–42.

Poussenkova, N. (2007) 'The Wild, Wild East', Working Papers, Carnegie Moscow Centre, pp. 4–15.

Sagers, M. (2007) 'Opportunities and Constraints: Prospects for Russian Oil and Gas Supply to Asia', Presentation in Washington, Cambridge Energy Research Associates, 29 April, slides 3–10.

Stepanyuk, V. (1995) 'Helium Production in Russia', *Chemical and Petroleum Engineering*, Vol. 31(1–2), pp. 67–9.

Theiss, L. (2007) 'The US Government's Role in Providing Helium to Worldwide Markets', *Cryogas International*, October, pp. 32–4.

Theiss, L. (2007) 'Consumers Feeling the Helium Squeeze', *Chicago Tribune*, 5 November.

Timoshilov, V. P., Mastepanov, A. M., Yu, D. and Shelekhov (2008) 'Gas Industry Formation, Additional Factor for Sustainable Development of Russia's Far East and its Integration into Economic Relations System of Asia Pacific', OAO Gazprom, pp. 1–10.

Udut, V. (1995) 'Present State and Development of the Helium Business in the World', *Chemical and Petroleum Engineering*, Vol. 31(1–2): 96–100.

US Department of Interior and Bureau of Land Management (2008) 'Helium Resources of the United States 2007', Technical Note 429: 7.

Vasenev, A. (2006) 'RFE Oil and Gas Projects', US Department of Commerce, slides 1–17.

Yakutseni, V. (1995) 'The Status of Natural Helium Resources, Geopolitics of Production and Use', *Chemical and Petroleum Engineering*, Vol. 31(1–2), pp. 61–3.

Yakutseni, V. (2001) 'Helium Resource Base: Current State, Prospects of Development and Use', VNIGRI, pp. 10–22.

7 India

Harnessing helium from the Earth's interior

Nisith K. Das, Rakesh K. Bhandari and Shri C. Mallik

Introduction

The extraordinary physical properties of helium and its ability to avoid chemical combination with other elements have made it an indispensable element in leading edge technologies involving space, nuclear energy and cryogenic technology, especially superconductivity, MRI and sensor development. Its low atomic mass, inertness, high specific heat and thermal conductivity, coupled with a very low boiling point (4.2 K) are extensively exploited in various practical applications such as inert blanketing gas for crystal growth and in gas discharge lasers for energy transfer to the lasing gases. Its low cross-section for nuclear reactions under neutron bombardment combined with its rather high thermal conductivity makes it an ideal choice for nuclear plant and fusion reactors. Its non-combustibility and buoyancy (second only to hydrogen) make it most suitable gas for balloons and other lighter-than-air craft. A mixture of helium and oxygen is often supplied as a breathing mixture for deep-sea divers and caisson workers and is used in decompression chambers; because helium is less soluble in human blood than nitrogen its use reduces the risk of caisson diseases. Helium is also used in electric arc welding, in growing semiconducting crystals such as silicon and germanium and in refining titanium and zirconium metals to maintain a non-reactive atmosphere. Temperatures in the millikelvins (mK) are routinely achieved using helium as the working fluid. These very low temperatures are utilized in studying smart materials, diluting magnetic alloys, elementary particle scattering, Fermi fluids and quantum computing.

Sources

Despite the high solar and cosmic abundance of helium, it is extremely sparse on the Earth. The atmosphere is the prime source of almost all the noble gases; however, the very lean abundance of helium (~5.2 ppm) in the atmosphere makes it uneconomical and unfeasible for large-scale extraction from this source (see Chapter 8). Natural gas reservoirs are the conventional source of commercial helium gas. It is also found in small quantities in radioactive rocks and in gases emitted from some mineral springs, volcanoes and fumaroles. A few attempts have been made across the world to extract helium from monazite sands through digestion processes, capturing spring emanations by cryocondensation followed by cryosorption (cryogenic adsorption) and from natural gas by adsorption or cryogenic methods. Natural gas containing helium concentration to the tune of 0.3 mole percent or above becomes economically promising for large-scale production. Currently, global commercial helium is supplied principally by four countries: Qatar, USA, Algeria and Russia. A small amount of helium is also produced in China, Canada and Poland. In India, it has been attempted to separate helium from three sources: monazite sand, thermal spring gas containing helium and natural gas. However, commercial production from these sources is yet to be realized.

Recovery processes

Helium-bearing natural gases are conventionally processed to yield refined helium, either by adsorption-based techniques or through cryogenic method, depending upon the level of helium concentration in the feed stream and site of operation. India has made substantial efforts over the past couple of decades to develop the cost-effective and energy-efficient helium separation technology that can be applied to gases containing traces of helium available in Indian soil. Cryocondensation, cryosorption and pressure swing adsorption methods have been employed so as to enrich and purify helium from raw field gases.

Methodology of helium separation

Alongside the traditional method of gas separation based on cryogenic distillation and liquefaction, non-cryogenic processes, like selective adsorption, also offer a promising means for the separation of gaseous constituents. However, where the concentration of the gas to be separated is very small, ~1.0 mole percent in the gas matrix, and the desired purity level of the gas is sufficiently high, serious difficulties crop up in the processing methodology. The separation of lean helium content present in some terrestrial gases is indeed a potentially difficult task to deal with, due to the fact that acquiring an isotherm for a particular gas component, constituting a heterogeneous mixture of reactive and non-reactive gases, is rather convoluted. We came across an analogous scenario in the wake of purifying helium from a spring gas containing 1.4 mole per cent helium mixed with other gases such as nitrogen (91.0 mole percent), argon (~2.0 mole percent), methane (2.5 mole percent), etc. Typical compositions of spring gas and natural gas are given Table 7.1. Even though the helium concentration from thermal springs is typically higher than that of natural gas, the feeble flow rate of spring gases as against natural gas makes them uneconomical for the large-scale extraction of helium.

In order to extract helium from hydrothermal gases, cryogenic and non-cryogenic approaches (hybrid process) were pursued. To make the raw gas ready for processing it was dehydrated and pretreated utilizing an adsorption technique.

Table 7.1 Composition of spring gas and natural gas

Gas	*Spring gas (mol.%)*	*Natural gas (mol.%)*
He	0.50–1.40	0.05–0.10
N_2	82.0–93.0	1.0–40.0
Ar	1.68–2.40	0.0–0.62
O_2	1.00–2.70	0.00–1.40
CO_2	0.03–1.50	0.35–2.90
CH_4	1.0–60.0	22.0–90.0
Flow rate (Nm^3/hr)	1.0–6.0	1,000–7,000

Absorption raises the helium concentration from 1.4 mole percent to 15.0 mole percent, followed by enrichment to a level of around 90.0 mole percent by means of cryocondensation. Finally the enriched helium is purified to Grade A (99.995 mole percent) through cryosorption on activated charcoal beds immersed at LN_2 temperature. An integrated scheme for helium extraction based on cryogenic condensation and adsorption is presented here (Das, 2003).

Enrichment of helium by cryocondensation

With reference to the flow diagram of the enrichment plant, after pretreatment the dry feed gas enters the cold box at a flow rate of 16 Nm^3/hr and passes through the main heat exchanger (HX-I), which is a brazed aluminium plate fin exchanger. This acts to reduce the thermal duty (precooling) of the incoming gas from 312 K to 106 K by counter-flow of the return LN_2 vapour from the gas liquid separator (S3) and from the LN_2 bath of the second heat exchanger, as shown in the flow diagram, Figure 7.1.

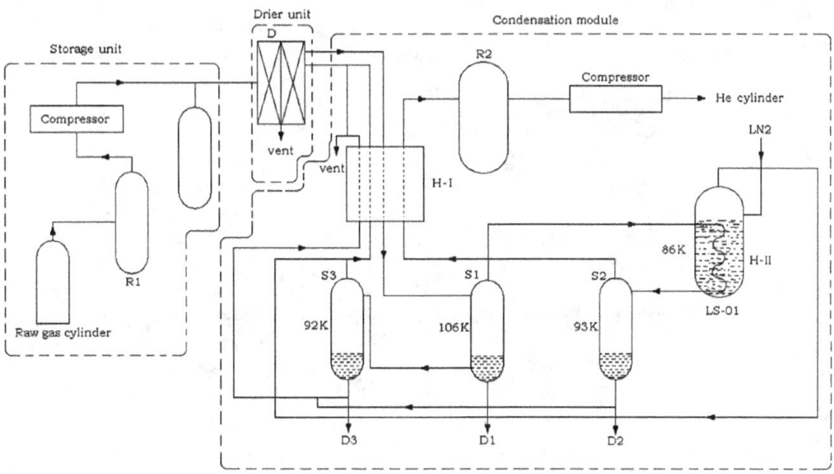

Figure 7.1 Flow diagram of enrichment plant based on cryocondensation.

At this temperature all the components like argon, methane, oxygen and partially nitrogen get condensed at a pressure of 15 bars. Resulting liquid gas mixture is partially flashed and separated in a G-L separator (SI), where the condensed liquid is collected. The exit gas (He+N_2) from this separator, containing 11.0–12.0 mole percent helium and reduced nitrogen at 88.0–89.0 mole percent, passes into the second coiled-tube heat exchanger (HX-II), immersed in a liquid nitrogen bath where the feed gas temperature is further lowered to 91 K. Consequently, at a pressure of 14.5 barg, most of the remaining nitrogen and traces of other gases become liquefied within this HX-II, and helium with small amount of nitrogen is bled out to another G-L separator (S2). This enriched helium (~94.0 percent) is then withdrawn as the product from the cold box and put into the receiver. Accumulated liquid from LN_2 bath of the HX-II heat exchanger is flashed into a separator (S3), wherefrom the gas is passed through the main heat exchanger, HX-I, and is vented out. This cooled gas is used up in cooling down and condensing out part of the incoming gas. The necessary refrigeration for the cold box is obtained from two sources: an internal refrigeration is provided by the expansion valve that reduces the temperature by 10–12 °C, and the pressure from 15 bars to 2–3 bars; and the external refrigeration is obtained from the LN_2 injection into the exchanger vessel (LS-OI). In this way an initial cooling of all the equipment and piping is achieved. LN_2 is used to make the plant function properly, and additional LN_2 injection supplements the loss of nitrogen vapour from the LN_2 bath and at the same time maintains the stability of the requisite temperature levels inside the cold box. Figure 7.3 displays the elemental composition of feed and product gases. The presence of a finite vapour pressure of LN_2 at S2 (93 K) is present as an impurity, which limits the helium purity to about ~90 mole percent.

Figure 7.2 Image of the helium enrichment plant based on cryocondensation.

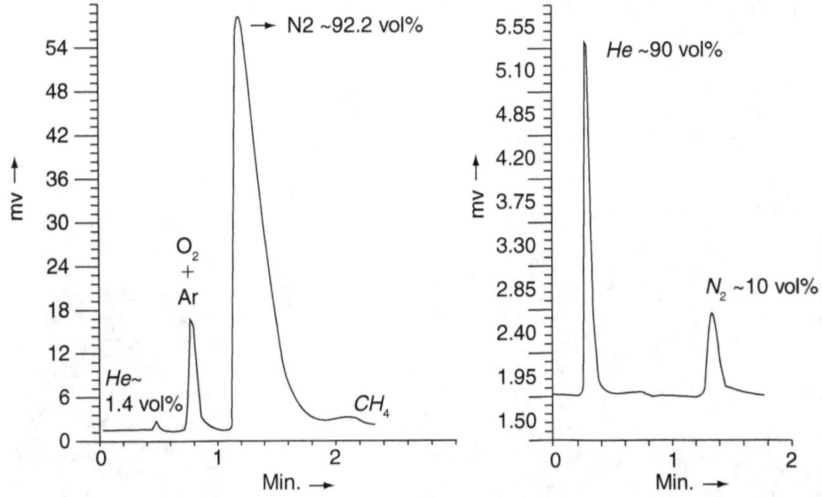

Figure 7.3 Elemental composition of feed and product gases. Presence of finite vapour pressure of LN_2 at S2 (93 K) comes out as impurities, which limits the helium purity to about ~90 vol%.

Final purification by cryosorption

Purification of enriched or semi-pure helium to Grade-A (99.995 mole percent) has been accomplished through selective cryosorption of unwanted gases, primarily nitrogen, contained in helium. Since the primary constituent of the impurities is nitrogen, adsorption takes place at a temperature close to the normal boiling point of liquid nitrogen. In the purification phase, the impure helium gas at an elevated pressure of 125–150 bars goes first through a counter-stream exchanger, where it is precooled by the purified helium gas resulting from the adsorbent bed, and the air (if any) condenses out to the water separation drum. Subsequently, the gas to be processed flows through four columns, where all gases except helium are adsorbed on active charcoal maintained at boiling temperature of LN_2 (–196 °C). This pure helium is then passed through the heat exchanger located inside the cold box before being removed as product helium. Since adsorption is a cumulative and reversible phenomenon, after each run of seven hours the adsorbents become saturated and are regenerated. This is carried out by heating the adsorbents to a temperature of 110 °C using an electric heater attached to the bed. The trapped impurities are extracted, thus reactivating the bed.

A sampling port provided at the outlet of the purifier enables us to analyze the product purity with the help of a high performance Quadrupole mass spectrometer (Pfeiffer Vacuum, model QMG-422). As is evident from the chromatogram presented in Figure 7.6, the product purity of the helium attained is better than 99.995 mole percent (Grade A).

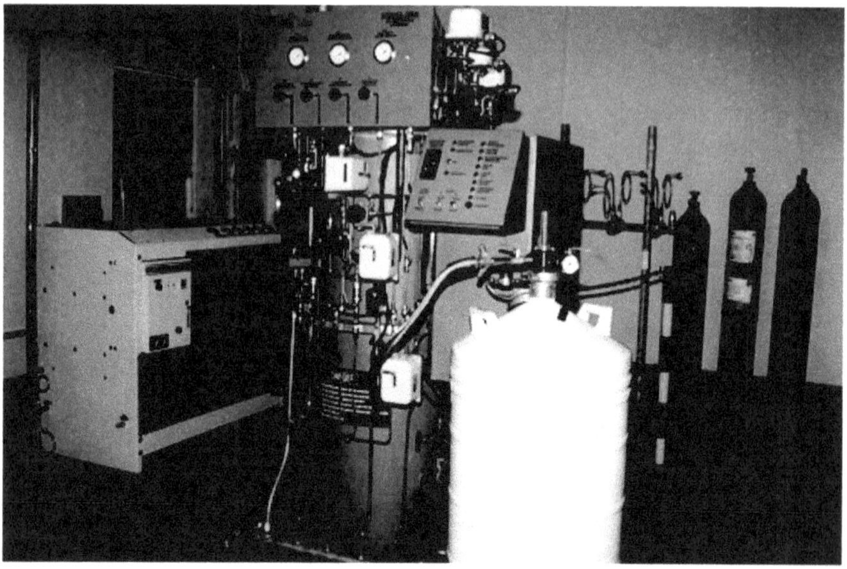

Figure 7.4 Image of the helium purification by cryosorption system.

The helium purification by cryosorption system presented in Figure 7.4 consists of the following components. The kernel of the purifier comprises two phase separators, four activated carbon columns, a vacuum-jacketed cold box, pre-filters and post-filters inclusive of process logic controller, with a local operator panel attached to the purification skid. Since nitrogen is the principal volatile impurity in the process gas, adsorption takes place at the boiling temperature of nitrogen. The carbon columns are, therefore, left immersed in a liquid nitrogen (LN_2) dewar within a vacuum enclosure. The technical data of the helium purifier is presented in Table 7.2.

The uniqueness of this purifier rests in its capability of tackling impurities as high as 40 percent, as against the traditional purifier that can only deal with impurities in the tune of 5–10 mole percent. It is worth mentioning that this plant is the first of its kind in India.

Table 7.2 Technical data for helium purifier

Flow rate	*17 Nm³/hr*
Operating pressure	125 bar g
Operating temperature	77 K
Input purity	40 vol%
Output purity	99.995 vol%
Reactivation	Thermal
Cooldown	Liquid nitrogen (LN_2)
LN_2 consumption	200 litre/7 hours

Operational details

As shown in the flow diagram (Figure 7.5), impure or enriched helium gas from the impure cylinder bank enters the gas bag, maintaining a maximum pressure of 1.75 bar via a regulator. A helium compressor draws the gas out of the bag and pushes into the first phase separator (PS-I) through the micron filter (F2). In PS-I the gross level of moisture, aerosols and oil droplets are congregated and are removed from the gas stream. This gas is then directed to the second phase separator (PS-II) held within the vacuum insulated space, but outside the LN_2 dewar.

Any remaining moisture, if present, is stripped off here and the condensates are blown down automatically with the aid of a pneumatically actuated solenoid valve. Subsequent to this, the process gas goes through a series of carbon adsorbers, maintained at a temperature of −196 °C, where all unwanted gases except the helium are adsorbed. Purification proceeds smoothly for seven hours and the pure helium, before extraction, flows through a post-filter (F1) present in the down stream of the adsorber unit to preclude the possibility of getting any dust through. At the same time it flows through a valve maintaining pressure at 200 bar toward the cylinders, pre-evacuated for filling with pure helium. With regard to Figure 7.3, the cold stream (80 K) of purified helium at the outlet of the purifier is passed through counter-stream exchanger, enabling the input gas to be pre-cooled at the expense of the temperature of the product helium. The chromatogram shown in Figure 7.6 reveals that grade-A helium is obtained as the product of the helium purifier.

Regeneration phase

As adsorption is a cumulative phenomenon, the activated carbon gradually becomes saturated by trapped impurities. At a given pressure, if the adsorbents are left in a low-temperature environment the rate of adsorption is scaled up. Again, the adsorption being a reversible occurrence, thermal reactivation is piloted to remove the impurities, thereby rejuvenating the adsorbents. The

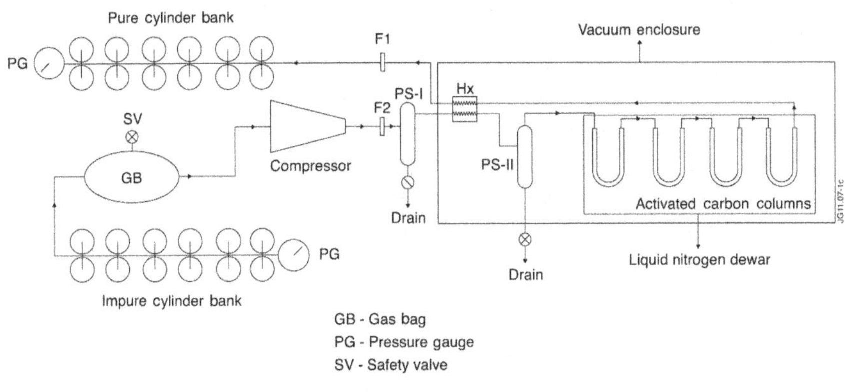

Figure 7.5 Flow diagram of purification unit by cryosorption.

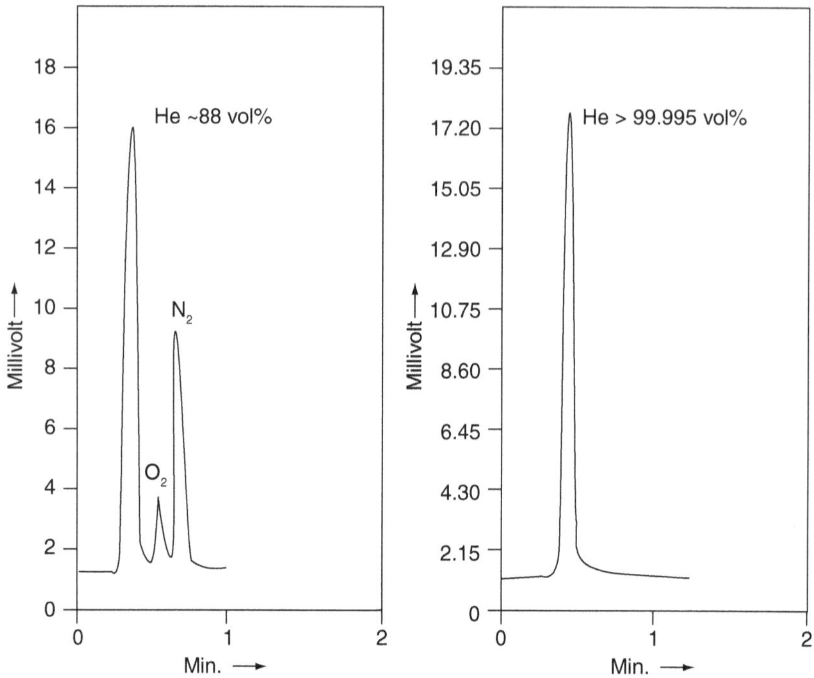

Figure 7.6 Chromatogram of product of helium purification.

carbon columns in the present case have been designed for a cycle time of seven hours with a feed having impurities less than 40 mole percent. In the course of plant operation, the contaminants adhere to the surface of the activated carbon, and as such reactivation is necessary to make it acceptable for purification. Regeneration is triggered by turning the purification 'OFF' and turning the heater 'ON' in order to decontaminate the adsorbent in the purifying columns. This is accomplished by depressurizing the columns to atmospheric pressure, followed by heating the adsorbents at a temperature of 110 °C for three hours; after drainage of LN_2 from the dewar, of course. A purging by pure helium of the entire system and gas lines is carried out toward the end of the regeneration. Furthermore, in order to exclude any moisture ingression within the purifier, the final pure helium is backfilled at a pressure of 5 bar. When the adsorption columns attain the ambient temperature, the requisite amount of LN_2 is poured into the dewar, preparing the system for operation.

Helium separation by adsorption

In India, the exploration of different sources has led to the discovery of small amounts of helium in natural gas, which is discharged through the natural gas wells in GCS Kuthalum, Nagapattinam District, Tamil Nadu. Besides the conventional method of helium recovery utilizing cryogenic separation, which is an energy-expensive process, a non-cryogenic helium purification system has been developed to separate out helium from the natural gas stream by means of the PSA process. The PSA plant produces a constant stream of helium with a concentration of around 98.5 mole percent from a natural gas feed stream with a helium content of around 0.06 mole percent, with an overall recovery rate of around 61 percent. The PSA system developed for helium purification involves four distinct components, as presented in Figure 7.7.

The multicomponent gas available from the wellheads of the natural gas field at GCS Kuthalam predominantly contains methane, besides heavy hydrocarbons, CO_2, nitrogen and helium. The raw natural gas is collected in a gas storage vessel at a pressure of 16 bar. The gas leading to the pilot plant is taken at a reduced pressure of 4–5 bar, and is routed through a refrigerated gas dryer to remove the bulk moisture content present in the feed stream. This is fed into the system at about a pressure of 4.5 bar. In contrast to the previous works (D'Amico *et al.*, 1996; Knaebel and Reinhold, 1998), where the feed gas contained about 2–4 mole percent helium, the present feed stream contains only a lean amount of helium, approximately 0.06 mole percent, which makes the system more challenging and involves four stages instead of two.

Process description

The pilot plant has four stages, with each stage consisting of three adsorber beds. The complete system comprises a network of piping and valves with associated instrumentation and control equipment that is integrated with the complete system. The number of adsorber beds used represents an economic balance between the capital cost of the pilot plant and the desired product quality.

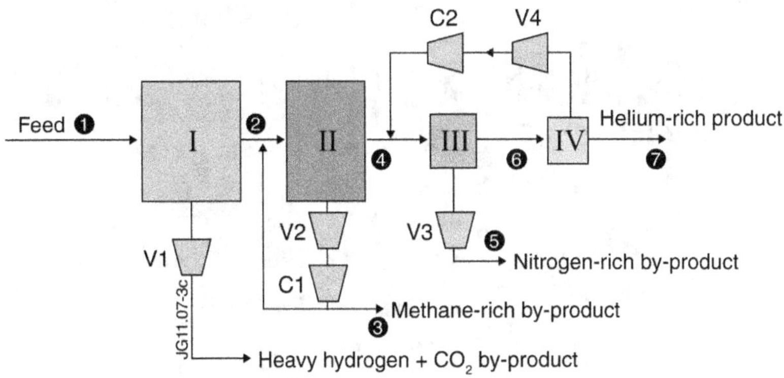

Figure 7.7 Schematic of the four stages of PSA system for helium purification: stage I – pretreatment of feed gas and removal of heavy hydrocarbons; stage II – recovery of methane (CH_4); stage III – elimination of nitrogen; and stage IV – final separation of helium.

Figure 7.8 represents the skid for stages I and II. Stage I is adopted to remove the heavier hydrocarbons and carbon dioxide from the feed stream, allowing methane, nitrogen and helium to flow and continue on to stage II. In stage I, the heavy components include carbon dioxide (CO_2), ethane (C_2H_6), propane (C_3H_8) and butane (C_4H_{10}), and are preferentially retained on the adsorbent, while the less absorbable effluent gas comes out through the exit end. Stage II adsorbs and rejects methane as the heavy product, allowing nitrogen and helium to flow and move on to stage III. The methane-rich heavy product from stage II is partially recycled for use in stage I as purge gas for the purpose of regeneration. Methane of high purity from the waste gas is recovered as the secondary product in this stage. Stage III adsorbs and removes most of the nitrogen, and allows a helium-rich mixture to pass through and advance to stage IV. The nitrogen-rich heavy product represents the third and final waste stream. Stage IV operates in a similar way to stage III; it adsorbs and casts off residual nitrogen, allowing purified helium to pass through as the final light product. The heavy product (which is a mixture of nitrogen and a little helium) from stage IV is compressed and optionally recycled to the inlet of stage III for the purpose of boosting the overall helium recovery. The entire process is controlled by microprocessor-based logic controllers (PLCs). Figure 7.8 shows some of the components of the helium pilot plant. In the present case, the absorbents used are silica gel in stage I, activated carbon in stage II, and zeolites in stages III and IV. The adsorption system operates by a sequence of steps, called a PSA cycle. Each stage in the purifying

Figure 7.8 Image of stage I and stage II machinery.

system has an associated cycle, as presented in Figure 7.7. The adsorber beds in each stage sequentially undergo the following seven steps: (1) adsorption; (2) co-current depressurization for pressure equalization; (3) blowdown to atmospheric pressure; (4) evacuation to the lowest pressure; (5) purge with product at the lowest pressure; (6) counter-current pressurization by pressure equalization; and (7) re-pressurization with product to a feed pressure.

Adsorption primarily occurs during the feed step. When the more strongly adsorbed components reach the top of the adsorber bed, and are about to contaminate the product, gas feed flow is discontinued and switched to the second adsorber bed. The first adsorber bed then undergoes regeneration by depressurization (called the blowdown and evacuation step), followed by the purge step. During the purge step, the enriched heavy component(s) are desorbed from the adsorber bed at low pressure by admitting the light components, instead of relying on evacuation (vacuum) alone (Das *et al.*, 2004). These steps induce desorption and release of the 'heavy' product. The heavy product will contain both of the more adsorbable components, which are enriched (relative to the feed) by the adsorber bed, and some of the less strongly adsorbed components released from the voids. The pressurization step, in which light product is used to repressurize and adsorb, can be thought of as regeneration, since it tends to displace the more strongly adsorbed constituents as it enters the adsorber bed. It can also be viewed as an adsorption step, since as pressure increases, the light components also get adsorbed to a greater extent. Pressurization with feed causes some of it to be adsorbed, while some feed remains in the gas phase to increase the column pressure. The process flow diagram showing the schematic flowsheet of the piping and instrument diagram of each stage is given in Figure 7.7. Stages I and III include equalization steps, during which one adsorber bed depressurizes into a parallel adsorber bed, partially pressurizing it. There are two equalization steps per adsorber per cycle, because each adsorber bed releases gas to one adsorber bed and receives gas from another adsorber bed. For example, stage I delivers gas to stage III, but stage I also receives gas from stage II.

Table 7.3 Description of the functions of the stages and steps of the helium pilot plant

Stage Step	*I*	*II*	*III*	*IV*
1	Feed	Pressurization with feed	Feed	Feed
2	Equalization	Feed	Equalization	Blowdown and evacuation
3	Blowdown and evacuation	Rinse	Blowdown and evacuation	Pressurization by light product
4	Purge	Blowdown and evacuation	Equalization	
5	Equalization	Pressurization by light product	Pressurization by light product	
6	Pressurization with feed			

Results and discussions

Figure 7.9 presents the chromatogram of the composition of the feed natural gas at the GCS Kuthalam site. It is observed that the major constituent of the feed stream is methane, while helium content is as low as 0.06 mole percent. The compositions of the feed and product gases in four stages of the PSA are shown in Table 7.3.

The product gas out of stage I after the removal of the majority of the heavy hydrocarbons and carbon dioxide is primarily methane with a marginal increase in helium and nitrogen concentrations, and it is also the feed gas of the stage II. The product gas of stage II, which is also the input gas of stage III, is essentially an admixture of methane, nitrogen and helium; whereas the output of stage III, which is also the feed gas of stage IV, consists of nitrogen and helium, accompanying a small quantity of methane. Any residual methane is completely removed from the gas stream and almost all the nitrogen is eliminated, essentially leaving purified helium as the yield of stage IV. Table 7.4 shows the operating parameters of the four stages of the PSA helium pilot plant. The plant intake is a

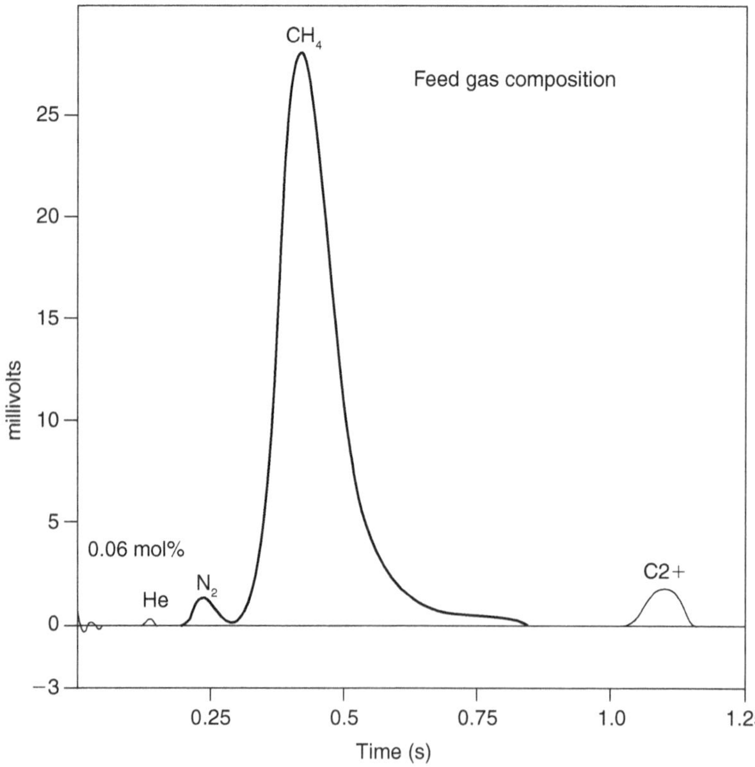

Figure 7.9 Chromatogram of the composition of the feed natural gas at the GCS Kuthalam site.

Table 7.4 Composition of feed and product gases at respective stages of the pressure swing adsorption (PSA) helium pilot plant

Composition	*Feed of stage I vol %*	*Product of stage I/feed of stage II vol %*	*Product of stage II/feed of stage III vol %*	*Product of stage III/feed of stage IV vol %*	*Product of stage IV vol %*
He	0.06	0.08	1.40	15.65	99.00
N_2	1.18	1.27	16.10	72.03	1.00
CH_4	88.50	98.42	82.50	12.32	0.00
CO_2	0.40	0.00	0.00	0.00	0.00
C_{2+}	9.86	0.23	0.00	0.00	0.00

Table 7.5 Operating parameters of the four stages of the PSA helium pilot plant

Parameter	*Stage I*	*Stage II*	*Stage III*	*Stage IV*
Feed pressure (bar)	4.50	4.50	2.25	2.00
Feed flow rate (slpm)	820.00	600.00	60.00	4.90
Product pressure (bar)	4.50	2.25	2.00	1.50
Product flow rate (slpm)	600.00	60.00	4.90	0.90
Regeneration pressure (bar)	1.3 (blowdown) 73 m bar (evacuation)	1.25 (blowdown) 75 m bar (evacuation)	1.1 (blowdown) 60 m bar (evacuation)	1.1 (blowdown) 60 m bar (evacuation)
Purge amount (slpm)	81	425 (rinse flow rate)	0.6	0.5
Helium in feed (vol%)	0.06	0.08	1.40	15.65
Helium in product (vol%)	0.08	1.40	15.65	99.00
PSA cycle time (min.) (three-bed cycle)	24.00	4.50	6.75	36.00
Bed composition (kg/bed)	Silica gel (~198)	Activated carbon (~38)	Zeolite 13X (~3)	Zeolite 13X (~3)
Column dimensions (H, height (cm); OD, outer diameter (cm))	H = 129.54 OD = 50.80	H = 104.14 OD = 30.48	H = 152.40 OD = 6.35	H = 152.40 OD = 6.35

constant feed of 50 Nm^3/h of natural gas containing 0.06 mole percent helium. The output is a constant stream of helium (>99.0 mole percent) at approximately (18 l/h). Practical implementation of this new system brings out the fact that loss of helium in stages II and III needs to be minimized through more precise cycle times and further fine-tuning to optimize the pressure ratios. Helium recovery from the four stages, as observed, is as follows: stage I, ~97 percent; stage II, ~87 percent, stage III, ~81 percent; and stage IV, ~87 percent. Overall helium recovery from the pilot plant is ~61 percent (Das *et al.*, 2008). The helium pilot plant presented above was installed at the GCS Kuthalum site during 2007 and commissioned in early 2008. The plant was formally inaugurated and dedicated to the nation on May 11, 2008. The yield of the PSA plant is given in Figure 7.10.

To the best of our knowledge it has been demonstrated for the first time that employing the non-cryogenic PSA technique, albeit on a pilot scale, helium has been purified to a level better than 99.0 mole percent starting from a concentration as low as 0.06 mole percent in the feed gas. The system is completely automated and separation has been achieved at room temperature. This enables considerable savings in energy and manpower compared to the low-temperature

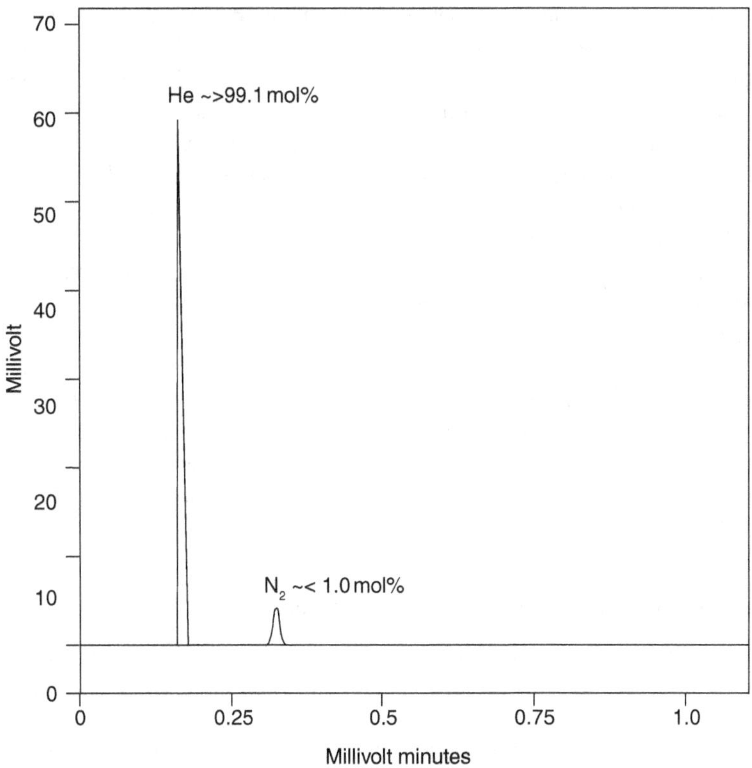

Figure 7.10 The yield of the PSA plant (product purity of helium from PSA plant).

separation process. We are of the opinion that extrapolation of existing natural gas reserves in India, on a commercial scale, would be able to meet the requirement for the domestic consumption of pure helium through the scaling up of such pilot plants.

Probing helium for major seismic studies

Helium concentration is regularly determined as a part of our helium separation program. In the course of helium measurement in spring emanations it was observed that occasionally the helium concentration scales up to a level significantly higher than its average ahead of an imminent major seismic tremor. Having noticed this, we were inclined to investigate this subtle time-dependent observation to decipher the physical implications of such anomalous fluctuations vis-à-vis major seismic events. To this end, an advanced experimental laboratory comprising a micro gas analyzer, to monitor radon and its decay products has been set up at the thermal spring site, Bakreswar, West Bengal. It has been noticed that prior to a major seismic event the concentration of helium and radon reach abnormally high levels ($>\pm2\sigma$, where σ is the standard deviation of the data set) for a considerable time period before returning to average concentration levels. Two out of several curious observations made are presented here. Figure 7.11 displays the anomalous fluctuations of radon concentration prior to the Pakistan quake (7.9 M) on October 8, 2005, and Figure 7.12 represents the abnormal variation of radon concentration ahead of the Sumatra quake and tsunami (9.1 M) that occurred on December 26, 2004. Repeatability of these distinct signatures at a tectonically sensitive spring site has led us to take up the activity, and it is found that helium and radon can serve as useful tools for seismic studies (Das *et al.*, 2006a).

Statistical analysis of the field data was pursued with a view to unravel the effects of the observed abnormal variations of gas concentrations with time. The recorded temporal data were found to be distinctly irregular, yet non-random. The governing mechanism underlying the dynamics of radon fluctuation is inherently non-linear and shows sensitive dependence on the initial conditions, capturing the impression of a non-linear, deterministic process. External perturbations on the under-earth fluid reservoir resulting from seismic events are accountable for the observed disparity in the statistical measures derived from two different geophysical conditions of the same system (Das *et al.*, 2006b).

Plausible mechanism for enhanced gas emission

The earthquake preparation process is characterized by the release of accumulated stress within the earth, causing a characteristic variation of lithospheric strain. It is established that the earthquake initiation process is marked by a substantial activation of the stress-strain conditions, while the strain rate is enhanced by orders of magnitude. A strain variation of the order of 10^{-8} is sufficient to allow liberation of trapped volatiles from the host rock into conducting channels

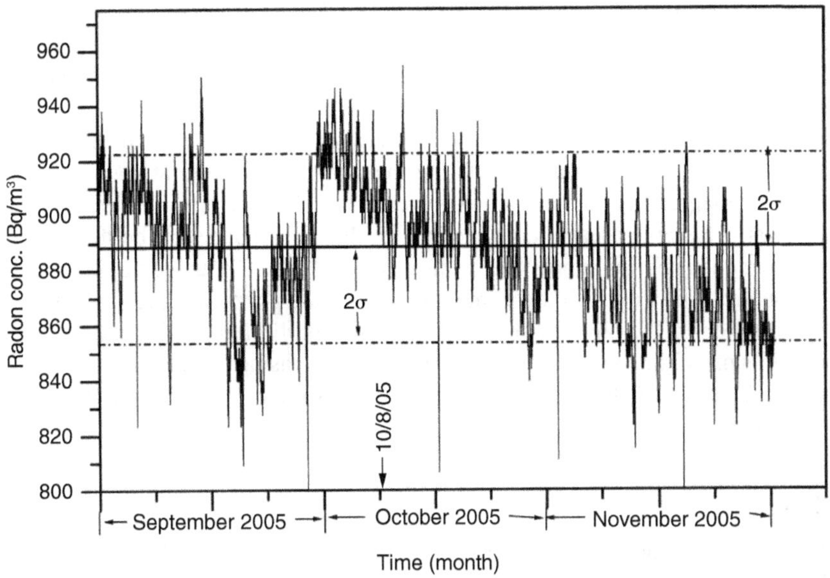

Figure 7.11 Radon concentration versus time prior to Pakistan quake (M=7.9).

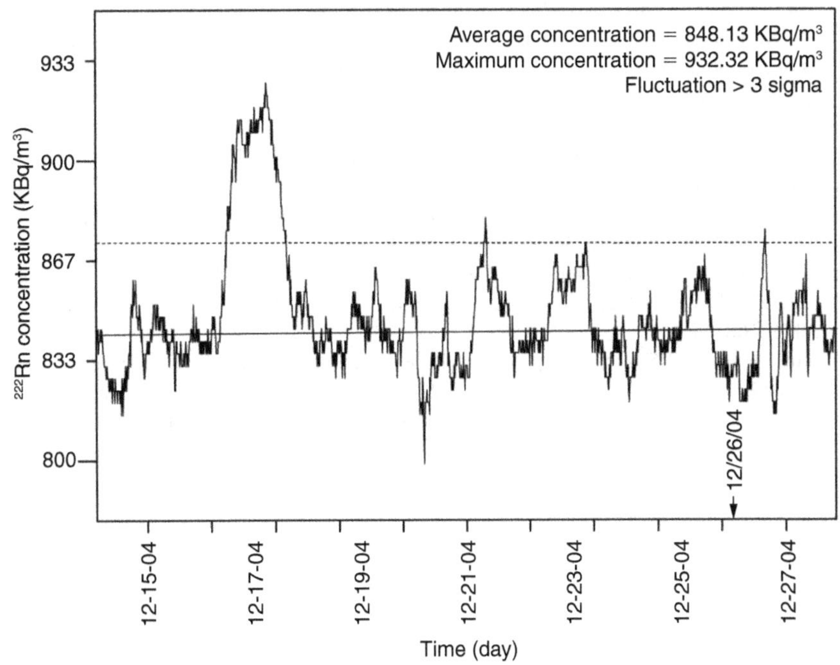

Figure 7.12 Radon concentration fluctuations versus time ahead of the great Sumatra quake (tsunami).

and migration up to the surface (Fleischer and Mogro-Campero, 1978). This event results in geophysical and geochemical anomalies of deep earth volatiles and is likely to be observed over a long distance, known as the precursory radius of seismic influence. Ahead of a major quake, the stress field approaches its critical value and emanates elastic waves in all directions. Consequently, the regional tectonic stress field is modified, enabling the strain energy alterations, particularly in an extensively faulted and seismically sensitive region like the one at Bakreswar. This effect leads to an increase in the permeability of the fluid conduits and a change in pore pressure, enhances the release of trapped volatiles and augments the fluid flow. All these mutually associated phenomena lead a temporary disruption of the physico-chemical dynamic equilibrium of the under-earth reservoir, amplifying volatile emission besides increasing concentration levels (Das *et al.*, 2009). A typical profile of helium variation ahead of the Pakistan quake is shown.

References

D'Amico, J. S., Reinhold, H. and Knaebel, K. S. (1996) US Patent No. 5542966, 6 August.

Das, N. K., Bhandari, R. K. and Sinha, B. (2003) 'Purification of Helium from Thermal Spring Gases', *Indian Journal of Cryogenics*, Vol. 28(1): 13–15.

Das, N. K., Bhandari, R. K., Sen, P. and Sinha, B. (2004) 'Proceedings of the International Seminar on Gas Technology', Kolkata, November 19–20: 252–6.

Das, N. K., Bhandari, R. K., Ghosh, D., Sen, P. and Sinha, B. (2009) 'Significant Anomolies of Helium and Radon Ahead of 7.9 M China Earthquake', *Acta Geodaetica Et Geophysica Hungarica*, Vol. 44(3): 357–65.

Das, N. K., Chaudhuri, H., Bhandari, R. K., Ghosh, D., Sen, P. and Sinha, B. (2006a) *The International Journal of Applied Radiation and Isotopes*, 64: 144–8.

Das, N. K., Chaudhuri, H., Bhandari, R. K., Ghosh, D., Sen, P. and Sinha, B. (2006b) *Lecture Notes on Physics – Modelling Critical and Catastrophic Phenomena in Geosciences*, Springer-Verlag, Berlin Heidelberg: 481–90.

Das, N. K., Chaudhuri, H., Bhandari, R. K., Ghosh, D. and Sinha, B. (2008) 'Purification of Helium from Natural Gas by Pressure Swing Adsorption', *Current Science*, Vol. 95(12): 1684–7.

Fleischer, R. L. and Mogro-Campero, A. (1978) 'Mapping of Integrated Radon Emanation for Detection of Long-Distance Migration of Gases Within the Earth: Techniques and Principles', *Journal of Geophysical Research*, Vol. 83(73): 3539–49.

Knaebel, K. S. and Reinhold, H. (1998) *Proceedings of Fundamental of Adsorption*, ed. F. Meunier, Elsevier, Giens, France: 763–7.

8 Helium from the air

The backstop

Richard H. Clarke and Roger Clare

Non-conventional helium

The major alternative source of helium, should helium-bearing natural gas not be available, would be its recovery from the atmosphere, which contains ~3.8 billion tonnes of helium.[1] Keesom (1942) considered air-sourced helium, believing that "helium reserves will become critical in a future, that is not too far remote". He estimated that the world's reserves (7 Bcf in 1942) "will do for about 280 years". Today, natural gas-related occurrences of helium represent about 1.7 Tcf (or about 8 million tonnes) of helium, according to the US Bureau of Land Management's data (Chapter 3). At a 2010 consumption rate of 6.15 Bcf per year this would also last 280 years, or 62 years if demand increases at the projected rate of 4.1 per cent a year (Garvey, 2011).

Helium and carbon dioxide also coexist, and some reserves have significant helium content, such as the Four Corners field in Wyoming. Also in the USA, at St Johns Arizona, the company EOR Inc. is preparing to sell its vast reserves of carbon dioxide through a pipeline to California for use in Enhanced Oil Recovery (hence the company name). The associated helium, perhaps 30 Bcf, will be recovered through a deal with Air Liquide (Lasker, 2008). On a smaller scale, the resources of the Sheep Mountain area in Colorado contain carbon dioxide and around 1 Bcf of helium (Gage and Driskill, 2004). On the face of it, if the last 70 years is anything to go by, helium reserves are keeping up with demand. Nonetheless, it is necessary to understand what the air has to offer.

In considering the Earth's atmosphere as a potential source of helium it is worth understanding where the helium comes from, and how long it is likely to stay in the atmosphere. Helium-4's sister isotope, the million-times rarer helium-3, is thought to be all of primordial "Big Bang" origin. Yet, perhaps surprisingly, around 7 per cent of the helium-4 is also primordial (Tolstikhin, 1975). The remainder has been – and continues to be – manufactured by the Earth itself through the radiogenic decay of uranium and thorium, in a long-lived chain that ultimately produces many helium atoms per parent atom: $^{238}U \rightarrow 8\ ^{4}He$, $^{235}U \rightarrow 7\ ^{4}He$, $^{232}Th \rightarrow 6\ ^{4}He$.

The rate of helium-4 outgassing from the surface of the Earth, on average, was estimated by Tolstikhin (1975) to be ~190 grams/second. This leads to two possible conclusions: (a) the Earth is young (<1 MY), as Vardiman (1990) suggests, since he concludes that there is no substantial escape mechanism from the upper atmosphere; or (b), if the Earth is aged ~4.5GY, then there must be other, non-Maxwellian, processes at work, such as the ion-outflow mechanism proposed by Lie-Svendsen (1996). This remains a subject of interest in atmospheric science but, for the purposes of foreseeing the future for helium, it seems reasonable to assume the atmosphere's helium will remain for millions of years, since atmospheric helium-4 is either accumulating or has reached a dynamic equilibrium. Anthropogenic helium releases (both by venting and through market mechanisms) have had a negligible impact on the helium concentration in the atmosphere (Holland and Emerson, 1987) and, to four decimal places, this remains the case.

Air separation

Currently, three atmospheric gases (oxygen, nitrogen and argon) are separated from air in major, *million tonne per day* quantities, and three – the rare gases neon, krypton and xenon – only in minor quantities. When air is separated by cryogenic fractional distillation into its components, it is the oxygen that liquefies preferentially (boiling point –183 °C). This is followed by argon (–186 °C) and then nitrogen (–196 °C), the "light component". The detailed composition of air is set out in Table 8.1.

Figure 8.1 shows the simplified schematic of a modern pumped liquid oxygen air separation plant (ASU). The points where rare gases would be drawn off are shown: at the outlet of the main condenser (helium and neon) and in the sump with liquid oxygen in the low-pressure distillation column (xenon and krypton). As oxygen is normally the principle product, this will determine the economics for the plant. There are exceptions, such as large nitrogen plants for EOR (enhanced oil recovery) or plants designed for argon (argon production influences capital cost strongly, as extra distillation column(s) are required to obtain high-purity argon).

The plant output is typically 99.6 per cent pure oxygen (the balance being argon, and almost no nitrogen because of the wide difference in the boiling points of oxygen and nitrogen). For customers with syngas (carbon monoxide +

Table 8.1 Typical dry analysis of air (everything below CO_2 is a trace element)

Gas	*Per cent by volume*	*Per cent by weight*	*Parts per million (by volume)*	*Chemical symbol*
Nitrogen	78.08	75.47	780,805	N_2
Oxygen	20.95	23.20	209,450	O_2
Argon	0.93	1.28	9,340	Ar
Carbon dioxide	0.038	0.0590	380	CO_2
Neon	0.0018	0.0012	18.21	Ne
Helium	0.0005	0.000073	5.24	He
Krypton	0.0001	0.00032	1.14	Kr
Hydrogen	0.00005	Negligible	0.50	H_2
Xenon	8.7×10^{-6}	0.00004	0.087	Xe

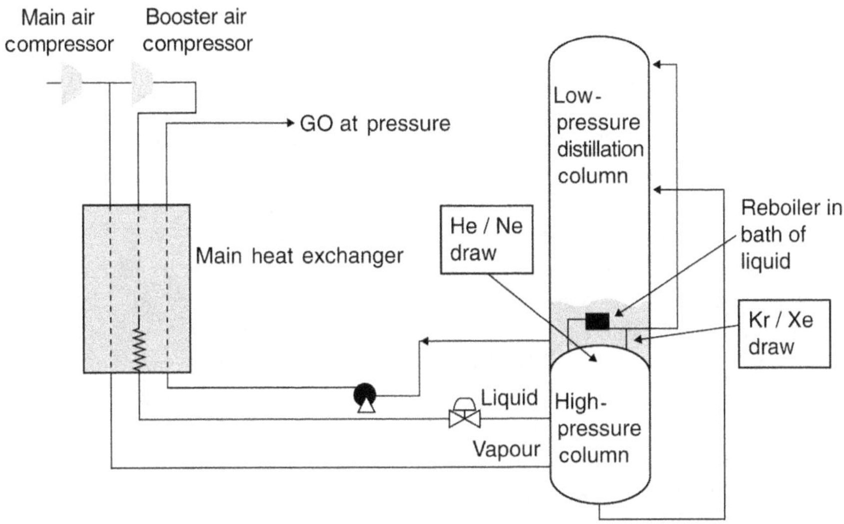

Figure 8.1 Simplified schematic of a modern ASU showing rare gas draw-off points.

hydrogen) plants used for power generation and chemicals, the impurity in the oxygen (although only 0.4 per cent or so) has a big impact on the application. Going from 99.6 per cent to 99.8 per cent halves the inert argon volume in the oxygen. It is relatively easier for the industrial gas provider to engineer a higher purity and help save the costs of the customer's inert gas removal system (which could also be a source for krypton and xenon).

Putting engineering considerations aside, the cost of power (electricity) is the dominant factor determining the economics of large plants. In smaller plants, particularly if argon is a product, the capital cost of equipment is much more significant. The concept of "capitalised power" (a function of contract term, discount rate and cost of power) is used to determine a plant's correct design to obtain the lowest lifetime plant cost. These values vary from ~US$2,000/tonne in the US to ~US$4,000/tonne in Western Europe. Since power is more critical in large and very large plants this is therefore of interest when neon might be extracted. While the surface area to volume ratio influences capital cost, which increases with (plant size)$^{0.65}$ as a rule of thumb, the power cost is directly proportional to plant size.

A particular factor that determines ASU design is argon production. The capitalised value of argon varies but is generally high – up to ten times the value of liquid oxygen. This depends on local market conditions. Argon is often associated with economically developing zones because of its use in welding. There are usually ways to increase or decrease argon production by altering the process cycle. The power implications of argon production are generally neutral compared to oxygen.

Rare gas markets are very specialised. They have high value and low volume. New technologies can emerge quite quickly and can create strong demand; for example, plasma TV screens – a neon application. Where neon is recovered in a side plant on large ASUs, using a nitrogen "wash column" as shown in Figure 8.2, the air separation process typically yields a crude neon stream containing ~30 per cent neon. The side column, if fitted, is largely independent of the main column except that it requires some nitrogen. This has a slight cost implication if the nitrogen is also a product. The reduced liquid nitrogen flow will also have some implication if argon is a co-product. As the flows of neon (and helium) are small, generally speaking, such columns would not be retrofitted to existing plants.

A typical process for neon purification with "wasteless" recovery of helium (and the change in concentration of the various gas components as they pass through the equipment) is shown in Figures 8.3 and 8.4. More detailed information is given in the book *Cryogenic Technologies of Rare Gas Extraction* (Bondarenko and Simonenko, 2009). Table 8.2 shows the typical specification of high-purity helium obtainable by industry: only the helium obtained in the Iceblick process comes from air; the normal source is natural gas.

The commercial uses of neon recovered from air currently include neon lamps (signs), neon-helium lasers, a cheaper refrigerant than once-through liquid helium (per kJ) if 27 K (–246 °C) is cold enough, infrared imaging, and (until LED screens prevail?) plasma TV screens. Neon's price is also very volume sensitive, similar to that of helium in the retail market, where party balloon gas costs more than the bulk helium used in MRI manufacturing (see Chapter 12). In a particular country or region, when demand grows to a certain level, the continuous operation of an ASU neon plant with its substantial investment may be considered economically viable.

The cost of helium produced as a by-product of neon will be much less than that of helium *exclusively* produced from air. Historically, helium was produced from air for special uses in nuclear engineering in some countries such as China, where in the 1960s to 1970s, other sources of helium were unavailable (see Chapter 17). Ongoing interest in this area is noted (Hu *et al.*, 2011). The option to produce helium from air is always open to the industrial gas companies. The problem associated with such production is its high unit cost resulting from the small production capacity. If all other things are equal, the additional cost per Nm^3 of further purification would be similar to that of crude helium from the BLM (Cliffside) reserve or crude helium coming out of the cold box of an LNG production process.

Will cryogenic air separation remain the primary means of industrial gas production worldwide? Ebbe Almqvist's (2003) book covers the early and modern-day processes in depth. Oxygen was first extracted from the atmosphere by a high-temperature chemical process in the 1890s. This was superseded over 80 years ago by the cryogenic (low-temperature) processes involving the liquefaction and distillation of air. The cryogenic process is still widely used for bulk or pure air products, but non-cryogenic techniques such as pressure swing adsorption and membrane diffusion, which were developed during the 1970s, have become increasingly significant for smaller or less demanding on-site applications. Cryogenic air separation is a mature and stable technology, although incremental technical advances are still yielding improvements in capital cost,

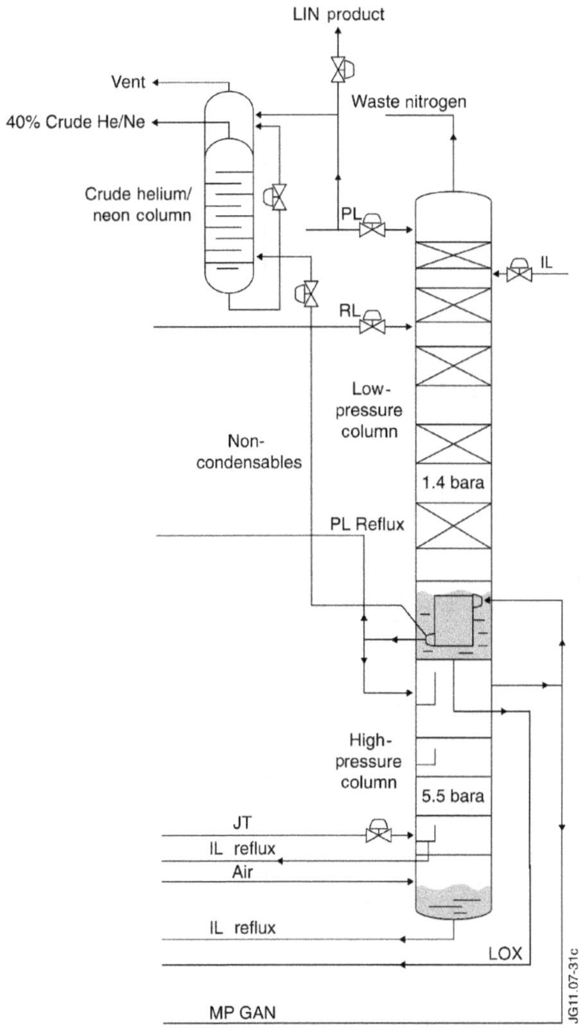

Figure 8.2 Additional equipment (the crude helium/neon "wash column", top left) is fitted to a large ASU for crude helium-neon (He/Ne) recovery. The He/Ne mixture is removed from the nitrogen side of the main reboiler, situated in the low-pressure column, and is fed into the bottom of the crude He/Ne column where nitrogen is by far the "heavy component" (the species with the highest boiling point). The sump liquid from the high-pressure part of the "wash column" is flashed into the top part, where it boils at a lower temperature. This causes condensate to reflux, or wash, across the distillation trays or packing in the main part of the He/Ne column. This strips out much of the nitrogen, leaving an enriched top vent of helium-neon mixture (approx 60 per cent nitrogen, 40 per cent He/Ne). Additional liquid nitrogen (from the LIN product stream) is needed to complete the energy balance. Waste nitrogen is vented from the top part of the column. It is possible to produce about 0.7 Nm^3/h of helium in a 1,000 tonne per day plant.

Table 8.2 Typical specifications of "Spectra" grade helium

	Helium producers	*Country*	O_2	*Ar*	N_2	*CnHm*	*Ne*	H_2O	*CO*	CO_2	H_2
			Impurity content in ppm (not more than)								
1	Linde	Germany	0.5		0.5	0.1		0.5	0.1	0.1	0.5
2	Air Liquide	France	0.5			0.1		0.5	0.1	0.5	0.1
3	Specialty Gases	USA	0.5		0.5	0.1		0.5	0.1**		0.5
4	Air Products	USA	1.0	0.5	1.0	0.1	0.5	0.2		0.1	
			1.0	0.5	1.0	0.01	0.5	0.2		0.1	
5	Airgas	USA	1.0	0.5	1.0	0.1	0.5	0.2		0.1	
			1.0	0.5	1.0	0.01	0.5	0.2		0.1	
6	Messer	Germany	0.5		0.5	0.1		0.5	0.1**		
7	Iceblick	Ukraine	0.1*		0.2	0.1	0.1	0.3	0.1**		0.1
8	NII KM	Russia	0.15*		0.45	0.1	0.15	2	0.1**		0.05
9	Monitoring Ltd	Russia	0.15*		0.45	0.1	0.15	2	0.1**		0.05

Source: Simonenko, Y.M. (2011).

Notes
* O_2 + Ar
** CO + CO_2

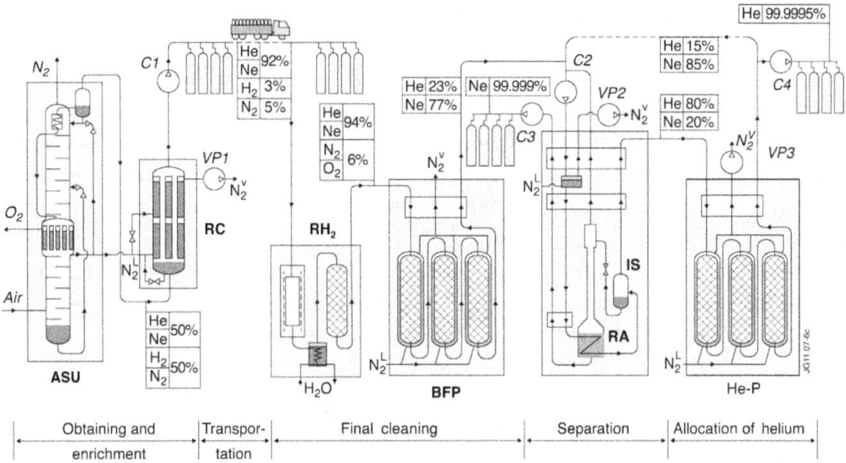

Figure 8.3 Concentration process for helium and neon, from the air separation unit through to the final product (Arkharov, 2003).[2]

operating cost, ease of operation, reliability and safety. The only significant "raw material", apart from air itself, is electricity, used to drive compressors, pumps and other equipment. In the authors' view, cryogenic air separation will continue to be the workhorse of bulk industrial gas production and so, potentially, a source of helium and the other rare gases.

A potential solution to technology's need for helium

The first thing to understand about using the atmosphere as a source for helium is that it would be a gargantuan task to supply the approximately 75 tonnes per day of helium the world now uses. At 5.2 ppmv (or 0.73 ppm by mass), over 100 cubic kilometres of air would need to be processed every day. The pumping power alone would consume the output of many power stations (Cai *et al.*, 2006). The results of several studies on the economics of helium production from air separation suggest that helium production from air *if none of the other components is used* would be about 60–100 times more expensive than helium production from natural gas (Beebe *et al.*, 2000). These studies are based on the logic that the energy cost is inversely proportional to helium concentration as shown in Figure 8.6, as derived from Earl Cook's work (Cook, 1979).

Taking conventional "economic" helium sources to be 0.3 per cent helium, Cook's chart indicates this requires ~600 kWh/Mcf of raw gas. Figure 8.6 suggests this would rise to ~150,000 kWh/Mcf (the lower of the "industry estimates") at 5 ppm. This ratio of energy requirements (~250:1) is consistent with the cost estimates, since in conventional helium purification plants the energy costs are about one-third of the total, giving a cost ratio of ~83.

Strictly speaking, while the above result may be about right, the logic is wrong. Keesom (1942) considered the theoretical minimum energy requirement[3] to separate: (a) 1.85 per cent (Cliffside) helium from natural gas and (b) 5.2 ppm of helium from air. The power ratio is about 2.7:1, not 250:1. However, the main reason for the higher cost is in the much larger plant that would be required to process air to give the same amount of helium product. As stated earlier, capital cost increases roughly with (plant size)$^{0.65}$. To extract helium from a volume flow rate 2,000 times larger (i.e. 1.85 per cent helium in natural gas/5.2 ppm helium in air) would involve a plant costing about 140 times as much. In addition there would be a considerable increase in pumping power. If these costs could be reduced, then perhaps technology of the type discussed in Chapter 7 might ultimately provide a more energy-efficient route to helium from air than dedicated cryogenic plants.

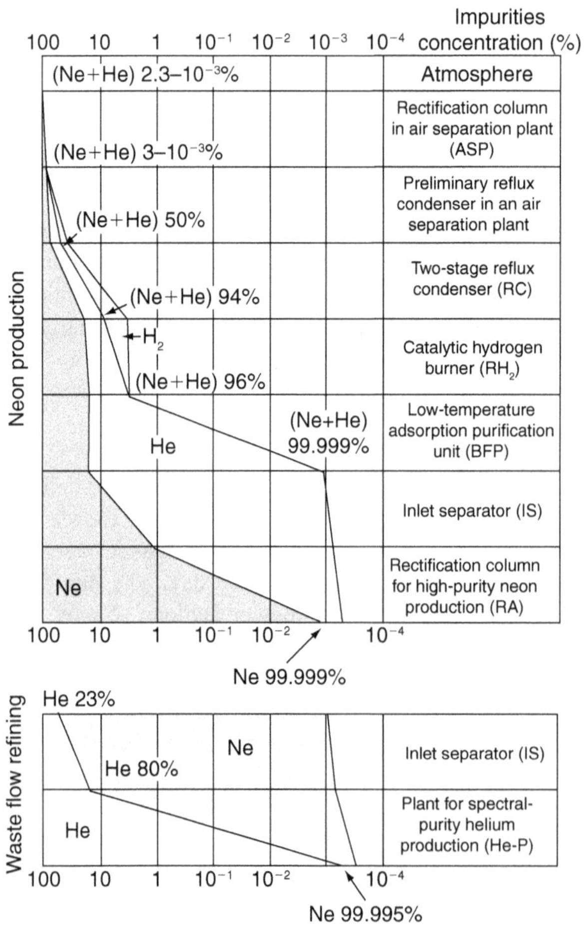

Figure 8.4 Technological sequence of the neon-helium concentration process: this particular process recovers both neon and helium (Arkharov, 2003).

(a)

(b)

Figure 8.5 Parts of a high-purity helium-neon production facility: (a) nitrogen-jacketed final cleaning column; (b) high-purity product compressors (courtesy: Iceblick Ltd).

However, in considering conventionally produced and LNG-derived helium versus air-derived helium, these conclusions may not be true for three reasons. First, current sources of helium are generally located in remote areas that are far from the end use markets, and this is why transportation can account for up to ~30 per cent of the bulk product cost. This portion may be saved because the air separation plant (ASU) may be located in, or close to, end-use markets. Second, the royalties of air are zero, since air is essentially infinite and free, yet in helium production from natural gas the feed gas typically accounts for ~20 per cent of the total cost. Third, and most important, helium can be produced as a relatively cheap by-product of cryogenic air separation when such technologies are already used to extract oxygen, nitrogen and argon. The key assumption here is that *the capital and energy costs of the main air separation plant are allocated solely to the main products* (oxygen, argon and possibly nitrogen). Plant retrofits for rare gas recovery are discouraged due to both the relatively high expense of modifications (when compared with designing a new plant) and also the potential disruption of supply or purity to the tonnage customer. The main cost that will be allocated to the rare gases – almost a fixed cost because of the precision control instrumentation required – is the cost of the neon (and helium) recovery and

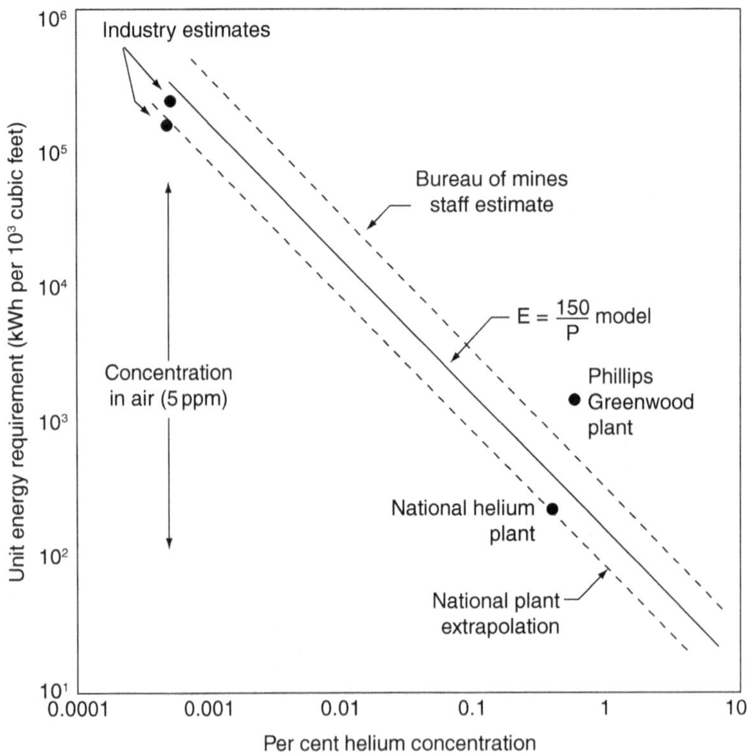

Figure 8.6 Energy requirements to extract helium from natural gas or air (Cook, 1979).

purification units, such as those shown in Figures 8.3 and 8.5. Depending on the purity required, typically these units will cost in the range €1.5–2 millon for use with a 1,500 tonne per day ASU.

Helium concentration in the residual gas after nitrogen and oxygen have been separated is much higher than 5 ppm; in the side plant, the helium concentration in the gas stream is 13 per cent mixed with neon at 27 per cent, the balance being nitrogen. If helium is produced as a by-product of neon then the capital and side-plant operating costs would be shared between the two products. Taking the above capital cost, and allowing for interest, maintenance, etc. over a 20-year period, the indication is that the helium would cost around US$200 per thousand cubic feet (Mcf) to produce in 2011.

On current trends the natural gas-derived helium price could surpass this level in the next few years, so will we see a switch to, or supplementation from, air-derived helium? This is unlikely to be the case in the near term, since the volume of air that would need to be processed in order to meet today's demand would be very large. Ultimately, helium-from-air production capacity would be limited by the size and nature of the industrial gas market, if helium is to enjoy a "free ride" with regards to energy costs. In the meantime, economies of scale favour the natural gas route, but it is clear that the cost of helium production from air need not be prohibitively expensive – certainly not "60 to 100" times as expensive as Figure 8.6 implies.

The key issue relating to the use of air separation to produce helium would be the realistic production capacity. Following a detailed discussion with John Oakey (a specialist at BOC-Linde's Guildford office in the UK) some points that will need considering are as follows:

1 Current oxygen demand. New oxygen capacity is expected to be added at about 75–100,000 tonnes per day each year for both "third-party" plants and industrial gas company plants. The current installed capacity, globally, is about 2,000,000 tonnes per day (at any given time, some of it will be idle). The cost of extracting a helium-neon mixture from the main stream (and compressing it) is about 1 per cent of the plant capital cost, i.e. small. The mixture could then be transported to a central purification plant in tube trailers. The cost of transportation depends on the distance and local infrastructures.
2 The cost of a large ASU can vary significantly depending on the economics of the local geography, local power costs, co-products and project scope. Having said that, the cost of a very basic ASU making 2,000 tonnes per day of high-purity, low-pressure oxygen (with no argon, no nitrogen at purity, no liquid sales), and including a fairly basic scope/infrastructure (cooling water, substation, minimal storage and back-up) and in a place with average power costs would be approximately US$65–80 million.
3 It is not wise, or not economically viable, to retrofit the existing plants to extract helium. One needs to redesign the system. So the capacity of helium production would be only from *new* plants.

4 An ASU's life cycle is about 25 years.
5 The cost of helium production from an ASU is becoming comparable with that extracted from natural gas if it is produced as by-product of oxygen (i.e. there must, in the first place, be a market for the oxygen, otherwise the full extraction costs *will* be passed onto the neon/helium market).
6 The primary justification for an ASU project is based on oxygen production only. The expectation of option value to produce by-products such as neon, argon and nitrogen very much depends on the economic conditions and requirements in the operating geographical region. Projects with future options to produce other gases have lower economic risk in economically booming areas such as China and India.
7 A few ASUs are devoted exclusively to producing nitrogen only (e.g. for enhanced oil recovery, such as the Cantarell plant in Mexico (Kuo, 2001).
8 The cost of making pure nitrogen from an ASU is marginal. In order to make pure oxygen, nitrogen is separated anyway, and so the cost to convert impure nitrogen into its pure form just involves some additional stages of separation in the distillation columns. If the nitrogen is required at some elevated pressure then, of course, there will be the associated cost of a compressor, but for the product itself as a pure gas at its natural pressure, the incremental cost would be less than 10 per centof the core plant cost. The cost associated with Kr/Xe production is approximately 3 per cent of the full plant cost as described above.
9 In order to produce argon a lot of additional distillation equipment is required, adding approximately 15 per cent to the capital cost. Although the quantity of argon that can be extracted is a function of the quantity and source of the nitrogen (i.e. which part of the main column system produces it) there is the possibility to make argon whether or not nitrogen is made.

A case study of helium production from air separation

In this section, a case study is presented of helium production from air to meet the demand of a hypothetical fusion reactor. In terms of power, there is a long-term scenario where approximately 1.5 kg/day helium per 1 GWe of fusion might be required (see Figure 15.10 in Chapter 15). This might also apply to the Generation IV fission reactors described by Richard Stainsby in Chapter 14, or possibly for maintaining the air-worthiness of future helium-filled airships that lose approximately 0.1 per cent per day of their helium.

To meet this demand, how much air should be processed and what power is consumed? Will it ultimately be possible to drive air separation plants using nuclear power in a "symbiotic" relationship, where the ASU provides the helium and the nuclear plant provides the electrical power? Table 8.3 illustrates the calculation process.

Helium-neon extraction from an ASU is not routinely carried out – it is entirely market-driven and usually led by neon. The economics lends itself to rare gas extraction from larger rather than smaller plants. Historically, there have

Table 8.3 Estimation of power required to produce 1.5 kg/day of helium from air

Air density = 1.2754 kg/Nm^3 and He density = 0.1785 kg/Nm^3 (at 1.013 bara, 0 °C)
He in air 5.24×10^{-4} by volume, or $5.24 \times 10^{-6} \cdot 0.1785/1.275 = 0.7$ ppm by mass
1 Nm^3 of air contains $0.00000524 \cdot 0.1785 = 0.000000935$ kg of He
1.5 kg of He is contained in 1.5/0.000000935 = 1,603,695 Nm^3 of air
1.5 kg of He per day is contained in 1,603,695/24 = 66,821 Nm^3/h of air
At 95 per cent He recovery (best possible), at a typical oxygen recovery of 98 per cent, 1.5 kg/day is recoverable from the air equivalent to oxygen of $(66821 \cdot 0.20956 \cdot 0.98)/0.95 = 14{,}445$ Nm^3/h = 495 tonnes of oxygen per day
The specific power for oxygen at low pressure with no co-products for a plant of about 500 tonnes per day would be about 0.4 kWh/Nm^3 of oxygen (Arkharov *et al.*, 2000). The total power would, therefore, be approximately 5.8 MW. A 22nd century fusion reactor might require about 1.5 kg/day He from air so, at 5.8 MW per GWe of fusion, about 0.5 per cent of the fusion or fission output power would be needed to sustain the reactor with helium (and make all the air products that help pay for the plant).

been more ASUs with krypton-xenon recovery than neon, and again that is market-driven. In a typical year new oxygen tonnage would be expected to be 40,000–60,000 tonnes per day. (This is "over the fence" gas supply as opposed to third-party plant supply – the latter being unlikely to have capital to spend on helium-neon extraction. In any case, it very much depends on the global economic cycle and also on industry-specific cycles, e.g. the steel industry).

If all new gas company plants were fitted for He/Ne extraction the new capacity per year might be expected to grow at 170–250 Nm^3/h of crude He/Ne. This is equivalent to 28–42 Nm^3/h of pure He (0.12–0.18 tonnes per day).

A significant portion of this new capacity would, however, come from small units where it would not be economically viable to extract He/Ne at today's prices. A more realistic capacity would produce 0.1 tonnes per day or ~37 tonnes per year; i.e. in about 20 years the ASU industry could build up about 2 per cent of today's global helium production.

Summary and recommendations

Detailed consideration has been given to the processes and costs associated with helium-from-air extraction as a means of meeting the needs of long-term, low-volume, high-technology users. It has been suggested (in Chapter 15) that in principle helium-efficient nuclear power plants might in the long term provide the power needed to drive cryogenic air separation units (ASUs) to make the helium required. The costs of deriving high-purity helium from air are likely soon to be similar to those of natural gas-sourced helium. However, the air-sourced helium (and neon) volumes are unlikely to be more than about 2 per cent of today's global demand, and so this will not in itself restrain the ongoing helium price increases. Additionally, the competitiveness of air-sourced helium depends critically on there being markets for the main air products – oxygen, argon and occasionally nitrogen. A watching brief should be kept on the development of industrial gases by the helium producers, which are often the same group of companies. The significance of air-sourced helium could increase if, for example, oxy-fuel combustion methods were used in carbon-capture schemes. Finally, we should not discount non-cryogenic methods of obtaining air-sourced helium: they may currently be more expensive but they would be more energy efficient than building dedicated cryogenic units. Of course, finding markets for the other air gases will again be important to make the economics work.

Notes

1 This mass, M_{He}, can be estimated from the mass proportion, x, of helium in the atmosphere ($0.73 \cdot 10^{-6}$ kg/kg); atmospheric pressure, P, at sea level (~10^5 N/m^2) and the surface area, E, of the Earth (~$5.1 \cdot 10^{14}$ m^2) from which $M_{He} = x \cdot P \cdot E / g = 3.81 \cdot 10^{12}$ kilograms (g = 9.81 m^2/s, the acceleration due to gravity).

2 Key to Figure 8.3: ASU – air separation unit; RC – reflux condenser; RH2 – section for catalytic purification from hydrogen; BFP – final purification block for the neon-helium mixture; RA – rectifying column (apparatus) for pure neon production; IS inlet separator; He-P – adsorption plant for pure helium production; VP1–VP3 – vacuum pumps for the evacuation of the nitrogen stream; C1–C4 – diaphragm type compressors.

3 Keesom (1942) states that the reversible separation energy – i.e. the energy needed to separate by a reversible process a mixture of two gases (supposed to be approximately in the ideal gas condition) into its components – is $-RT\ [y \cdot ln(y) + (1-y) \cdot ln(1-y)]$ per mole of the mixture, if y is the molar composition with regard to one of the components. To get a mole of this component the energy needed is $-RT\ [ln(y) + (1-y)/y \cdot ln(1-y)]$. In this expression T is the absolute temperature of the surroundings (assuming the process is reversible). As the gas constant R = 8.315 joules per mole, and taking T = 293 K, we obtain for y = 0.0185 – i.e. the molar helium content in the Cliffside natural gas – that to prepare 1 m^3 of helium, 0.141 kWh is needed. For atmospheric air, y = 1/185000. To extract 1 m^3 of helium from the air, 0.375 kWh is needed.

References

Almqvist, E. (2003) *History of Industrial Gases*, Kluwer Academic/Plenum Publishers, New York, NY.

Arkharov, A. M., Bondarenko, V. L., Arkharov, I. A., Savinov, M. Y., Volinskiy, B. I., Simonenko, Y. M., Rura, V. N. and Losyakov, N. P. (2003) "Highly productive technologies of obtaining neon and helium from atmospheric air", Twenty-first International Congress of Refrigeration, Washington, DC, 17–22 August.

Arkharov, A. M., Marfenia, I. V. and Mikulin, I. (2000) *Cryogenic Systems*, Bauman Moscow State Technical University Press, Moscow.

Bondarenko, V. L. and Simonenko, Y. M. (2009) *Cryogenic Technologies of Rare Gases Recovery*, Izdatelskiy centr, Odessa, Ukraine (in Russian).

Beebe, R. R., Reppy, J. R., Goldman, A. M., Lander, H. R., Macauley, M. K., Miller, M. A., Rose, A. Z., Siewert, T. A., Weisskoff, R. M., Shapero, D. C., Ehrenreich, R. M., Aylesworth, K. D. and Morgan, D. F. (2000) *The Impact of Selling the Federal Helium Reserve*, National Academy Press, Washington, DC.

Cai, Z., Clarke, R. H., Nuttall, W. J., Ward, N. and Glowacki, B. A. (2006) "Is helium from air a future source for helium production?", CR06–271 ICEC-21, Prague, July.

Cook, E. (1979) "The Helium Question", *Science*, 206: 1141–7.

Gage, B. D. and Driskill, D. L. (2004) "Helium Resources of The United States 2003", US Department of the Interior.

Garvey, M. D. (2011) "Tight Supply Reins in the Worldwide Helium Market", *Cryogas*, October: 30–2.

Holland, P. W. and Emerson, D. E. (1987) "A determination of the helium-4 content of near-surface atmospheric air within the continental United States", *Journal of Geophysical Research*, Vol. 92(B12): 12557–66.

Hu, Z. J., Qui, Y. N., Li, Q. and Zhang, N. (2011) "Helium Extraction and Production Techniques from Air Separation Plant Purge Gas", Eighth International Conference on New Theories, Discoveries and Applications of Superconductors and Related Materials, Paper P-VI-05, Chongqing, PR China, 8–11 June.

Keesom, W. H. (1942) *Helium*, Elsevier, Amsterdam, London, New York, Princeton.

Kuo, J. C, Elliot, D., Luna-Melo, J. and De León Pérez, J. B. (2001) "World's largest N2 generation plant", *Oil and Gas Journal*, 12 March.

Lasker, B. (2008) Enhanced Oil Resources overview website (accessed May 2011): www.enhancedoilres.com/presentations/eorppt0308.pdf.

Lie-Svendsen, Ø. and Rees, M. H. (1996) "Helium escape from the terrestrial atmosphere: the ion outflow mechanism", *Journal of Geophysical Research*, Vol. 101(A2): 2435–43.

Simonenko, Y. M. (2011) "Optimisation of temperature level of Spectral pure helium production by method of absorption" (private communication from Iceblick Ltd, Odessa, Ukraine).

Tolstikhin, I. (1975) "Helium isotopes in the Earth's interior and the atmosphere: a degassing model of the Earth", Earth Planet. Sci. Lett., Vol. 26: 88–96.

Vardiman, L. (1990) *The Age of the Earth's Atmosphere*, Institute for Creation Research, San Diego.

9 Helium demand

Applications, prices and substitution

Zhiming Cai, Richard H. Clarke, William J. Nuttall

Introduction

Helium, owing to its unique electronic configuration, is endowed with many extraordinary properties, such as the lowest boiling point, which renders it a valuable commodity. Apart from party balloons and perhaps in scuba diving, ordinary citizens do not come into contact with helium that often. Helium use remains the province of scientific technologies.

Helium demand is one of the key components to influence helium market dynamics and the depletion path through its interaction with supply, government policies and production technologies.

Historically, helium has been regarded as a national strategic material by the US government. Many high-technology industries are developing new commercial applications that depend on the unique physical properties of helium. In addition, the expansion of the global economy is also a strong driving force for helium demand, especially in areas of rapid economic growth such as China and India. All these factors bring about an increasing demand for helium. The increase in helium demand, coupled with under-investment in helium production capacity, results in dynamic changes in market prices and helium availability.

What are the qualities that make helium useful

Helium has an extremely low boiling point of –269 °C, 4.2K above absolute zero. It is the coldest of all liquid refrigerants. Helium is colorless, odorless, tasteless, non-flammable, non-reactive and non-toxic at all practical temperatures and pressures. It is the second lightest gas after hydrogen, which makes it useful for balloon or lifting applications. Helium molecules are the second smallest after hydrogen and are able to diffuse through many materials and minute leaks, hence its use in leak detection. Helium has a high thermal conductivity, exceeded only by that of hydrogen. If cooled enough its viscosity is close to zero, which means it becomes a superfluid. Helium has relatively high specific heat (C_p and C_v) compared with other common gases such as air, argon and nitrogen, as shown in Table 9.1, making it useful for heat transfer applications.

Table 9.1 Specific heat of various common gases at 300 K

Gas	*Molar mass, kg/kilo-mol.*	*Cp, kJ kg^{-1}K^{-1}*	*Cv, kJ kg^{-1}K^{-1}*	$\gamma = Cp/Cv$
Air	28.97	1.005	0.718	1.400
Argon	39.94	0.5203	0.3122	1.667
Helium	4.00	5.1926	3.1156	1.667
Nitrogen	28.02	1.039	0.743	1.400

Source: Kyle, 1999.

Helium has two stable isotopes in nature: ^{4}He (Helium-4) and ^{3}He (Helium-3). ^{3}He is primordial, and much rarer than ^{4}He. Figure 9.1 shows the phase diagram of helium. Liquid helium remains in the liquid phase under its own vapour pressure and would do so right down to absolute zero temperature. High pressure is required to produce solid helium (25 times atmospheric pressure, or more), because of the small mass and extremely weak forces between the helium atoms. When liquid helium is cooled to 2.17 K, it undergoes a phase transition. The terms He I and He II are used to distinguish the liquid above and below T_{λ}[1], respectively.

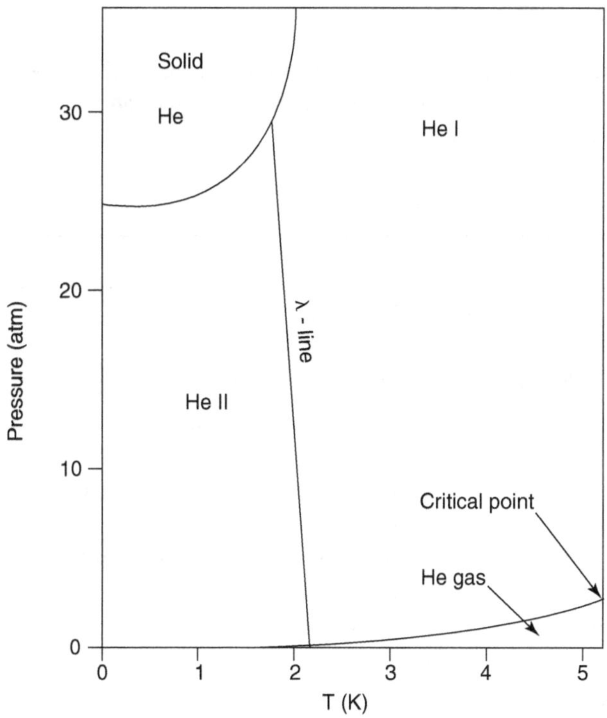

Figure 9.1 Phase diagram of liquid helium.

Because of its atomic configuration, helium exhibits a number of extreme physical and chemical properties that give it unique advantages in a number of industrial and technological applications. Table 9.2 lists some of these properties and key applications derived from each or combinations of them.

Of these properties, some result in significant commercial opportunities. Examples include the following:

- Helium has the highest ionisation potential of any atom, which makes it useful for electric arc welding of exotic metals such as titanium, magnesium and aluminum alloys used in aerospace and shipbuilding.
- Helium's extremely low solubility in water (and therefore blood) has lead to substitution of helium/oxygen mixtures for nitrogen/oxygen breathing mixtures in order to avoid nitrogen narcosis ('the bends') when deep-sea diving.
- The extremely high speed of sound travelling in helium has led to development of some very interesting alternative processes for producing metal coatings, which rely on spraying metal powders at great velocities through a high-pressure helium stream.
- Finally, helium exhibits the curious property of superfluidity at 2.17; below this temperature the transport properties of helium (He II) are excellent and are used in big physics experiments such as the Large Hadron Collider (LHC).

Helium applications

The applications of gaseous and liquid helium are listed in Tables 9.3 and 9.4 respectively. In the rest of this section, a number of main or important applications are singled out for discussion in more detail.

Table 9.2 Properties and applications of helium

Properties	*Applications*
Lowest boiling point; does not solidify at atmospheric pressure and 0 K	Liquid cooling of low Tc superconductors; purging liquid hydrogen systems
Second lightest element after hydrogen	Lifting medium for balloons, airships
Smallest molecular size, with the greatest permeation rate of any substance	Leak detection; supersonic wind tunnels (high Reynolds number)
Chemically inert	Carrier gas; analytical uses; semiconductors
Very high specific heat and thermal conductivity	Faster cooling in vacuum furnaces; gaseous cooling – fibre optics; nuclear reactors
Radiologically inert (no radioactive isotope caused by neutrons)	Heat transfer medium in nuclear fission or fusion reactors
Highest ionisation potential	Metal-arc and plasma-arc welding
Extremely low solubility in water	Deep-sea diving gases
Very high sonic velocity	Metal coating

MRI

The MRI helium market can be subdivided into two key segments: the original equipment manufacturers (OEMs), which precool and test the magnets prior to shipment; and hospitals and other field locations, which need to replace helium lost through evaporation in service. The remaining helium in the superconductor sector is used for NMR devices, SQUIDs (superconducting quantum interference devices), for particle accelerators and for myriad other, smaller applications.

Table 9.3 Gaseous helium applications

Medicine	"Heliox" medical therapies: ^{4}He (asthma and neonatal distress treatment); ^{3}He (lung function MRI imaging) "Heliox" breathing mixtures for deep-sea diving
Protective atmospheres	Detection gas for leak detection Optical fibre manufacturing Semiconductor manufacturing Rare metal smelting Heat treatment of metals or components Shielding gas for cutting and welding Working gas for laser applications
Heat transfer medium	Heat exchange in gas-cooled nuclear reactors
Transportation	Balloon and airship inflation Inert gas in space sensors Airport neutron detectors (^{3}He)
Laser technology	Resonator gas component
Metrology	Carrier and purging gas in gas chromatography Calibration and test gas in analysis

Table 9.4 Liquid helium applications

Metrology	Nuclear magnetic resonance analysis (NMR) Superconducting magnetic field probes Cooling of sensors
Medicine	Magnetic resonance imaging (MRI) Magneto-encephalography
Research and development	Nuclear fusion and particle physics Low-temperature physics Superconducting energy storage Superconducting materials and sensors Ultra-low-temperature material research Materials testing
Technology and industrial applications	Strong-field magnetic separator Magnetic melt stabilisation Particle accelerators Infrared detectors Josephson junction devices

Development of high-temperature superconductors (HTS) cooled by liquid nitrogen or other means is not expected to blunt helium demand significantly in existing low-temperature superconductor (LTS) markets. Instead, this technology is likely to extend the use of superconductors into new areas, such as electrical energy storage and transmission, and magnetic levitation transport.

MRI scanners are used for medical diagnostics. The patient is moved into the centre of a large and powerful cylindrical magnet and subjected to a varying frequency magnetic field in the presence of a radio frequency pulse. Certain atoms preferentially absorb the magnetic energy compared with others, and high definition cross-sectional images are produced that can clearly show tumours and other abnormalities (see Chapter 13).

Superconductivity is the phenomenon shown by certain materials to have near zero resistance to the flow of electric current at temperatures close to absolute zero. Superconductivity can provide an extremely powerful magnetic field, as required for MRI imaging. Liquid helium is required to cool these magnets. Demand for helium in new machines is much reduced as they have mechnical refrigerators to recondense evaporated helium. Some manufacturers are moving towards liquid helium-free designs because an MRI machine that warms up is unusable, and they perceive that their customers (hospitals) fear the consequences of helium shortages.

Semiconductors

The main users of helium in the semiconductor industry is related to the supply of silicon wafers,[2] plasma etching and vacuum pumping. Helium-cooled superconducting magnets are used in the process of pulling 300 mm crystals to stabilise mechanically the hot boules of semiconductor material. In vacuum pumping, liquid helium-cooled surfaces cryopump gas contaminants (Galasiewicz, 1971).

Since helium is crucial to the semiconductor processing industry, an increase in helium price could be accommodated. As a result, the demand for helium in the semiconductor industry is rather inelastic.

Carrier gas

Helium is the most commonly used carrier gas for gas–liquid and gas–solid chromatography. Chromatography is one of a range of methods for analysing the presence of minute quantities of elements and compounds in complex mixtures, by separating the components on liquid or solid absorbents (Heftmann, 2004). Helium is chemically inert and minimally dissolved or absorbed by the absorbent. Its low viscosity also minimises pressure drop through the apparatus. Furthermore, helium's very low molecular weight gives it physical properties greatly different from the materials being analysed. The industries that use helium in analytical applications (principally gas chromatography) and leak detection (mass spectrometry, discussed in the next section) are moving towards more rigorous quality control and compliance with more demanding environmental standards. As a result, helium demand in this sector is growing steadily, and continuous improvement of recovery systems helps to mitigate the rising cost of helium.

Leak detection

The small size of the helium molecule, its high diffusivity, and its ease of detection in a mixture of gases make it an ideal gas for detecting minute leaks in vacuum and other gas-tight systems. A helium leak detector, which is a small mass spectrometer tuned to detect helium, is connected to the piping or assembly to be tested. The piping is evacuated by a vacuum pump and helium directed to the outside of the piping. Any helium entering the system is soon detected. Because its molecules are so small, helium is superb for detecting and locating minute leaks in pressure and vacuum systems or for confirming the absence of leaks. Its leak detection capability is invaluable to the aerospace, semiconductor, nuclear, cryogenic vacuum, refrigeration, food and automotive industries (Hucknall, 1991).

Leaks, caused by special types of defect, can be important in systems where they have influence on safety and performance. In a helium leak test, helium is used as a search gas. Principally, there are two methods applied for leak testing and localisation of leaks, the 'vacuum method' and the 'overpressure method'. In the vacuum method, the object to be examined for leaks is evacuated and enclosed in a plastic bag containing helium, the search gas. The gas enters through any leaks present in the object and is detected by a sensor connected to the leak test instrument. In the overpressure method, the object to be examined for leaks is filled with helium, under slight overpressure. The search gas escapes through any leaks present to the outside and is detected by a detector probe, which is called a 'sniffer', acting as a gas-sampling probe. Leaks can be detected in good time to prevent safety or operational hazards and expensive unplanned shutdowns of operational systems (Rasmussen and Jeppesen, 1998). Some substitution to low-percentage hydrogen mixtures is being considered but helium will remain the gas of choice for demanding applications.

Optical glass fibre manufacture

Helium is used in two of the processing stages in the manufacture of optical glass fibres used in the telecommunications industry. The initial stage of the process is to form a glass 'soot' preform by depositing fine glass particles onto a silica rod. The glass 'soot' is formed by reacting oxygen with glass-making reagents such as silicon tetrachloride and silane in a natural gas flame. The 'soot' is next sintered to a solid preform in a helium-chlorine atmosphere. The chlorine removes moisture whilst the helium removes gas bubbles. The preform is then transferred to the drawing machine where it is heated to be drawn into a fibre. The fibre is rapidly cooled by a high-velocity stream of helium-nitrogen gas mixture. Such a gas mixture is used because it provides excellent heat transfer.

Vacuum furnace quenching

Vacuum furnaces, used for the hardening of aerospace alloys and tool steels, use a gas quench to achieve the desired metallurgical characteristics. The gas mixture, often based on helium, is passed through the furnace chamber and the heat transferred to the gas is dissipated in an external heat exchanger. The ability of gases to transfer heat is dependent on a combination of their molecular weight and heat capacity. Low molecular weight gases such as helium and hydrogen have high heat capacity and thermal conductivity but lack mass. Higher molecular weight gases such as nitrogen and argon have lower thermal conductivity and heat capacity. The highest heat transfer is achieved by a mixture of helium-nitrogen, but helium is not essential.

Welding (inert atmosphere)

The first patent for welding in an atmosphere of helium was granted to Henry Hobart in the 1930s (Wheeler, 1955). An early application of this was made during World War II by Northrup Aircraft to repair and salvage faulty magnesium castings. Large quantities of helium were used from the late 1950s into the 1960s as a protective atmosphere in the industrial production of titanium, zirconium and hafnium. In this application, helium serves mainly as a shielding gas, preventing atmospheric contamination of the molten metal and stabilising the arc. In gas metal arc welding (GMAW), when compared to conventional argon-CO_2 mixtures, helium-enhanced mixtures provide increased arc stability at greater wire-feed speeds, thus leading to the rise of weld metal deposition rates. With the help of helium's greater heat conductivity, the performance over mild base metal can be improved and the need for costly rework can be avoided.

Argon cover gas could be substituted for helium in most welding applications, but for the space industry, in which high quality of welding is required, helium is essential. Therefore, helium used in welding is desirable rather than essential in most applications except for aerospace. The elasticity of helium demand in welding is relatively low in aerospace applications but relatively high in the others.

Diving

Helium was first experimented with as a replacement gas for nitrogen by the US Navy beginning in 1912, and its first big success came in 1939 when it was integral to the diving salvage of the downed submarine *USS Squalus* at a depth of 243 feet (74 m). Because nitrogen causes narcosis and central nervous system toxicity, it was recognised early that a different mix of breathing gases would provide advantages, allowing divers to stay down longer and with greater safety.

Helium is, however, a very efficient conductor of heat, which means that if the diver is not kept warm it can hasten the reduction of core body temperature.

Another problem is, of course, the effect it has on the diver's speech, which has resulted in the development of many special communications devices over the years to correct the change in pitch that is caused by rapidly vibrating vocal cords.

Deep-water divers using compressed air for breathing start to suffer nitrogen narcosis (the bends) at a depth of about 40 m. Narcosis is caused by the dissolution of nitrogen in the bloodstream at the elevated pressures experienced, and the formation of nitrogen bubbles in the bloodstream. As the diver ascends this impairs oxygenation, which causes the narcotic effect and leads to unconsciousness. Helium and oxygen are mixed to form artificial air, which safely allows dives to depths in excess of 350 m. Deep-sea divers use gaseous helium as a carrier gas for oxygen in their breathing mixtures. This makes possible pioneering undersea scientific investigation, as well as the exploration and development of offshore oil and gas resources. Hydrogen is being considered as a substitute, but there are safety concerns.

Party balloons, blimps and airships

With a density relative to air of 0.138, helium has considerable lifting power and is the only safe gas for inflating balloons for most purposes. Pure helium has a lift of about 65 pounds per thousand cubic feet compared with a lift of about 76 pounds for hydrogen. Hydrogen has greater lifting power but as a highly flammable gas its use involves the danger of fire and explosion. It is used only for meteorological balloons deployed from remote locations, often by the military.

The use of helium in party balloons[3] has created a sizeable market. The use of very large helium-filled balloons and blimps, or airships as they are more commonly known, continue to be developed for various applications from communication satellites to cargo carriers, and though there has been little commercial success to date, some think the 'tipping point' is close (Limb, 2009). These various 'lifting' applications are the second-largest market for helium. Today, blimps represent a fairly small part of this use. The greatest volume in this sector is for party and promotional balloons, a market that is heavily concentrated on the United States. In the long term, some interesting projects are under way to develop blimps for heavy-lift transport of loads measured in hundreds of tonnes and for telecommunications platforms in the upper atmosphere. Neither of these applications, however, is expected to have an appreciable impact on the demand for helium in the next decade. Some people have found the use of helium in balloons contentious, yet this use provides a stimulus to recover helium that might otherwise be vented to atmosphere. It can be argued that this helps stabilise prices and supply to other users.

Helium in balloons also lifts important cargoes ranging from weather forecasting instruments, television equipment and radar stations to communications relays, blimps and aerostats.

Purging and pressurising

Since the liquefaction temperature of helium is well below that of hydrogen, it is used to purge and pressurise liquid hydrogen tanks and piping systems prior to use. Any other inert gas such as nitrogen or argon would freeze on contact with liquid hydrogen and plug pipes and valves. Helium's low solubility and low boiling point combine with its inertness to purge and pressurise the liquid hydrogen fuel systems of rockets and spacecraft. It is the only inert substance that remains a gas in such low-temperature environments. For this application there is no substitute for helium. This is very significant, as liquid hydrogen is the rocket fuel of choice and most of the helium cannot be recovered.

Typically, helium is used as pressurising gas because of its low molecular weight, lack of chemical reactivity, and relative insolubility in propellants. Either pump-fed liquid propellants or solid propellants are utilised in most rockets. In liquid-fuelled rockets, typically, propellants are stored in tanks which are pressurised with helium. The pressurised propellants are then fed from their pressurised onboard tanks to the rocket engine. Although pressure-fed liquid propellant rockets are typically heavier than pump-fed rockets, and hence lack certain performance characteristics, they are reliable, and well adapted to missions that require stopping and starting the engine several times. Because of this advantage, the Space Shuttle used pressure-fed liquid rockets in its orbital manoeuvring system, and its attitude control system. Pressure-fed liquid rocket engines are also used in interplanetary spacecraft for course correction, orbital insertion and other important functions that must be reliably performed (Helium Study Committee, 1978).

When the pressurised propellant is fed to the rocket engine, the pressure in the propellant tank declines due to expansion of the gas. During the rapid withdrawal of a substantial fraction of the propellant, there is little time for heat transfer from the tank walls and pressurised propellants to the gas. In this event, the gas expands adiabatically rather than isothermally, and the gas cools as it expands. This cooling further exacerbates the decreasing pressure in the tank. A separate tank of high-pressure gas may be provided in order to counteract the effect of decreasing pressure caused by propellant withdrawal from the pressurised tank. Helium from this separate storage tank is then introduced into the propellant tank to regulate the pressure on the propellant (Dunn, 1999). The launching, primarily for peaceful purposes, of reusable spacecraft through the 'Shuttle' programme required 200,000 cubic metres (7.2 MMcf) of helium (for each launch) (Andreev, 1995).

Helium does not become radioactive when exposed to radioactive materials or to neutron bombardment. Also helium is chemically inert, so the system does not corrode. The combination of these two characteristics, together with relatively high thermal conductivity, makes it a feasible to use high-pressure helium as the heat-transfer medium in gas-cooled nuclear reactors for power generation. Unlike fusion, the dependency of near-term nuclear reactors on conventionally sourced helium is probably rational.

HELIUM DEMAND

Historically, helium's demand structure has been constantly changing. It needs to be pointed out that the structure of helium consumption and use varies with the passage of time as a function of the development of science, engineering and industry. This section presents an overview of historical helium uses.

Global demand overview

Figure 9.2 indicates the uses for helium in various industries throughout the world. Over the past 20 years, demand has mainly been driven by the development of MRI and the need to cool the magnets to near absolute zero to become effective. MRI became, and still remains, the major use of liquid helium. The other major drivers of demand in the past were the use of helium as a heat-transfer agent and carrier gas for fibre optic production, followed by pipeline leak detection, the leisure industry, electronics, specialised welding and airships.

Demand: MRI

Figure 9.2 shows that MRI accounted for about 20 per cent of the worldwide use of helium in 2006. It is the largest end-use sector. This is expected to remain an important sector but future growth will be very much related to the continued shift in technologies from low-temperature superconducting to high-temperature superconducting magnets. Demand for MRI machines is still growing, although

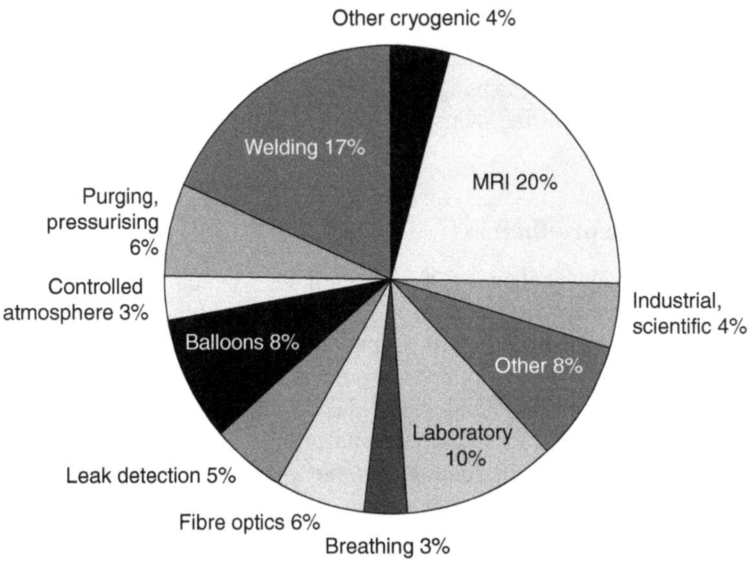

Figure 9.2 Global helium demand in 2006 (*Gasworld*, 2006).

growth has slowed from the recent highs. MRI use is driven by strong growth in emerging nations investing in such technology and the replacement of older units in developed regions. In the short term there will be stagnation in demand for liquid helium in MRI, and in the medium to long term, demand will fall.

A technology shift in the way MRI magnets are produced and maintained has had a significant impact on the amount of liquid helium used to cool them. The use of closed-cycle systems in the magnet has generally resulted in liquid helium use being reduced by 75 per cent in new magnets. Manufacturers have also installed recovery systems, leading to more efficient use of liquid helium. The technology shift would have been a major concern to helium producers and suppliers, but this decline in the use of helium per magnet is offset by the recent rapid rise in the production of MRIs to meet global demand – especially in the Far East. This means that the overall demand for liquid helium in MRI applications will remain stable to slightly positive as a result of the technology shift.

The global market for new MRI installations continues to show a healthy growth, but advances in MRI design help reduce the need for increased liquid helium supply to field locations.

Demand: welding

Welding represents the second-largest application for helium. It accounts for 17 per cent of the total global market demand. This use is heavily concentrated in the United States, where helium historically has been somewhat less expensive and more freely available than in other parts of the world. Of all the major applications for helium, its use for welding is judged to be the most susceptible to substitution as helium costs rise. In fact, in parts of the world where helium has traditionally commanded a higher price, alternative, more sophisticated shielding gas mixtures already fulfil many of the same needs that helium meets in the United States. Resistance to changing construction codes in aerospace, shipbuilding and other manufacturing industries, however, will slow the move away from helium in US welding.

Demand: fibre optic production

Demand reached a peak in about 1999. Up until then helium use had been growing at about 5–6 per cent per annum. With the collapse of the telecommunications market in 2001, however, helium demand in this segment dropped by at least 50 per cent. The slump in fibre optic production in 2001–2 and the slowdown in the electronics industry and major developed economies after 2001 caused a reduction in demand for helium, as shown in Figure 9.3. In the meantime, these significant declines in volume, however, have been partly offset by growth in other major demand categories, namely MRI and helium for speciality gases, leak detection and welding (including laser welding).

Figure 9.3 also demonstrates that 2003–4 saw a strong rise in demand from the decline seen in 2001–2. Most industrial gas companies witnessed a rise in

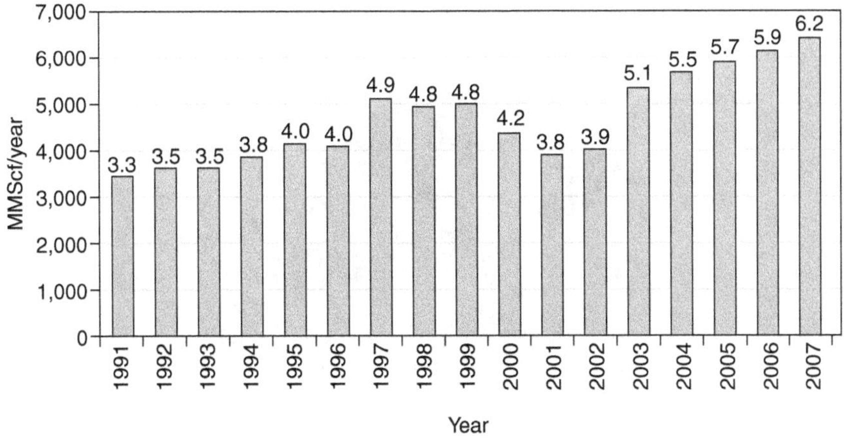

Figure 9.3 Global helium consumption, 1991–2007 (Pacheco, 2008; Air Products, 2007).

demand of 4–5 per cent – some experienced even higher sales due to market share penetration. The short-term boost was driven by the high demand for helium in the electronics sector – especially in flat-panel display manufacturing for LCD and plasma televisions. A number of major manufacturing plants have been established in China, Korea and Taiwan. As a result, helium demand steadily increased. A slight reduction in demand during the 2008 downturn has now been reversed, as demand from developing nations increases.

The UK market

Figure 9.4 indicates that in 2005 the main drivers of helium demand are MRI/NMR and balloons. From 2002 to 2005 the UK helium market experienced significant growth at Siemens, which uses helium for manufacturing MRI machines. GE Manufacturing departed from the UK, which led to the reduction of helium supplied by BOC and Air Products Ltd (part of APCI). Price increases in this period resulted from helium shortages and rising costs. During the period from 2005 to 2008, the demand for helium continued to grow in the MRI sector and universities. Some price erosion may have been expected due to a helium surplus arising from the 2008 global turndown, although the weakening US dollar has offset this effect and prices have continued to rise.

Geographical movement of helium demand worldwide

The trend in the movement of helium demand is that demand in the US has been declining, while in economically emerging areas such as Asia demand has been increasing rapidly. To quote news released on 28 June 2006, from the China-TechNews website (2006):

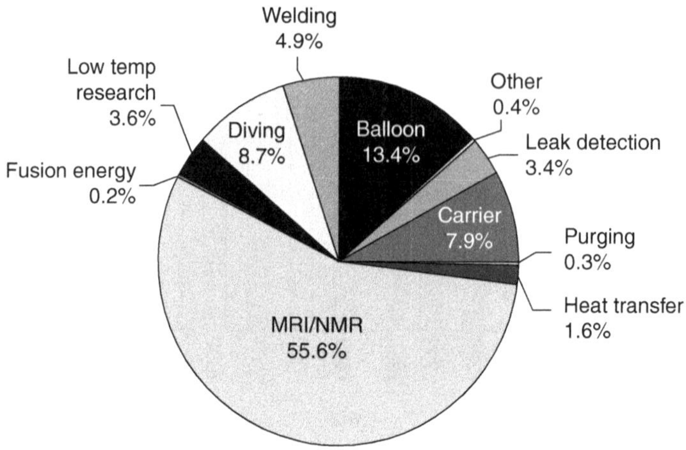

Figure 9.4 Helium market demand structure in the UK market, 2005 (data provided by Linde-BOC).

> China is one of BOC's fastest growing helium markets. Demand has grown at the rate of 16 percent during the last five years, driven by scientific and medical applications, such as magnetic resonance imaging (MRI) and nuclear magnetic resonance (NMR), as well as high-tech applications including optical fibre and semiconductor manufacturing.

Figures 9.5a, 9.5b and 9.5c show the distribution of demand worldwide in 2001, 2006 and 2010 split by geography. Due to the historical concentration of helium sourcing and the nature of the principal helium applications, the United States still accounted for nearly 60 per cent of global demand in 2001. The next largest market for helium was Europe (22 per cent), followed by Japan (11 per cent). Asia and the rest of the world (ROW) accounted for just 4 per cent and 2 per cent of helium demand respectively.

Moving to Figures 9.5b and 9.5c, the market share of the US decreased significantly from 2001 to 2010, while demand for helium increased in Europe, the main change was in the Middle East and the economically booming Asian area. This is highlighted in the stacked bar chart shown in Figure 9.5d.

Historical demand: helium consumption 1917–60

In this period, helium was predominately used for military (aeronautical) purposes. A very small amount of helium was used in a mixture with oxygen for deep-sea diving and as lifting gas for toy balloons (Macy's Thanksgiving Parade balloons started in 1927) and also as an inert atmosphere in metallurgical operations. Therefore, helium demand in this period depended almost entirely on

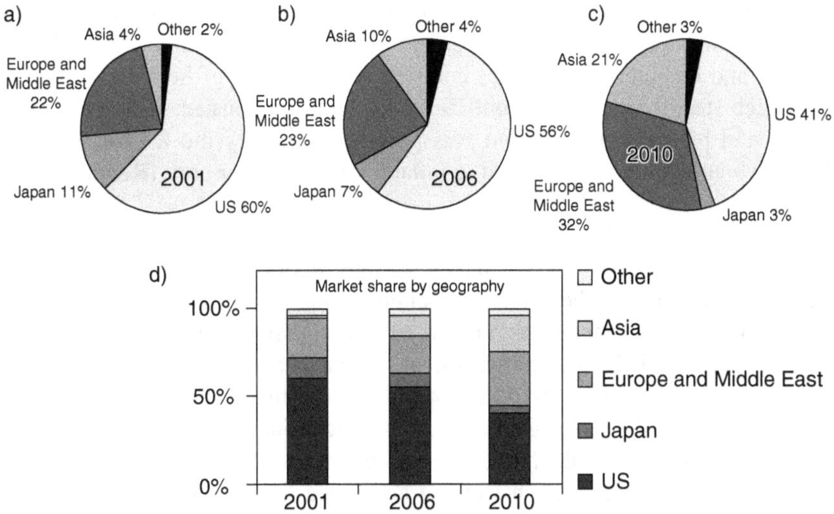

Figure 9.5a, b, c and d Global helium demand distribution by geography 2001, 2006 and estimated for 2010 (Air Products, 2007).

advancement in the use of airships. Table 9.5 shows that during World War II, helium was principally consumed by the federal government, directly or through defence contracts. The US Navy was the principal user of helium among the federal agencies. The US Weather Bureau's use of helium continued to grow. A very small portion of helium was for private use, particularly for welding. After World War II, helium used by the US government dramatically decreased, and stayed relatively stable until the outbreak of the Korean War in 1950.

Table 9.5 Helium consumption during WWII, MMcf/year

	Navy	*Weather Bureau*	*Army and other federal*	*Federal total*	*Non-federal*	*Total shipments*
1938				5.47	0.0	5.54
1939				5.12	1.07	6.19
1940	3.53	2.62	1.88	8.04	1.51	9.55
1941	9.19	4.41	5.31	20.91	1.25	22.16
1942	25.40	5.09	4.79	35.28	0.78	36.06
1943	107.24	5.63	2.44	115.31	1.32	116.63
1944	19.08	7.04	2.44	120.56	3.01	123.56
1945	38.09	8.01	9.76	57.86	2.93	60.79
1946	15.74	9.71	9.29	34.73	9.19	43.92
1947	26.51	6.35	4.49	37.35	14.97	52.32
1948	21.53	6.48	6.87	34.88	16.04	50.92
1949	20.40	6.40	8.34	35.13	16.37	51.50

Source: Mullins and Wheeler, 1950.

After the enactment of the Helium Act of 1937, the expected wider use of helium materialised because the regulations authorised the Bureau of Mines to sell helium that was not needed for government use. Helium used for medical, scientific and commercial purposes contributed to the rise of helium consumption, which steadily increased from 1949 to 1960. The constant increase in the demand and production of helium was partially driven by the Korean War in 1950 in which helium was required to inflate lighter-than-air craft (Kennedy and Mullins, 1951).

Another important factor that pulled up the demand for helium in this period was the lower price, as directed by the regulations of 1937. This provided a strong incentive to encourage the R&D of helium-related technologies such as the MASER (microwave amplification by stimulated emission of radiation), the precursor of the LASER (light amplification by stimulated emission of radiation).

Furthermore, the application of helium as an inert shielding medium increased rapidly in shielded-arc welding of aluminum, magnesium, stainless steel, copper and other metals. In 1955, the first survey of helium use conducted by non-federal users shows that helium used in welding accounted for 66 per cent of the total helium used in private industries, followed by titanium and zirconium production and fabrication, leak detection and controlled atmospheres, as shown in Figure 9.6.

The breakdown of federal helium use was available for the first time in 1959, as shown in Figure 9.7. This shows that 21 per cent of helium was used for guided missiles, the second-largest user was atomic energy, then welding, accounting for 18 per cent. Other uses included medical applications, titanium and zirconium production, controlled atmospheres for the growth of germanium and silicon transistor crystals, leak detection in industry and low-temperature research.

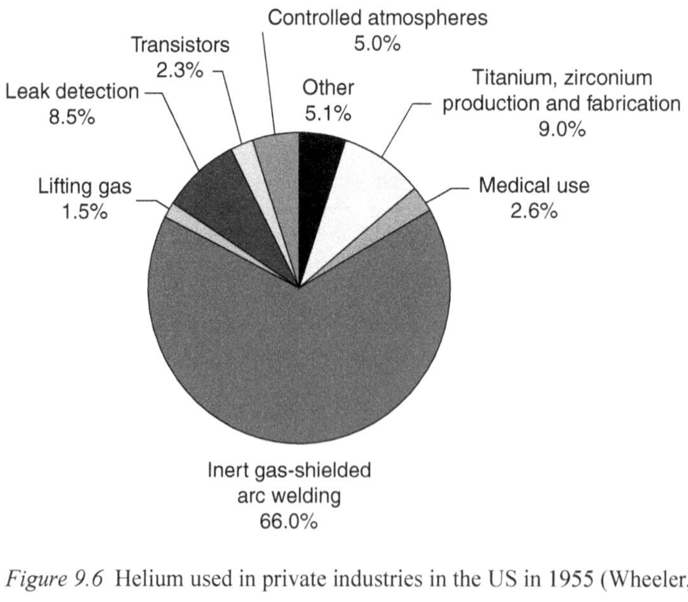

Figure 9.6 Helium used in private industries in the US in 1955 (Wheeler, 1955).

Helium pricing 1925–60

From 1925 to 1938, the price of government helium sold to government agencies was determined by the operating costs, including the labour and maintenance cost, without taking into account the depreciation of the capital cost of production plants and depletion of gas resources. This notional pricing became a part of the administrative tradition for government helium activity.

Figure 9.8 shows the helium price for federal users and the triple-price scheme for non-federal use – i.e. medical use, commercial use, and scientific use. The term 'triple pricing' arose during the period 1938–46, where different users were charged different prices per unit of helium. The Helium Act 1925 was further amended to adopt a dual pricing scheme for helium sold to

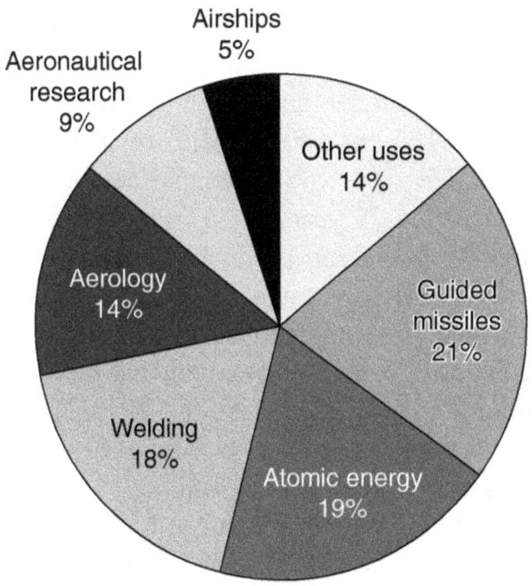

Figure 9.7 Estimated uses of helium in 1959 (Wheeler, 1962).

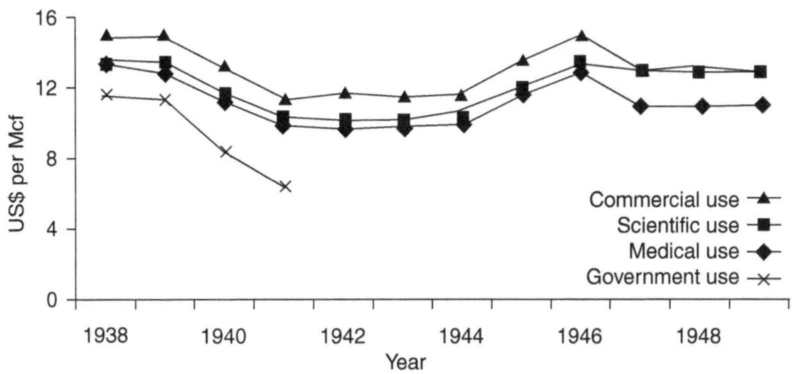

Figure 9.8 Helium market prices 1938–49 (Mullins and Wheeler, 1950).

non-federal users and was put into effect in February 1947. Commercial and scientific users were charged the same price ($13/Mcf) and medical users were charged $11/Mcf.

Figure 9.9 shows the helium price for US government agencies and non-government purchasers between 1950 and 1964, indicating that the price for government use was lower than that for non-government purchasers. The change in the market helium price was made through an amendment of regulations made effective on 27 February 1950 (Wheeler, 1955). Another amendment of the 'Regulations Governing the Production and Sale of Helium' (30 CFR 1) became effective on 13 March 1954. This amendment directed a price-setting mechanism. As shown in Figure 9.10, it can be seen that from 1955 to 1958 the helium demand exceeded its supply, and in 1959 and 1960 helium supply was in excess of demand, but the price did not respond to this change and stayed constant during the period 1955–60. This is because the US government was the *sole producer* in the US helium market, and had the market power to set prices. A new pricing policy was prescribed in the 1960 Helium Act Amendments. It aimed to implement the conservation programme on a 'self-liquidating' basis. The costs to be paid for included several elements: loan borrowings from the Treasury with an annual compound interest rate on unpaid balances (to be paid within 25 years or 35 years if necessary); the net capital and retained earnings of the helium production fund (as of September 1960) and any further investment in the Keyes plant, together with the annual compound interest rate (to be paid within 25 years); and for funds borrowed to acquire or construct helium plants and facilities, with annual compound interest rate on unpaid balances (to be paid within 25 years of the date of borrowing). Under this new policy, the initial price was calculated at $35/Mcf and charged to both federal and non-federal users. This was made effective in November 1961.

To sum up, in the period 1925–60 the helium price was regulated by the US government, via the Helium Act 1925, Helium Act 1937 and the 'Regulations Governing the Production and Sale of Helium', amended in 1947, 1949 and

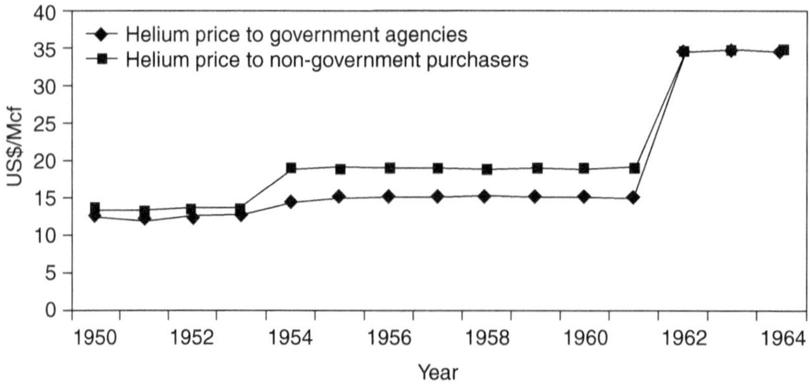

Figure 9.9 Helium market prices in the US, 1950–64 (Thomason, 1965).

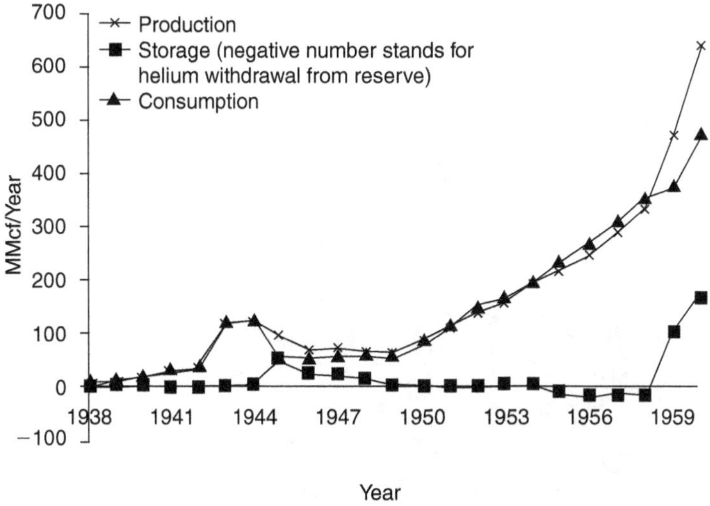

Figure 9.10 Helium production, consumption and storage, 1938–1960 (Kelly *et al.*, 2007).

1953 respectively. The US government had a monopoly on helium, being the sole producer and the major user. It completely controlled helium production and pricing through the regulations with the aim of achieving security of helium supply in the public interest. There were basically two reasons for the US government's dominance in the helium market before 1960. Economies of scale was the key justification in the helium supply, since the rated capacity of the Amarillo plant was 24 MMcf/year yet the demand for helium was not more than 15 MMcf/year before 1938.

Helium consumption after 1960

As shown in Figure 9.9, from 1961 to 1966 helium consumption increased steadily. During this period, helium was mainly used in pressurising, welding, controlled atmospheres and research, as well as purging, airships and balloons, as a heat-transfer medium, cryogenics and leak detection. Figure 9.11 shows the main uses of helium in 1963. These totalled 630 MMScf.

In 1967, consumption of Grade-A helium from Bureau of Mines plants was about 607 MMcf, a decrease of 202 MMcf or about 24 per cent compared with total sales of 809 MMcf in 1966. This decrease in sales was attributed to three factors (Thomason, 1967): (1) decreased activity in the US space programmes following the Apollo 1 disaster at Cape Kennedy in January 1967; (2) the general cutback in all governmental programmes not related to the Vietnam conflict; (3) the growth of the private helium industry which, in 1967, supplied virtually all Grade-A helium to the civilian market.

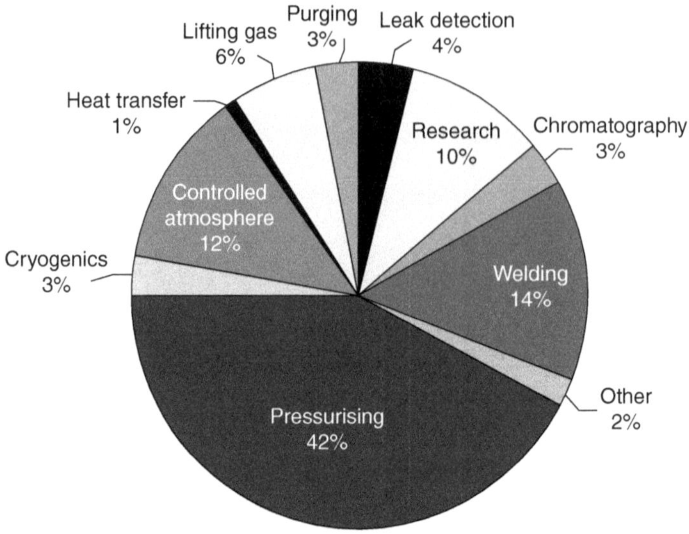

Figure 9.11 Helium end uses in the US in 1963 (Lipper, 1964).

> The decline in total helium sales in the United States follows a trend first established in 1967. Peak sales occurred in 1966, when 948 MMcf of helium was sold (used). Most of the decline in helium sales (usage) was believed attributable to the reduced level of activity in the Nation's space and missile programs and the shift in these programs from development to operational status. Also, the overall slowdown in the Nation's economy has undoubtedly contributed to the reduction of helium sales in the past few years.
>
> (Thomason, 1970: 569)

Helium redistribution continued satisfactorily under contracts between the General Services Administration and private companies. The private companies purchased helium in bulk from the Bureau of Mines, repackaged it in smaller containers and distributed it to the helium-using federal agencies. These contracts made relatively small quantities of helium readily available to the agencies with reduced freight charges for small purchases. The largest user of helium in 1970 was again the US space and missile programme. Private industry and research organisations also continued to use large quantities of helium. Table 9.6 shows the shipments of Grade-A helium from Bureau of Mines plants to various customers (in MMcf) during 1963–70.

Although the US space and missile programme continued to be the largest user of helium, their demand was declining while private use of helium increased steadily. This trend continued until 1972, when domestic sales of refined helium started to rise. Much of the increase was attributable to the overall improvement

Table 9.6 Shipments of Grade-A helium from government plants to various customers (MMcf)

Federal Agencies	*1963*	*1964*	*1965*	*1966*	*1967*	*1968*	*1969*	*1970*
Department of Defense	337.3	347.3	313.2	232.9	226.1	279.9	191.6	119.5
Atomic Energy Commission	75.5	64.8	59.9	52.9	39.3	28.5	23.3	17.1
National Aeronautics and Space Administration	52.6	79.7	99.9	214.7	147.0	86.6	63.8	37.4
Weather Bureau (ESSA)	7.8	7	5.6	5.8	5.5	4.5	4.5	3.1
Other	0.5	0.6	0.5	0.6	0.7	0.9	0.8	0.4
Subtotal	473.7	499.4	479.1	506.9	418.6	400.4	284.0	177.5
Non-federal customers	153.6	168.0	219.5	301.7	188.4	78.0	78.7	53.2
Total	627.3	667.4	698.6	808.6	607.0	478.4	360.7	230.7

Source: Koelling, 1971.

in the US economy and increased requirements for helium in research and in breathing mixtures.

From 1972 to 1983 helium demand in the US domestic market increased slowly. From 1984 to 1999, helium demand in the domestic market grew more quickly, largely due to the active space exploration activities. For instance, in 1984, five more successful Space Shuttle launches were made by the National Aeronautics and Space Administration (NASA) using Bureau helium. Each Space Shuttle launch required about 7.5 MMcf helium for purging and pressurising.

Demand declined during the period of 2000–2. This flattening of growth in global helium demand was caused by several factors. One of the main factors is the fibre optics 'bubble', which had propelled helium growth rates and set expectations of ever expanding demand. It 'burst' in 2001–2 with the telecoms/internet downturn.

Price elasticity of demand

Price elasticity of demand is a key variable in the system dynamics model, and it was necessary to take a close look at it. Helium is preferred in most cases because it is technically superior. For instance, helium is preferred as a lifting gas because of the danger of explosion related to hydrogen. Nitrogen can be substituted for helium as a pressuring medium, but only where weight is not a problem. Because helium is preferred in most applications, its price tends to be inelastic at the lower end of the price range.

Although helium demand may be inelastic at lower prices, little is known about what effects future price increases may have on helium demand. Attempts have been made to estimate the price elasticity of demand for helium using econometric methods. Those studies agree that, within a range of prices for which experience exists, elasticity is from about –0.5 to –0.63 (Uri, 1987).

The price of helium can be very inelastic for technologies where helium is a small component of the total investment. Time also matters in the price elasticity of demand. In the longer term, the price elasticity of demand tends to be higher because time is needed for customers to respond to price changes. For example, recent soaring prices (2009) did not significantly reduce demand since a period of time (perhaps three to five years) is required for users to seek substitution, either through recycling or through the development of new technologies that are free of helium or less helium intensive.

To sum up, the following insights and propositions from the critical discussion in this chapter can be used in the system dynamic model presented in Chapter 10.

- Helium trading is basically governed by long-term contracts with take-or-pay clauses. They slow down the 'clock speed' of price adjustment, delaying adaptation to the changing market conditions. The law of supply and demand is still valid in helium pricing since pricing typically 'reopens' once a year in the upstream market and price adjustments are more frequent in the downstream market in response to changing market conditions.

- Evidence shows that price increases result from two elements: rising costs and the existence of excess demand.
- Price elasticity of helium demand is estimated to be within the range of –0.5 to –0.63, in the realm of relatively low prices. It remains to be seen what will happen to price elasticity as prices continue to increase significantly.

Notes

1 The temperature at which the transition (from gas to liquid) takes place is called the lambda temperature (T_λ).

2 What is a wafer? A wafer is silicon disc with a diameter of 200–300 mm. Wafers provide the active structure of the electronic chips. In clean rooms, 300 different operations involving over 50 gases create multiple thin films in a dozen or so levels on the wafers. These layers create the interconnection to and from the individual transistors, which are tiny electrical switches that allow or prevent the flow of electricity. When the manufacturing is completed, the wafer is cut into individual chips. Each is then packaged and installed in everyday products such as mobile phones, laptops, TV sets, etc.

3 *Helium balloon safety information*
The British Compressed Gases Association (BCGA) advises as follows:

> "Helium is a very light and inert, non-toxic gas, but if it displaces oxygen it can be fatal. The 'fun' to be found in the squeaky voice helium trick is far from funny when people, often youngsters, die trying this. It does not take many breaths of helium to fall unconscious and die this way. Youngsters in particular should be made aware of the danger and BCGA would advocate a warning hazard being attached or given with every helium filled balloon and with every cylinder of helium balloon gas supplied to members of the general public who are unlikely to read the full detail contained in safety data sheets." Further safety information on helium and other gases can be found at the BCGA website www.bcga.co.uk.

References

Air Products, 2007, 'Wide Helium Shortages', in CryoUsers Meeting, Oxford, UK.

Andreev, I. L., 1995, Helium Industry in Russia and World Experience with the Development and Operation of Helium Equipment, *Chemical and Petroleum Engineering.* Vol. 31(2): 73–85.

ChinaTechNews, 2006, www.chinatechnews.com/2006/06/26/4321-boc-opens-opens-distribution-center-in-suzhou.

Dunn, B. P., 1999, 'Method and Apparatus for Pressurized Feeding of Liquid Propellants to a Rocket Engine', US Patent No. 5,961,074, 5 October.

Galasiewicz, Z. M., 1971, *Helium 4.* Pergamon Press, Oxford, New York.

Gasworld, 2006, 'Helium, Changing Faces, Technology Shifts and Tight Supply – Challenges Facing our Industry'. *Gasworld Magazine*, Vol. 17.

Heftmann, E., 2004, *Chromatography: Fundamentals and Applications of Chromatography and Related Differential Migration Methods.* Elsevier, Amsterdam, New York.

Helium Study Committee, 1978, 'Helium, A Public Policy Problem'. National Academy of Sciences, Washington, DC.

Hucknall, D. J., 1991, *Vacuum Technology and Applications.* Butterworth-Heinemann, Oxford.

Kelly, T. D., Bleiwas, D. I. and Pacheco, N., 2007, 'Helium Statistics', in 'Historical Statistics for Mineral and Material Commodities in the United States' (online). US Geological Survey.

Kennedy, H. S. and Mullins, P. V., 1951, *Minerals Yearbook 1951*. United States Government Printing Office, Washington, DC.

Koelling, G. W., 1971, 'Helium', in *Minerals Yearbook: Metals, Minerals, and Fuels 1971*. Bureau of Mines, Washington, DC: 577–83.

Kyle, B. G., 1999, *Chemical and Process Thermodynamics*. Prentice Hall PTR, Englewood Cliffs, London.

Limb, D., 2009, 'Airship Transportation System – Helium Handling And Storage', Fifth International Airships to the Arctic Symposium, Calgary AB, 7–9 October 2009, www.airshipstothearctic.com/docs/pr/19_David_Limb.pdf (accessed February 2011).

Lipper, H. W., 1964, 'Helium', in *Minerals Yearbook: Mineral Fuels 1963*. http://minerals.usgs.gov/minerals/pubs/myb.html.

Mullins, P. V. and Wheeler, H. P., 1950, 'Helium', in *Minerals Yearbook 1950*. Bureau of Mines, http://minerals.usgs.gov/minerals/pubs/myb.html.

Pacheco, N., 2001, 'Helium', in *US Geological Survey Minerals Yearbook 2001*. US Geological Survey.

Pacheco, N., 2008, 'Helium', in *Mineral Commodity Summaries*. US Geological Survey, http://minerals.usgs.gov/minerals/pubs/mcs/.

Rasmussen, H. H. and Jeppesen, L., 1998, 'Industrial applications of Helium Leak Test'. *The NDT Database and Journal*, Vol. 3(12), http://www.force.dk.

Thomason, E. M., 1965, 'Helium', in *Minerals Yearbook 1967*. Bureau of Mines: 457–64.

Thomason, E. M., 1967, 'Helium', in *Minerals Yearbook 1967*. Digital Collection Center, University of Wisconsin: 757–64.

Thomasson, E. M., 1970, 'Helium', in *Minerals yearbook: Metals, Minerals, and Fuels 1970*. Bureau of Mines, http://digicoll.library.wisc.edu/cgi-bin/EcoNatRes/EcoNatRes-idx?id=EcoNatRes.MinYB1970v1.

Uri, N. D., 1987, 'Helium Conservation: Supply and Demand Projections in the USA Reconsidered'. *Energy Economics*, Vol. 9(2): 93–8.

Wheeler, H. P., 1955, 'Helium', in *Minerals Yearbook 1955*. Bureau of Mines, http://minerals.usgs.gov/minerals/pubs/myb.html.

Wheeler, H. P., 1962, 'Operation Foresight – Bureaus of Mines Program of Helium Conservation and Research'. *Advanced in Cryogenic Engineering*, Vol. 7: 25.

10 The dynamics of the helium market

William J. Nuttall, Zhiming Cai, Bartek A. Glowacki, Nikolaos Kazantzis and Richard H. Clarke

Introduction

Chapter 2 has described the interesting and unusual history of the helium industry, which has been dominated for so long by the United States. In Chapter 3 the continuing role of US production and storage was discussed. We have seen how the past, present and future of helium production are intimately connected to the natural gas industry, and trends in that industry have been considered in Chapter 5. In this chapter we seek to gain insight into possible long-term future developments in helium prices and supply.

In 2005 a desire to improve understanding of the long-term trends in helium supply led to the formation of a collaborative group comprising a helium supplier (BOC Limited), major helium user (Culham Centre for Fusion Energy (CCFE)[1]) and academic researchers at the University of Cambridge (from Judge Business School, the Engineering Department and the Department of Materials Science and Metallurgy).

The project was constructed as a PhD research programme for a student, Zhiming Cai, with a background in Nuclear Engineering and Sustainable Development. The work was led by Zhiming Cai with support and advice from the wider collaborative group. The core results of that PhD research have been published in the journal *Resources Policy* (Cai *et al.*, 2010). In this chapter we seek to elaborate on the choice of research methodology, to summarise the main findings, and to discuss the limitations of the approach. We pay particular attention to demand-side assumptions. At the end of the chapter we shall consider recent developments and make suggestions for further work.

Helium as a non-renewable co-product

Within the research team it was soon appreciated that the supply of helium is a complex matter involving market forces, public policy intervention, resource depletion, and demand growth ahead of that seen for other commodities, which generally track overall economic activity.

The helium industry has two relatively unusual aspects. First, helium is a co-product of another industry – the natural gas industry.[2] While helium is used in a variety of diverse industrial sectors and for a wide variety of purposes, natural gas has rather different demand and supply dynamics as it is primarily an energy commodity (although some natural gas is used as a chemical precursor in, for instance, fertiliser production). Second, helium that is not used is not necessarily conserved. Because helium is found mixed with natural gas, any helium not extracted from the natural gas supply chain remains mixed with the natural gas. Helium is an inert gas occurring in small proportions in commercial natural gas (typically less than 0.3 per cent) (Cook, 1979), and as such it is unproblematic for the natural gas industry. In those cases the helium typically remains mixed with the natural gas throughout the supply process until the end use of the natural gas in combustion. At that point the helium is simply lost to the atmosphere along with the gaseous combustion products. Such lost helium is said to be 'vented' and the work led by Zhiming Cai has shown that venting rates are of central importance to the medium and long-term sustainability of the helium industry. Helium and nitrogen are impurities affecting the energy content of natural gas as measured for a given supply pressure by the Wobbe Index.

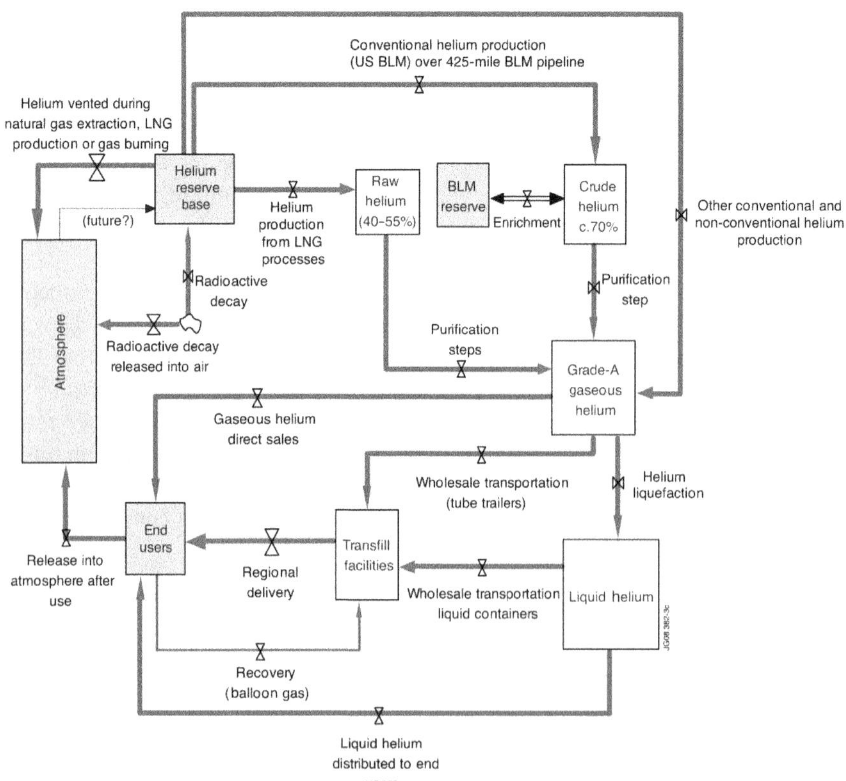

Figure 10.1 Stocks and flows of helium in the industrial supply chain.

Noting these complexities the research collaboration decided as a first step to focus on insights that might be obtained by deploying the methodology of system dynamics pioneered by Jay Forrester and colleagues (Forrester, 1961) at the Massachusetts Institute of Technology in the late 1950s. In those early days it was known as 'industrial dynamics' (Sterman, 2000).

System dynamics is a methodology for studying and managing complex feedback systems, for investigating the dynamic behaviour of sociotechnical systems and their response to policy. It makes use of three distinct fields of endeavour to analyse economic-social-technical systems: feedback control theory, organisational behaviour and information technology.

The basic processes of the system dynamics methodology comprise: problem identification; development of a dynamic hypothesis explaining the cause of the problem; building a computer simulation model of the system at the root of the problem; testing the model to be certain that it reproduces the behaviours seen in the real world; and finally devising and testing in the model policies that may mitigate helium supply problems. This is an iterative process, however, because the starting hypothesis is always provisional and subject to revision. Such revisions occur as we learn from the modelling process and from new 'real world' data (Sterman, 2000).

The origin of natural resource economics is widely recognised as Harold Hotelling's seminal article, 'The Economics of Exhaustible Resources' (Hotelling, 1931). Hotelling's paper laid a solid foundation for an economic theory for an exhaustible resource. Hotelling established the well-known 'Hotelling rule', stating that the price of an exhaustible resource must grow at an interest rate, as shown in Equation 1 below.

$$P(t) = P_0 \exp(rt) \tag{1}$$

Where $P(t)$ is price in period t, P_0 is the initial price, and r is the real interest rate. The Hotelling rule is based on three key assumptions: first, the firm's objective is to maximise the net present value of its current and future profits; second, the resource market is always perfectly competitive; and third, the interest rate stays constant. Hotelling further analyses the effects (on price and output paths) of various factors, including: monopoly, demand, cost dependency on cumulative production and uncertainty with special reference to externalities.

Hotelling further observed that a monopoly-based industry will produce at roughly half the 'optimal' rate. The optimal rate is that which would maximise the current value of the resources extracted. Oligopolistic production leads to a 'waste' of resources when compared with a fully competitive market.[3] Hotelling's perspective is that of the firm and his work predates by several decades notions of sustainability and the sense that society's long-term interests are not necessarily best served by today's producer profit maximisation. In the early 1930s challenges to Hotelling's implicit recommendations would have been more Marxist in origin; recommending monopoly in production as part of a rebalancing of power between labour and capital. Whether motivated by

Marxism or by a sense of sustainability, concern for the depletion of primary resources, such as oil, is generally premised on the notion that any unused resource remains for use by later generations. In the case of helium, however, the situation is interestingly different. Helium is a co-product of natural gas and if it is not separated for immediate use then, in effect, it is lost forever to the atmosphere. Hence for helium, any argument that there might be long-term sustainability benefits in market inefficiency does not hold true.

There have been relatively few economic studies by academics of the helium industry. One interesting set of observations came in the 1980s from Alice Mae Hughey, when she considered the co-product issues of helium for her 1984 PhD dissertation at Columbia University. She later published her helium work in the Eastern Economic Journal (Hughey, 1991). Hughey developed an economic model of jointly produced natural resources. She further analysed the effects of different policies such as tax and subsidy of helium. Demand uncertainties were given special consideration. Her model was derived on the basis of several assumptions, one of which is that the market is perfectly competitive. This assumption made it possible for her to achieve an analytical solution to her model. We suggest, however, that for the real helium industry such an assumption is unduly optimistic. While we see no evidence for cartelisation in the helium industry, it appears clear that there are instances of oligopoly in the supply chain and that for certain territories the helium market is far from competitive. These realities lead us to conclude that attempts at analytic solutions should be replaced by techniques better suited to handling emergent complexity, such as system dynamics. Despite the limitations inherent in her approach, Hughey observed and recognised the important role of helium venting.

The sustainability, or otherwise, of helium is driven by other factors, mostly relating to the growth of the natural gas industry and helium demand growth for high-technology applications.[4] Such complex interdependencies are well suited to investigation using system dynamics, further motivating Zhiming Cai's choice of system dynamics as his research methodology.

System dynamics of the helium industry

In the course of his PhD research Zhiming Cai constructed a system dynamics model comprising six interconnected feedback loops (labelled B1–B6 in Figure 10.2).

System dynamics and intermediate microeconomics represent complementary and sometimes competing approaches to the study of natural resource depletion. For system dynamics the core of the approach is a concern for quantity (via stocks and flows influenced by causal loops), whereas for microeconomics the central determinant is price. Some of us, in our earlier work, have sought to integrate these two approaches to develop an economically valid system dynamics model of helium supply and resource depletion (Cai *et al.*, 2010). While it is possible to comment on the price behaviour in the various system dynamics

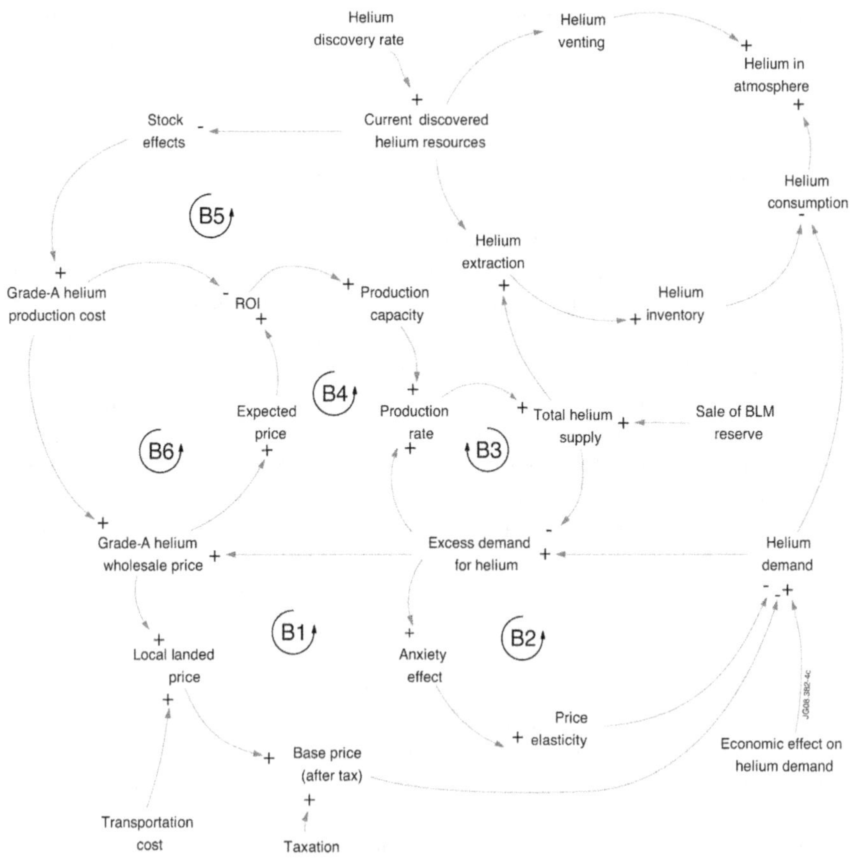

Figure 10.2 Helium systems dynamics – the causal loop model.

loops it is arguably more helpful, and more conventional, to regard the polarity of each loop solely in quantity terms.

In the model developed by Cai and colleagues the growth of helium demand is mainly driven by an exogenous force: the growth in demand driven by macro-level economic growth adjusted for its particular effect in helium-dependent technology sectors. The various feedback loops tend to militate against such growth in supply – i.e. these loops provide a negative feedback. The relevance and relative importance of each feedback loop varies with time depending on prevailing conditions. For example, the resource depletion loop (B5) only becomes strongly influential at later dates in the range of time modelled, i.e. many decades from now.

It is interesting to note that the combined loop B1 and B2, via excess demand for helium (a quantity term), appears to represent a negative feedback, and indeed it is in terms of the elements presented. However, the overall role of B1 and B2 together is to link helium price to helium demand. Most of its elements are price quantities. Any feedback loop that is determined by price-related quantities and that has a negative polarity could be regarded essentially as having an opposite effective polarity in quantity terms. It would, however, not be conventional to regard loops B1 and B2, as presented, to be anything other than negatively reinforcing.

A core objective of Zhiming Cai's PhD research was to understand better the supply-side issues of the helium industry. It was noted from the start that the demand side is of equal importance to the industry's dynamics and it is arguably of greater complexity, but it was agreed that such a depth of understanding could only be attempted once the supply side was better understood. For his PhD Zhiming Cai represented the demand-side issues in a simplified way, relying on a set of key assumptions: all industrial sectors requiring helium were lumped into a single user profile; all diverse territories with different local (landed Grade-A) wholesale helium prices were lumped into a single notional territory; and growth was driven by a single exogenous parameter (Cai *et al.*, 2010). As regards the single lumped user approximation, we note that the real demand side is multifaceted and dynamic. At any instant the demand for helium includes both users whose needs are increasing and users whose needs are diminishing. Despite such individual fluctuations, overall demand has tended to grow steadily at between 4 per cent and 6 per cent, at least until the global recession of 2008–9.

The various demand-side approximations and simplifications adopted in the work of Cai and colleagues might in future be unpacked in an expanded system dynamics model. Rather than relying on an exogenous growth parameter, growth in these expanded models might be driven more by endogenous demand-side dynamics rather than by exogenous factors. Detailed expansion of demand-side modelling could in principle be matched to the system dynamics model as developed thus far. In taking such a step, however, one would need to recognise the risk inherent in all system dynamics modelling: that a move to greater detail and complexity can represent a move away from simplicity and hence away from true understanding. We note that more detailed and refined system dynamics models endowed with a higher degree of complexity are frequently developed as a consequence of a desire to enhance the model's descriptive power. The price paid is to introduce more drivers of uncertainty into the model, however, which risks undermining the model's performance and utility in policy terms. Such concerns have helped to shape the level of approximation implicit in the model of Cai and colleagues.

Observations concerning helium supply

Figure 10.3 presents our assessment of the future trends for helium supply emerging from Zhiming Cai's system dynamics model for the long term. Looking in particular at the 50-year period 1990–2040, we see an S-shaped or 'logistic' curve that has become regarded as an archetype in system dynamics modelling, as illustrated in Figure 10.4. Such S-shaped or 'logistic' behaviours typically correspond to a transition or tipping from an initial situation in which the dominant effects are driven by positive reinforcing loops and an exogenous driver of global demand growth. While in some sense the combined loop B1 and B2 may be regarded as a negatively reinforcing loop, we note that it is a loop mostly determined by price-related quantities, as discussed previously. In time, parameter values can change to bring into force negatively reinforcing loops and over this timescale we suggest that the most important negative feedback will come from loop B1, denoting demand-side response. In the event of supply scarcity,[5] wholesale prices rise, local landed prices rise and in turn (linking to loop B2) retail prices rise, leading to a reduction in demand tending (via part of loop

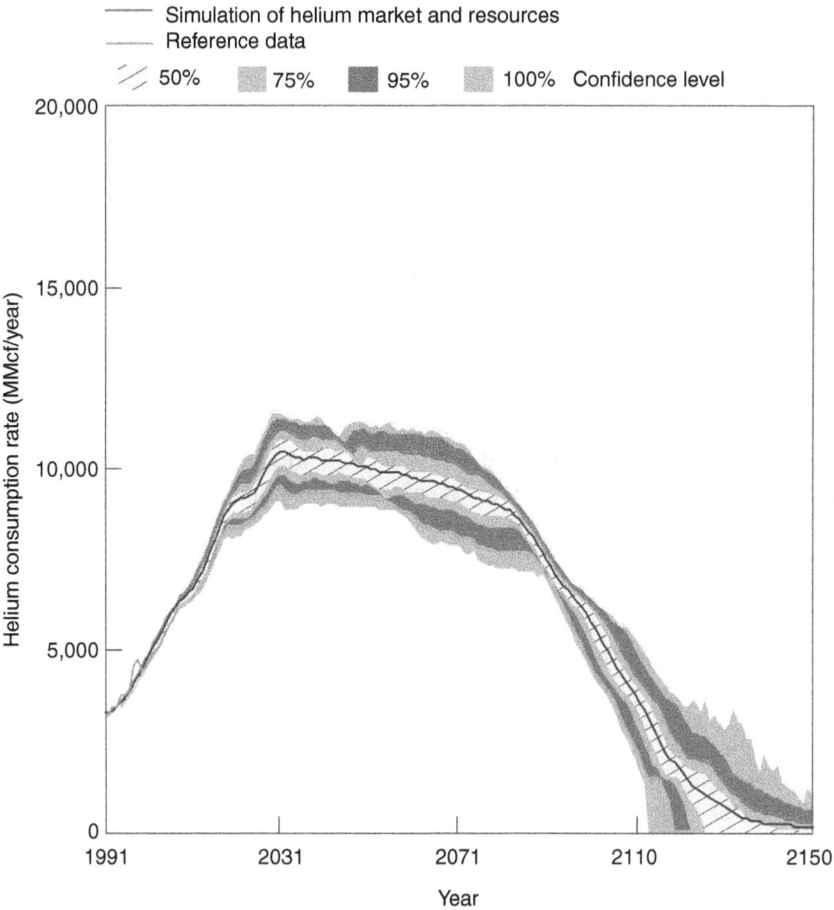

Figure 10.3 Zhiming Cai's system dynamics simulation of helium supply (probabilities derived from Monte-Carlo simulation in the Vensim professional system dynamics).

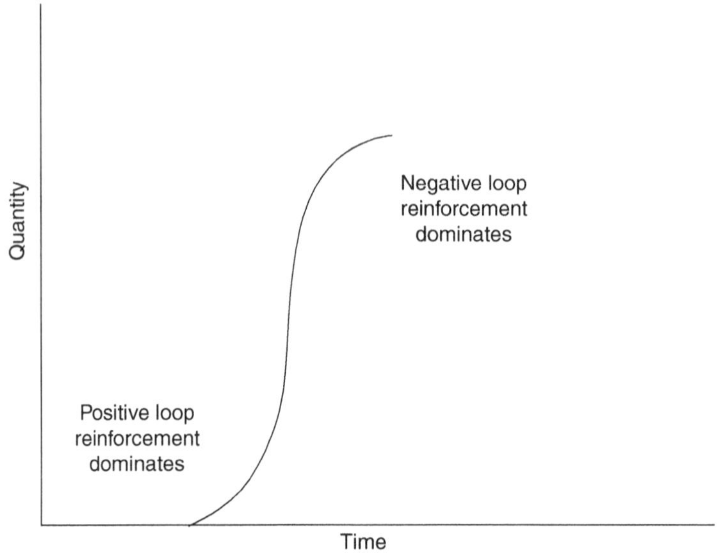

Figure 10.4 S-shaped curve. Curves of this type are widely seen in system dynamics and they occur when, at a certain point in the evolution of the model, negative feedbacks come to dominate over positive drivers of growth.

B3) to a reduction in the quantity supplied. Rising prices have two dominant consequences: first, on the supply side there is a greater incentive to develop new supplies (i.e. to fit extraction capacity to natural gas facilities and thereby to reduce venting); and second, on the demand side to increase efficiency and in some cases substitute away from a dependence upon helium. The search for new supplies and the desire to reduce demand are in constant tension within the model, but with time the relative importance of the two effects changes, such that around 2030 demand-side reductions start to dominate over the supply-side production expansion, turning the quantity supplied through an inflexion point and into an S-shaped curve. As we consider possibilities for future investigation we suggest that a sensitivity analysis should be performed to assess the robustness of the 2030 date at which it is predicted that a shift from supply-side production expansion to contraction occurs.

If, over a period, the exogenous demand growth rate were to be reduced, then the inflection point would occur earlier. High-technology and mission-critical users, such as magnetic resonance imaging equipment manufacturers, are also particularly sensitive to disruptions in supply such as the one that occurred in 2007 when a collective failure to invest in new helium production facilities resulted in shortages in some regions when maintenance and start-up issues impacted a supplier's ability to meet demand. The anxiety this created may weaken demand over time, particularly if such shocks are found to recur.

Richard de Neufville at the Massachusetts Institute of Technology terms such disruptions 'trend breakers' and he observes that they can make forecasting particularly challenging. The existence of trend breakers in the helium supply industry leads us to note that this sector may be well suited for a 'real options'-style assessment of the value of investment in designed-in supply chain flexibility. Such concerns fall beyond the scope of the research discussed here.

It is noteworthy that the 2030–70 helium production plateau predicted by Cai and co-workers is at a level roughly double today's helium supply rate. As such, any comment to the effect that the helium industry will decline in the next two (2010–30) decades is completely at odds with the predictions arising from the system dynamics simulations. It also appears that the considered opinion of the industrial and speciality gases industry is that, while helium supply will remain

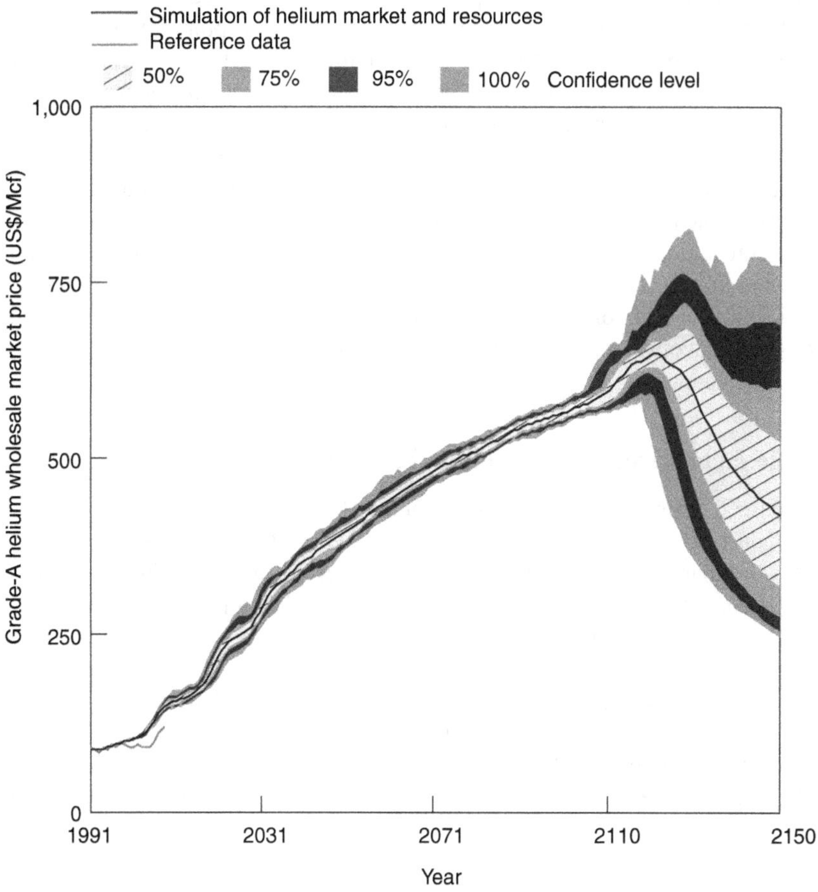

Figure 10.5 Future trend in the real price of Grade-A wholesale helium (1990 normalised to US$100 – in fact this nominal benchmark is close to the actual US BLM Minerals Data helium price for 1991). Modelling as per Figure 10.4 (Cai *et al.*, 2010).

an issue for decades to come and while one expects occasional short periods of difficulty, at a fundamental level helium will be available via the industry for at least 50 years ahead (*Gasworld*, 2006).

One clear prediction from the model is that the real price of helium will rise steadily in the coming decades, until the extraction of helium from natural gas enters its end-game in the early twenty-second century.

Limitations of the approach

As noted in Section 3, our approach has neglected to consider various important demand-side subtleties. Furthermore, on the supply side the modelling led by Zhiming Cai does not include within it an endogenous process for generating acute market shocks and supply chain crises as have been seen and must be expected in future. As such, the approach is arguably ill suited to modelling the 2008–9 economic downturn. A slowing of production is captured by the model, but not the slump encountered in real life. The model might have tracked reality better through that period if the exogenous growth parameter had been based upon a NYSE stock market index. Indexing in that way would also have captured well the 'dot-com' bubble that burst after 2000. However, reflecting longer term behaviours, the model relied upon a technology sector parameter $g(t)$ as described in the paper by Cai and colleagues (2010).

Last, the model developed thus far is ill suited to handle major structural changes in the natural gas industry, such as the move to liquefied natural gas, as discussed in Cai and colleagues (2010) or the recent US innovations concerning shale gas (for consideration of the natural gas industry, see Chapter 5 and the concluding chapter).

Specific factors shaping the helium industry

As discussed elsewhere in this book and in the 2010 US National Council of Sciences report, the mandated sale of US federal helium had a price-making influence in the market. This important effect is expected to be transitory and does not lie at the heart of the system dynamics model, as developed.

We note especially the impact of liquefied natural gas investments, such as 'Qatar-2': Air Liquide is commissioning a second phase of helium production in Qatar. This is expected to add about 17 per cent to global helium production with a single-train helium liquefaction process producing 8,000 litres per day, making Qatar the source of around a quarter of the world's helium (Dauguet, 2010). The gas reserves associated with this extraction unit are expected to last for around 40 years.

Recommendations for future system dynamics studies

The previous work by Zhiming Cai and co-workers placed the bulk of its emphasis on supply-side considerations (Cai *et al.*, 2010). A detailed system dynamics model was developed bringing together exploration, extraction, the supply chain, end use and losses via venting. The demand side is assumed to be adequately represented by a single lumped aggregate demand at a single notional location. We note, however, that the actual landed price of helium varies greatly between different parts of the world such as Europe, Latin American and Australia, as shown in Table 10.1.

Typically final retail costs have numerous components (USA wholesale price, transport, taxes, etc.) and these generate the retail price differences from one territory to another, as do levels of retail market competition. In addition, in all territories a major factor in price differentiation is the quantity ordered. The larger the amount of helium ordered by an individual customer the lower the unit price available.

Cai and co-workers adopted a standard system dynamics iterative feedback approach to the determination of price and quantity, which has been described in detail previously (Cai *et al.*, 2010). As noted earlier, and at the risk of over-simplifying somewhat, central to any economic assessment of the helium industry must be the notion of price (rather than quantity), and we start from that premise. We assume the existence of a meaningful single global wholesale market price P, on which all real retail prices are based in some way. We further assume that price determines consumer behaviour, and as such we do not assume that consumers have needs measured simply and solely in terms of quantities of helium.

We shall assume that each individual consumer has a willingness to pay, which determines the quantity of helium to be purchased at a given price. So assuming that the local retail price faced by that ith consumer is P_i and that the quantity they are willing to purchase is Q_i, then simply: $Q_i = f(P_i)$. We assume that aggregate market demand is given simply by a sum over all n consumers, where all consumers have the same inverse demand function f.

$$Q = \sum_{i=1}^{n} f(P_i) \tag{2}$$

This relates to the stronger assumption made by Cai and colleagues that

$$Q = f(P) \tag{3}$$

Table 10.1 Retail prices of a 200-cubic-foot helium balloon cylinder rental, US$ (an informal collection of data from online retailers and online consumer comments, September 2010)

Brazil, São Paolo	*US, California*	*US, Alabama*	*US, New York*	*UK*	*Singapore*	*Australia*	*NZ*
\$1.14/cf	\$0.785/cf	\$0.35/cf	\$0.90/cf	\$0.70/cf	\$1.85/cf	\$1.80/cf	\$3.60/cf

Now for simplicity rather than considering individual consumers parameterised by i we shall consider notional aggregate consumers in different territories denoted by j, where all consumers in territory j may be approximated as paying an average price $\langle Pj \rangle$. Hence we assume:

$$Q = \sum_{j=1}^{n} f\left(\langle P_j \rangle\right) \tag{4}$$

We may now consider an inverse demand curve in territory j as being given by $f(\langle Pj \rangle)$. We only assume the function f to be monotonically decreasing, but we make no assumption of linearity or otherwise. The monotonically decreasing function f can be either concave or convex. If we follow the single lumped demand assumption of Cai and colleagues (2010), then our assumption of monotonically decreasing f is simply saying: $\mathrm{d}Q/\mathrm{d}P = \mathrm{d}f(P)/\mathrm{d}P < 1$.

While equation 2 has general validity, much of the canonical economics literature assumes linear local (and aggregate) demand curves (at least as a valid approximation). In the case of the helium industry we see no particular basis for the validity of any assumptions in the linearity of $\mathrm{d}P/\mathrm{d}Q$, $\mathrm{d}P_i/\mathrm{d}Q_i$ or $\mathrm{d}\langle Pj \rangle/\mathrm{d}Q_j$.

In Figure 10.6 we show the special case of linear inverse demand curves. In such a case the slope of aggregate inverse demand is simply the sum of the slopes of the component curves and is independent of price. Market system dynamics are most strongly affected by functional forms (e.g. $f(p)$) than by absolute values. In those (linear) cases, where $\mathrm{d}f(P)/\mathrm{d}P$ is independent of P differences in local territorial pricing will have no affect and hence a lumped approximation will always be robust.

Many microeconomics textbooks introduce demand curves via linear approximations, but for real commodities such as helium such assumptions are almost certainly incorrect.

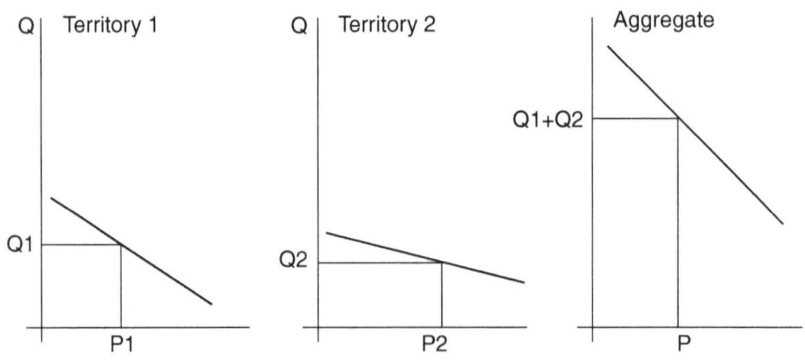

Figure 10.6 Demand aggregation under an assumption of linear local territorial inverse demand curves. Under this assumption the total global demand curve is also linear. Global price P is assumed to lie above P_1 but below P_2.

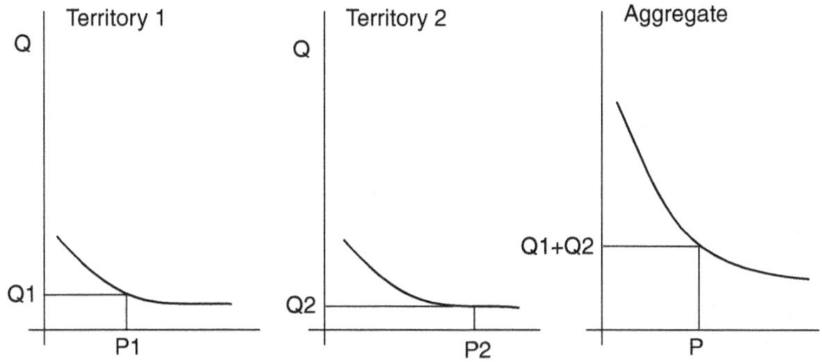

Figure 10.7 Non-linear territorial inverse demand curves are in this case assumed to be equal in territories 1 and 2. Aggregate demand has the same functional form $f(P)$, albeit rescaled – the functional form is self-similar. Critical for system dynamics is the price response denoted by $df(P)/dP$. For non-linear $f(P)$, price response will be price dependent.

Figure 10.7 illustrates non-linear inverse demand curves in two territories. In this case it is assumed that the customer base in each territory is identical and the local demand curves in the two territories are the same. The only difference assumed is a difference in local helium prices (such as a difference in the local landed price or in local taxes). This places consumers in different territories at different points on the demand curve and hence exhibiting different levels of willingness to pay.

Aggregate demand is simply the sum of local territorial demands and the slope of the aggregate inverse demand $df(P)/dP$ is simply the sum of local slopes at the global price P. Consumers in the individual territories do not see the price P, however. Let us assume that they each see a local price offset by a simple constant displacement ΔP_j and allowing for average behaviour in any territory to be representative of that territory, then:

$$\frac{dQ_j}{dP} = \frac{df(P + \Delta P_j)}{dP} \tag{5}$$

Which by substitution is equivalent to:

$$\frac{dQ_j}{dP} = \frac{df(P_j)}{dP_j} \tag{6}$$

Now let us assume a set of 'n' territories, each with a random ΔPj but subject to the condition that they are equally distributed such that the average ΔPj is zero. Then, as is illustrated in Figure 10.7 we see that the aggregate demand response in case of non-linear demand is such that:

$$\frac{dQ}{dP} = \sum_{j=1}^{n} \frac{df(P_j)}{dP_j} \neq \sum_{j=1}^{n} \frac{df(P_j)}{dP} \tag{7}$$

It is this discrepancy that illustrates the approximations inherent in the lumped approximation adopted by Cai and co-workers (Cai *et al.*, 2010). For concave non-linearity of the type shown in Figure 10.7 the inequality in equation 7 would actually take the form:

$$\frac{dQ}{dP} = \sum_{j=1}^{n} \frac{df(P_j)}{dP_j} > \sum_{j=1}^{n} \frac{df(P_j)}{dP} \tag{8}$$

In several respects this is reminiscent of the famous 'flaw of averages'. In order to elucidate these ideas further it is best to adopt an empirical approach and to construct a system dynamics model without the lumped demand approximation. In so doing there is no need even to adopt the simplifying assumptions adopted above (that inverse demand curves are self-similar in all territories, that all territories have price structures simply displaced from the global price by an offset and that quantity needs have a similar distribution in all territories). A disaggregated approach to demand could seek to model the realities of key global markets for helium in a more realistic manner.

The approach of Zhiming Cai and colleagues represented a sensible approximation, especially given a lack of proper data set $\{\langle Pj \rangle\}$, let alone $\{P_i\}$. As noted previously, there remains much scope for improved demand-side modelling.

In an Appendix we present a further discussion of the formation of aggregate demand from the behaviours of individual agents. These ideas flow from the realm of intermediate microeconomics (Varian, 2009). The Appendix illustrates that even in a situation of linear demand curves, kinks and discontinuities in aggregate inverse demand curves can be observed.

Figure 10.2 as presented in this chapter has been updated slightly since being published by Cai and colleagues (2010). First, the impact of excess demand on helium production rate has been corrected to show a positive influence. Second, the version presented here alters the treatment of price elasticity in loop B2. Formally elasticity is defined as (d*Q*/*Q*)/(d*P*/*P*) and, as such, is always price dependent even in the case of linear demand curves. We therefore use the term 'response function' to denote the demand curve slope (d*Q*/d*P*). In the case of a linear demand curve the response function is price independent. Figure 10.2 is based on a generalised premise of a non-linear demand curve, which we shall assume to be monotonic function; either concave or convex. In such a case, a change in base price (after tax) leads to a different look-up point on the demand curve, but it does not change the shape of that curve (or the 'elasticity') in any profound way. That is why we now show a direct link from base price to helium demand as it is simply mediated by a static demand curve. The anxiety effect, however, can in principle alter the very shape of the demand curve. In proposing this change we are keen to distinguish cases where certain factors, such as 'anxiety' cause a shift of the entire demand curve (a 'shift in demand' as it is known in microeconomics) from those effects which simply alter the point of inspection on a static demand curve. Shifts in demand alter, at a fixed price, the willingness to buy and the quantity demanded. Such shifts can arise as a result of the action of variables such as market expectations, prices of related goods, income effects and anxiety effects. Depending on how the above factors change with time, one might usefully regard the whole demand curve as essentially 'parameterised' or 'driven' by them. Complex (with multiple loops interacting, and information to stock/flow links) system dynamics models may simultaneously incorporate both cases – price-led changes in demand and deeper demand effects. As such, system dynamics can be the subject of methodological concern from microeconomists troubled by the mingling of two such different concepts. In redrafting Figure 10.2 we have tempted to make our thought processes more explicit in this regard.

Conclusions

System dynamics is a methodology well suited to the complex realities of the helium industry. In particular it does not rely on analytical tractability and as such is better suited to the oligopolistic realities of the helium industry. The modelling work led by Zhiming Cai extends resource utilisation system dynamics to the particular challenges of co-product production. While numerous insights have already been gained, we do not suggest that the work we have presented represents a comprehensive or definitive system dynamics assessment of the helium industry going forward. In particular we note the need to improve the detailed handling of the demand side. We look forward to a future in which more sophisticated insights can emerge from system dynamics for this unusual industrial material that is so important for the advanced energy, medical and communication technologies of the twenty-first century.

Appendix

On market demand curves: computation and representation

The ensuing notes aim at a more accurate mathematical representation of the market demand curve in the case of linear individual demand curves and represent a quantification effort of an interesting pertinent remark made by H. Varian (2009). As a starting point, consider two linear individual demand curves:

$$p = g_1(q) = -c_1 q + d_1$$
$$p = g_2(q) = -c_2 q + d_2$$

with c_1, $c_2 > 0$. Due to the monotonicity of the demand curves and since q is always a positive quantity an inverse demand curve is defined as follows:

$$q = f_1(p) = \max\{b_1 - a_1 p_1, 0\} = \begin{cases} 0 & \text{if} \quad p \geq p_1^* \\ b_1 - a_1 p & \text{if} \quad p < p_1^* \end{cases}$$

where $a_1 = \frac{1}{c_1}$, $b_1 = \frac{d_1}{c_1}$ and p_1^* is the price above which demand vanishes, i.e. satisfies $b_1 - a_1 p_1^* = 0$. Notice that the inverse demand curve is a piecewise linear function rather than a linear one, as Varian also observes (2009). In a similar spirit, an inverse function can be defined for the second consumer by substituting 2 for 1 above.

Without loss of generality, let us assume that $p_1^* > p_2^*$. Then, the above piecewise linear inverse demand curves can be equivalently represented as follows:

$$q = f_1(p) = \begin{cases} 0 & \text{if} \quad p \geq p_1^* \\ b_1 - a_1 p & \text{if} \quad p_2^* \leq p < p_1^* \\ b_1 - a_1 p & \text{if} \quad 0 \leq p < p_2^* \end{cases}$$

and

$$q = f_2(p) = \begin{cases} 0 & \text{if} \quad p \geq p_1^* \\ 0 & \text{if} \quad p_2^* \leq p < p_1^* \\ b_2 - a_2 p & \text{if} \quad 0 \leq p < p_2^* \end{cases}$$

Furthermore, the market demand curve can now be readily computed by forming the horizontal sum of the above inverse demand functions:

$$q = f(p) = (f_1 + f_2)(p) = \begin{cases} 0 & \text{if} \quad p \geq p_1^* \\ b_1 - a_1 p & \text{if} \quad p_2^* \leq p < p_1^* \\ (b_1 + b_2) - (a_1 + a_2)p & \text{if} \quad 0 \leq p < p_2^* \end{cases}$$

Please notice that the market demand curve is also piecewise linear exhibiting a kink at p_2^* mathematically realised by a discontinuity in the first-order derivative (market demand response) $\frac{dq}{dp}$ as shown below:

$$\lim_{p \to p_2^{*+}} \frac{f(p) - f(p_2^*)}{p - p_2^*} = \lim_{p \to p_2^{*+}} \frac{b_1 - a_1 p - b_1 + a_1 p_2^*}{p - p_2^*} = -a_1$$

and

$$\lim_{p \to p_2^{*-}} \frac{f(p) - f(p_2^*)}{p - p_2^*} = \lim_{p \to p_2^{*-}} \frac{(b_1 + b_2) - (a_1 + a_2)p - b_1 + a_1 p_2^*}{p - p_2^*} = -(a_1 + a_2)$$

The above analysis quantifies and elucidates Varian's remark (2009) on the manifestation of an inevitable kink (discontinuity in market demand response) on the market demand curve. As a result, it would be more accurate to use the characterization piecewise linear when describing market demand curves derived from individual demand curves as shown above rather than linear.

Notes

1 At the time of the research CCFE was known as UKAEA Fusion.
2 To be precise, helium is a co-produced by-product, or a 'tail gas'. There is perhaps a weak coupling between natural gas and helium production dynamics associated with the input energy requirements at the extraction site. That cost signature should in principle be present in both helium and natural gas supply profiles.
3 Oligopoly refers to a situation in which supplies are sourced from a sufficiently small group of market players that price convergence occurs without any collusion. Typically a free market requires more than five vendors if oligopoly is to be avoided. The helium industry is on the cusp of oligopoly globally, but verges on monopoly in some territories.
4 It is important to note that even if venting could be entirely eliminated the world would not be able to establish a sustainable helium industry. We estimate that in the absence of venting helium depletion would occur roughly 50 years later than in a business-as-usual scenario.
5 Supply scarcity is rarely a consequence of an absence of geological resource, more typically it is a consequence of a supply chain problem. Zhiming Cai's system dynamics model includes within it delays inherent in supply-side infrastructure investment; for instance, in areas such as extraction and production of Grade-A product.

References

Cai, Z., Clarke, R., Glowacki, B. A., Nuttall, W. J. and Ward, N. (2010) 'Ongoing Ascent to the Helium Production Plateau – Insights from System Dynamics', *Resources Policy*, Vol. 35: 77–89.

Cook E. (1979) 'The Helium Question', *Science*, Vol. 206: 1141–7.

Dauguet, P. (Air Liquide) (2010) Private communication.

Forrester, J. W. (1961) *Industrial Dynamics*, MIT Press, Cambridge, MA.

Gasworld (2006) 'Helium Changing Faces, Technology Shifts and Tight Supply – Challenges Facing Our Industry', *Gasworld*, September–October.

Hotelling, H. (1931) 'The Economics of Exhaustible Resources', *The Journal of Political Economy*, Vol. 39(2): 137–75.

Hughey, A. M. (1989) 'Uncertainty and the Joint Extraction of Helium and Natural Gas', *Resources and Energy*, Vol. 11(1): 65–79.

Hughey, A. M. (1991) 'Joint Natural Resources and Government Policy: Helium and Natural Gas', *Eastern Economic Journal*, Vol. 17(1): 80–8.

National Academy of Sciences (2010) *Selling the Nation's Helium Reserve*, National Academies Press, Washington, DC.

Sterman, J. (2000) *Business Dynamics: Systems Thinking and Modeling for a Complex World*, Irwin/McGraw-Hill, Boston, MA.

Varian, H. (2009) *Intermediate Microeconomics*, Eighth Edn, WW Norton and Company.

Zhiming Cai, Richard Clarke, Bartek A. Glowacki, William J. Nuttall and Nicholas Ward (2010) *Ongoing Ascent to the Helium Production Plateau – Insights from System Dynamics*, Resources Policy, **35** pp. 77–89 (2010).

11 Closed-cycle refrigeration

Minimizing helium demand in cryogenic applications

Thomas W. Bradshaw and Trevor Miller

Motivation

The price of liquid helium has risen over the past few years due to several factors including the reserves at the Bureau of Land Management in the US becoming depleted. Any given refrigerator design includes a cost benefit analysis over the lifetime of the equipment. In recent years reliable 4 K coolers have become available and this has led to the situation where the use of closed-cycle refrigerators (or "cryocoolers") is more likely to be considered.

Factors that have led to the more widespread use of closed-cycle refrigerators

1. Elimination of manpower costs to "top up" helium systems – the most notable example of this is the use in MRI systems to produce "zero loss" cryostats. A side benefit of this is a reduction in the total liquid helium inventory. This means that a more compact and lighter product can be offered.
2. The ability to site MRI magnets and complex cryogenic systems in remote locations away from the infrastructure that can supply bulk liquid helium. These include MRI magnets and astronomical telescopes which often require detectors operating at cryogenic temperatures.
3. With cryogenic temperatures required in a wide variety of applications, the use of closed-cycle coolers in a cryogenic application means that relatively unskilled people can operate the equipment. This has led to a more widespread acceptance of cryogenic technology.
4. The elimination of liquid cryogens can be a positive consideration when making safety cases, particularly where the equipment is sited in a confined space.
5. If there is a requirement for "round-the-clock" operation then it is often easier to achieve this with closed-cycle refrigerators as the equipment is not dependent on the security of helium supply.

Factors that govern the uptake of closed-cycle refrigerators

1 Competitiveness compared to conventional "wet" systems that are seen as the safe option in cryostat design.
2 Reliability is a key issue.
3 Large, life-limited systems on spacecraft can be replaced by long-life, low-mass and often low-vibration cryocoolers.

Availability and cost of helium

Depending on geographical location, current market demand and local laboratory conditions, high-purity helium may not be easy to obtain in every case. Supply chain disruption issues are predicted to continue to cause some market pressures over the years ahead and prices are likely only to continue to rise.

Conservation of helium

Helium is a finite and irreplaceable resource (with any currently envisaged technology). Whilst global quantities are calculated to last potentially for some hundreds of years, venting helium to the atmosphere is not a desirable practice as each event loses a little more of the total available.

History

The first use of closed-cycle Stirling refrigerators was by Alexander Carnegie Kirk (Kirk, 1874). At the time he was engaged at the paraffin oil works of Young, Meldrum and Binny at Bathgate in Scotland. The firm was involved in the production of oil from the coal (of a type called torbanite) found in the area. The motivation was the separation of the highly prized paraffin wax from oil. This had previously been achieved in the early days of the factory by waiting until the winter cold weather and putting the oil out on trays, and later by using a refrigerating device based on ether. The first method had economic difficulties and could only be used at certain times of the year; the latter had obvious dangers. Considerations like this, of safety and economics, are still relevant today. Because of the shortcomings of the available technology, Kirk developed a cooling engine based on the Stirling cycle. He realised that "Every machine for producing cold is simply the negative solution of a motive power heat engine" (Kirk, 1874). This laid the foundations for much of the later work on closed-cycle refrigeration.

Commercial cryocoolers

A cryocooler may be defined as an electromechanical refrigerating machine to produce cryogenic temperatures. Usually the working principle is by compression and expansion of a fluid (very often helium gas) around a closed-loop circuit. Cryocoolers begin cooling from room temperature (or just above) but typical applications are <100 K, normally in the temperature regime between 4 K and 100 K (–269 °C to –173 °C).

That today's cryocoolers can routinely cool from room temperature to 4 K, or less, within an hour or two of being switched on, is either a source of amazement or taken for granted, depending on one's point of view or familiarity with the current technology. Nevertheless, today it is a fact that temperatures of a few degrees above absolute zero are easily reached in very many thousands of installations in countries all over the world, and even in satellites in space applications.

To reach 4 K the medium used is high-purity (typically 99.99$\underline{9}$ per cent) Helium-4 gas. Thus a typical cryocooler uses the very element that is most often used in liquid form to achieve the same temperatures, simply because everything else freezes at the very low temperatures required. However, a cryocooler uses relatively small amounts in a hermetically sealed closed-loop circuit.

Cryocoolers can often replace liquid cryogens completely in so-called "dry" systems, except where the requirement is to remove large amounts of heat from a system quickly. Even then a cryocooler often maintains the temperature initially achieved by boiling off a quantity of cold liquid. Later it is used to recondense, or at least substantially to reduce, boil-off.

Base temperatures in machines typically operating between 10 K and 100 K can be reached by machines using helium, or towards the higher end of the range by using nitrogen or various gas mixtures to maximize the cooling power.

Generally such cryocoolers consist of a compressor connected by gas lines to a cold head which has a warm flange (as will later demonstrated in the section "Applications") to connect to a vacuum vessel and one, two or occasionally three stages of cooling providing some Watts at each heat station. Cryocoolers operating below about 50 K are usually built with more than one stage of cooling to reduce the heat load on the subsequent stages. This is particularly important as the radiative heat load increases as the fourth power of the temperature difference. After compression the gas expansion therefore occurs in one to three stages. Heat can be extracted from each of these stages by the use of thermal links.

Some of the terms for closed-cycle refrigerators, otherwise known as a cryocoolers, are often used loosely. The term "Stirling", for instance, is mainly applied to a piston refrigerator operating on a Stirling cycle.

Stirling cycle

A typical Stirling cycle is presented in Figure 11.1

Gifford-McMahon cryocooler

At the end of the 1950s Gifford and McMahon developed a concept to separate the compressor from the expander essentially by introducing valves between them which allowed control of the pressure wave to the cold head, as presented in Figure 11.2, and G-M technology was born.

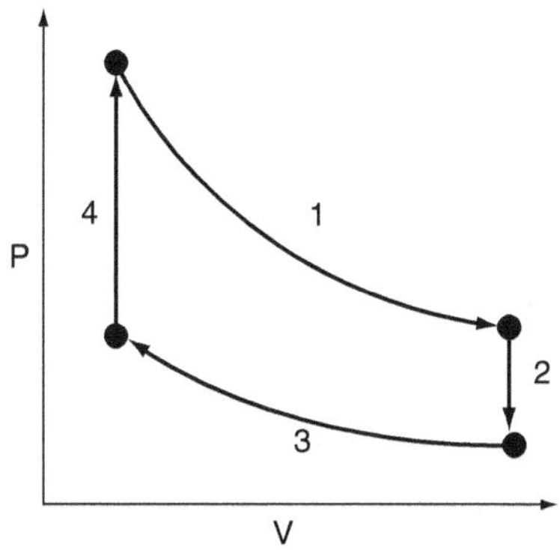

Figure 11.1 An ideal Stirling cycle presented on the pressure–volume diagram: (1) isothermal expansion – the expansion-space is heated externally, and the gas undergoes near-isothermal expansion; (2) constant-volume (known as isovolumetric or isochoric) heat removal – the gas is cooled by passing through a regenerator, and transfers heat to the regenerator for use in the next cycle; (3) isothermal compression – the compression space is intercooled, so the gas undergoes near-isothermal compression; (4) constant-volume (known as isovolumetric or isochoric) heat addition – the compressed gas flows back through the regenerator and picks up heat on the way to the heated expansion space.

Pulse tube closed-cycle refrigerator

This type of closed-cycle refrigerator (CCR), is different from all the other systems. Instead of using the Gifford-McMahon process shown in Figure 11.2, it uses a radically different approach. In an effort to reduce the amount of maintenance and vibration in a standard CCR, a pulse tube in the cold head uses no moving parts. This means there is no displacer, seals or control hardware to wear out and require maintenance. Instead the system uses an acoustic standing wave in a pulse tube, hence its name. There are other differences in the actual process than simply the mechanics, or lack thereof, but that is beyond the scope of this discussion (De Waele, 2000).

Pulse tube refrigerators can in fact be of either G-M or Stirling type depending on the type of compressor technology used to drive the cold head.

Figure 11.3 shows various types of commercial cryocoolers: (a) G-M (Gifford-McMahon); (b) GM-JT (G-M with Joule-Thomson stage); (c) PTR (pulse tube refrigerator); (d) Stirling. A review of some of the techniques behind the cryocoolers presented is given in Walker and Bingham (1994).

Cryocooler technology justification

Cryocooler technology has been commercially available a nd has achieved a steadily increasing presence over several decades but commercial applications have rapidly expanded in recent years and this trend looks set to continue.

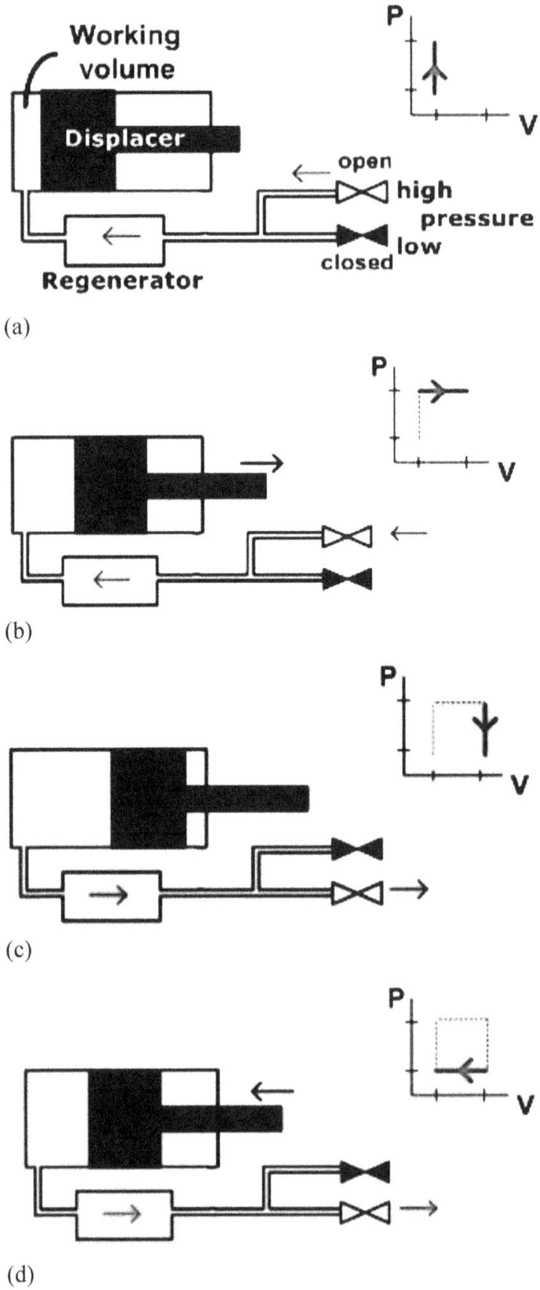

Figure 11.2 Schematic of the four steps of the Gifford-McMahon cycle, representing: (a) pressurise expansion cylinder at minimum volume with regenerated (precooled) gas; (b) maximise V and maximise P (more precooled gas added); (c) depressurise and exhaust through the regenerator (cooling); (d) minimise V at minimum P (more exhaust and regenerator cooling). (Adapted from NCNR-NIST).

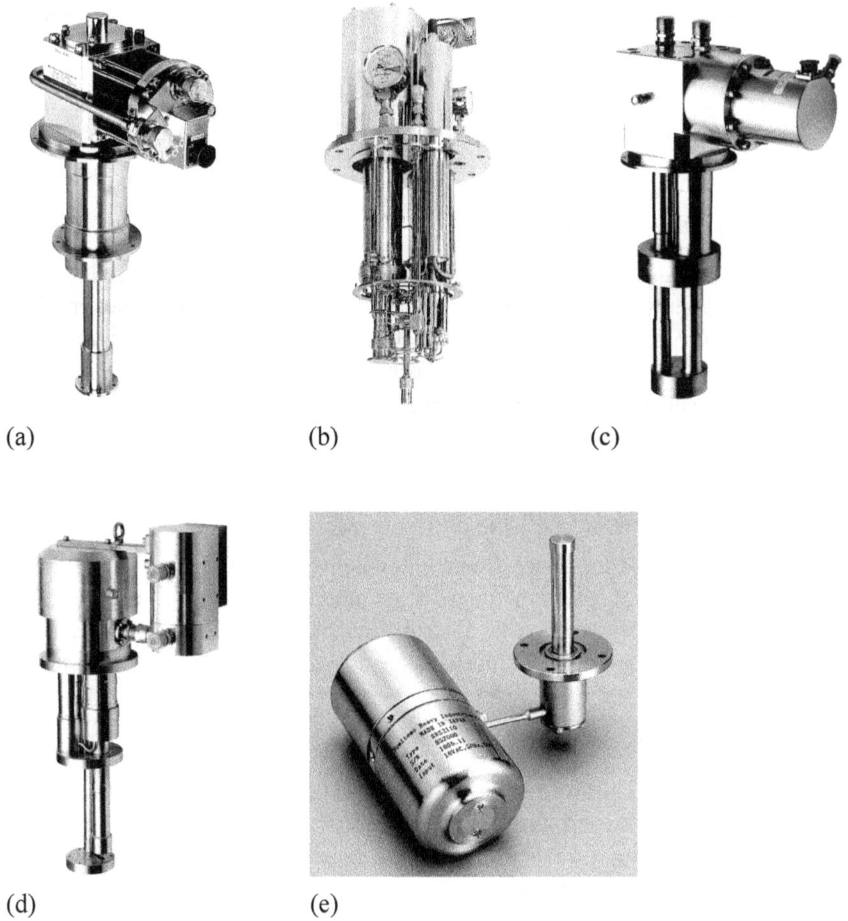

Figure 11.3 Different types of cryocoolers: (a) 1 W at 4 K G-M cryocooler; (b) GM-JT cold head providing 3 W at 4 K; (c) The 1 W at 4 K PTR cold head; (d) The 0.5 W at 4 K PTR cold head with separable valve; (e) Stirling cold head and compressor (note close coupling typical of this type of device).

There are various reasons for this trend, as follows.

Convenience

Anyone who, as a research student, has had to get up in the middle of the night to top up the liquid helium in a cryostat will understand this one. Being able to push a button and have your system cool down whilst you go and do something else provides the user with much greater flexibility.

Health and safety

The risks of transporting and handling liquid cryogens mean that special training is required, both for awareness and practice. These extra precautions often lead new users to choose a cryocooler as a safer alternative. In some countries the law relating to use of cryogens has been tightened to an extent, and this has also significantly driven the use of cryocoolers.

Reduction of operating costs

This can be significant driver in many applications. Cryocoolers reduce the quantities of helium or other liquid cryogens required to operate a system. The liquid saving is part of the rational. Cryocoolers can also reduce the requirement for a system to be attended by service personnel for cryogen top up or other reasons. The reduction in operating costs and extension of system hold times means the capital cost of the cryocooler is readily offset.

Simplification of system design and build

This is another significant driver. Large liquid helium dewars require an intermediate layer to shield them from room temperature, reducing the heat load. By enabling reduction of some parts, systems can be designed smaller and with fewer components, making them more compact, simpler and cheaper to build.

Reduction of system cost

The lower design and build costs have been passed on in price savings to the end customers, resulting in reduced system prices and increased performance. At least in part the system price reductions have been enabled by the use of cryocoolers for low-wattage systems.

Reliability

The relative unreliability of early cooler designs in the past has been replaced over recent decades by machines that routinely run without problems provided they are supplied with clean gas, clean electrical supply and appropriate ambient cooling. It has therefore become normal in many thousands of installed systems running continuously to have at least one year between services, and in many cases much longer. In the case of research-type systems, which are often used discontinuously, whilst the overall run time between servicing may be reduced compared to continuous use, the reality is that they may be serviced only once or twice during a lifetime of many years, or in some cases never at all. For many system types when service is required it is available from the manufacturer from one of their global centres. Therefore the servicing and reliability aspects of ownership do not generally produce any barrier to cryocooler use.

Downside

The thermodynamic efficiency of most cryocoolers is very low (typically 1 percent or so). Therefore, the main downside to the use of cryocoolers is the electrical requirement, which for a typical 1.5 W, 4 K machine is in the region of 5.5 kW. In other words, the CO_2 footprint $[CO_2]$ from using a cryocooler is about twice the CO_2 footprint using liquid helium, as shown below.[1]

$$\frac{[CO_2]_{Cryocooler}}{[CO_2]_{LHe}} \approx 2.1$$

A proper assessment would need to be taken in the round and compared to the energy consumed to produce and transfer liquid cryogens to the point of use, keep them topped up and so on. Also, as mentioned previously, cryocoolers may enable designs that save energy or material in other parts of the system. Such a comparison is therefore somewhat complex and can give no absolute answer as to which is more or less energy efficient. Nevertheless, reduction of energy consumption is a worthy future goal for cryocooler development and some systems are already coming to market with this in mind.

To illustrate this point further, the theoretical maximum specific input power – that is the input power per watt of refrigeration at T_{cold} – is given by the relation:

$$\frac{InputPower[W]}{Refrigeration\ at\ T_{cold}[W]} = \frac{(T_{ambient} - T_{cold})}{T_{cold}}$$

This is plotted in Figure 11.4. The specific input power required rises dramatically as the temperature decreases. This theory was developed by the French engineer Sadi Carnot and represents the performance of an ideal engine cycle.

The situation is far worse when the performance of real refrigerators is analysed. Strobridge (1974) has analysed the performance of real refrigerators (also see Kittel, 2006). The best achieve around 30 per cent of the Carnot performance and these are very high-capacity machines at the higher temperatures. Low-capacity refrigerators (1 W and below) tend to be dominated by losses and can achieve only 1 per cent of the Carnot performance. Most of the research work on miniature refrigerators is involved in the minimisation of these losses; e.g. improving the performance of the regenerator, increasing motor efficiency, etc.

Applications

History teaches us that it is easy to underestimate the development times of new applications as they emerge. However, in general the number of new applications is increasing and therefore the market for cryocoolers is projected to continue to grow over the decades ahead. Cryocoolers can now be found in the

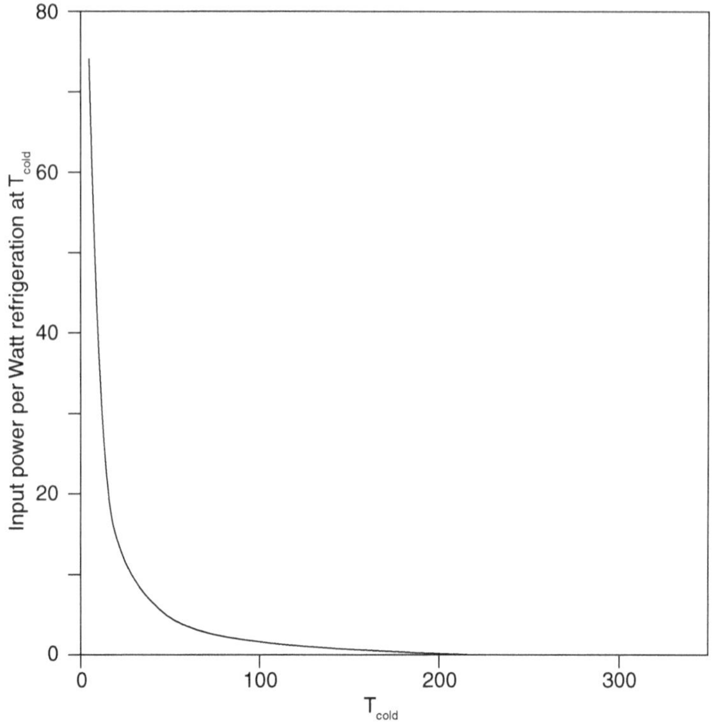

Figure 11.4 Theoretical maximum specific input power vs cold end temperature, T_{cold}.

laboratories across the low-temperature physics community, in industrial applications such as superconducting induction heaters and in hospitals where MRI systems are installed and where simplicity and non-reliance on the availability of liquid cryogens is of concern. Today's most common cryocooler applications are briefly summarised below.

Magnetic resonance imaging

Magnetic resonance imaging (MRI) is a radiological technique used in medicine, mainly for detailed analysis of internal organs and other parts of the body. MRI can produce high-contrast images in soft tissue, which is particularly useful for instance, in oncological (cancer), neurological (brain), cardiovascular and musculo-skeletal imaging. Powerful magnetic fields are used rather than ionising radiation such as the X-rays used in computed tomography (CT). Today 1.5 T or 3 T fields are typically used (Campbell *et al.*, 2008c). Previously 0.5 T has also been popular. Up to 7 T fields are used in some whole-body systems. Superconducting magnets are generally chosen to produce these fields because of their performance advantages, not least of which is removing the necessity to provide the magnet with power continually to generate the field. Additionally, radio frequency (RF) fields are used to alter the alignment of nuclear magnetisation in the tissues of interest, very often of hydrogen molecules in water. Manipulation of the rotating magnetic fields produced allows sophisticated 3D images to be constructed, which allow physicians to visualise medical problems and plan suitable treatments. After the earliest studies in humans in the mid-1970s, the technique gained popularity from the mid-1980s and today is a mainstream non-invasive diagnostic tool used routinely in hospitals around the globe. Current installations run into the tens of thousands.

Early systems used a cryostat containing both liquid nitrogen and liquid helium, LHe, in separate annuli. Heat losses in the cryostat mean that this helium would normally boil off in quite a short space of time, requiring frequent system servicing. So for economical reasons cryocoolers were soon adopted. This also enabled cooled shields to replace the liquid nitrogen. The first-stage cooling is used to intercept the heat load from room temperature and to keep the thermal radiation shields at low temperatures and reduce the load further at 4 K. This is invaluable where the magnets are being operated in a hospital environment, which can be a long distance from the source of helium supply. The use of this technique leads to a reduction in the helium inventory, reduced servicing costs and greater system availability, which keeps overall costs low. Today a few systems are directly cooled by cryocoolers but most modern systems continue to use a superconducting magnet kept cold in a bath of LHe, which is conserved by use of a cryocooler condensation system as described in Chapters 12 and 13. The cooler is used in various ways, as shown in Figures 11.5, 11.6 and 11.7.

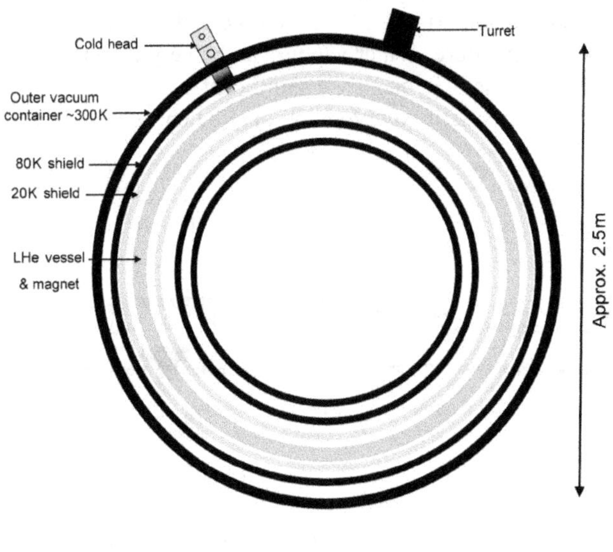

Figure 11.5 10 K shield cooler cooling a magnet (see also Chapter 12).

A vacuum cryostat is essential to reduce thermal losses. Shield temperatures shown in Figure 11.5 are nominal only. Superinsulation (not shown) is also used to reduce losses. The cryocooler maintains the system temperature as low as possible. Helium boil-off is very significantly reduced, but here there is no recondensation.

The cryocooler in Figure 11.6 is thermally linked to the shields and connected to the room temperature cryostat via a room temperature flange. The temperature gradient from room to cryogenic temperatures occurs over a few centimetres. There is an outer vacuum cryostat. However, it is not essential to surround the cold head itself with a vacuum and helium gas just above atmospheric pressure may surround the cooler, depending on system design.

A cryocooler capable of achieving 4 K, Figure 11.7, may be directly linked to the magnet in a system without liquid helium or used as a recondensor in a near-lossless design. Future developments in magnetic resonance imaging could lean towards the use of high-temperature superconductors. In particular, MgB_2 operated in liquid hydrogen at 20 K has been proposed as a viable replacement for conventional low-temperature superconductors (see Chapter 17).

Cryopumping

In applications where a fast pumpdown that is free from contamination is required, cryopumps are very often the best choice. These consist of a cryocooler with the cold stage attached to an extended surface that is usually covered by a highly absorbent material such as activated charcoal or zeolite. These materials are able to absorb large amounts of gas at low temperatures and can be regenerated by heating.

To provide a high vacuum, nearly all gases can be pumped at cryogenic temperatures by cryocondensation. Figure 11.8 shows the vapour pressure curves for various gases versus temperature. Water vapour can be very effectively pumped at 100 K. Since water vapour is frequently a significant contributor to the partial pressure in a high-vacuum system, a cryopump may be chosen for its very high

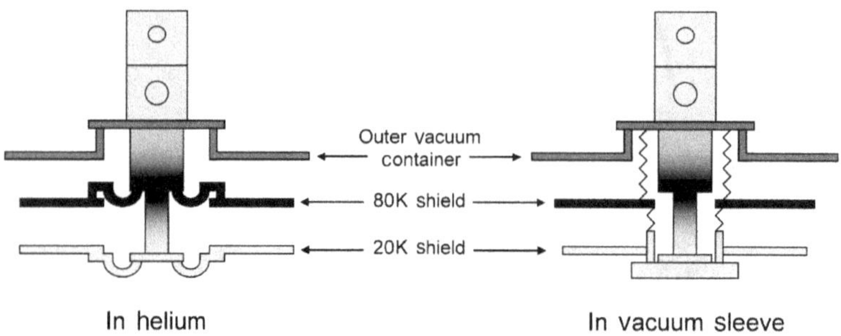

Figure 11.6 Methods of cooling MRI and research electromagnets using 10 K Gifford-McMahon shield cooler.

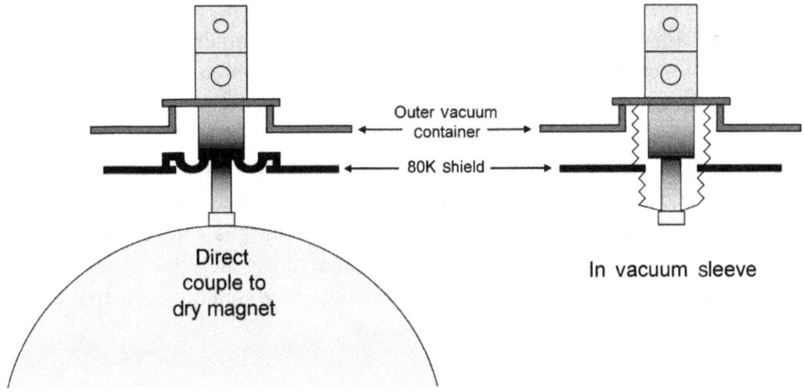

Figure 11.7 Methods of cooling MRI magnets for zero He inventory (left figure) or "zero" loss (right figure).

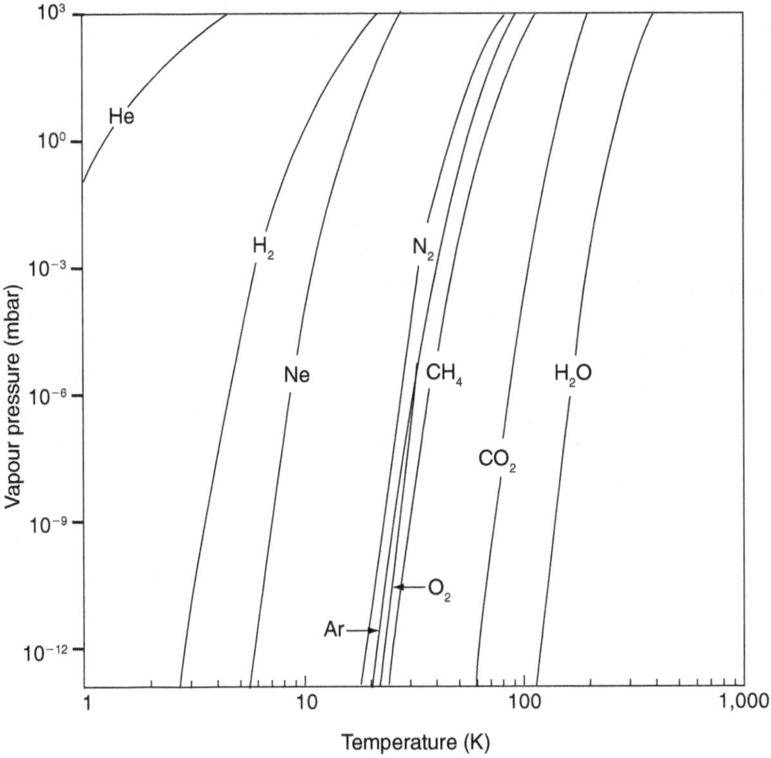

Figure 11.8 Vapour pressures of "common" gases.

water vapour pumping speed. All gases except hydrogen, helium and neon can be pumped using cryocondensation by the time the temperature is below 20 K. At standard pressure, STP, helium is liquid at 4.2 K. Even at 1 K He has a significant vapour pressure. The very common type of G-M cryopump in use today therefore employs activated charcoal at the second stage of the cryopump to cryosorb hydrogen, helium and neon. This is achieved with temperatures between about 4.2 K and 20 K. Cryopumps are capture pumps and eventually reach their capacity limit or safety (flammability) for a particular gas and require regeneration. Modern designs can accommodate automated regeneration cycles that can be fitted into the running schedule of the machine to which they are applied.

Detector cooling

Infrared systems are commonplace in military systems, for applications such as night vision goggles and imaging systems as well as for missile targeting. These are generally described under the name FLIR (forward-looking infrared) systems. For normal operation in the 3–5 micron range the detectors are generally uncooled. For longer wavelengths, where the resolution of hot parts at long distances is possible, then the detectors need to be cooled to low temperatures. The cooling is either by the use of thermoelectric solid-state elements, miniature mechanical coolers or powered by gas bottles. A wide range of small rotary devices have been developed by various manufacturers, and these are low powered with a lifetime in excess of one year.

Space – astronomy and Earth observation

The use of closed-cycle refrigeration in space has opened up a new era in long-term global monitoring and observation of the Earth. The economics of launching instruments into space means that mass is at a premium, and closed-cycle refrigerators are lightweight and not life-limited compared to carrying liquid or solid cryogens. From space, temperature measurements of the Earth's surface can be made using sensors that detect in the infrared. These detectors have to be cooled to typically 80 K to achieve the required sensitivity. The number of space-based infrared instruments using closed-cycle refrigerators that have been turned on the Earth are too many to list here, but there are several good reviews (Bhatia 2002; Glaister *et al.*, 2000; Jewell, 1996). These have been used to monitor the weather, pollution, sea surface temperature, crops and ice formation. In particular, the global measurement of sea surface temperature has been used to further our understanding of the effects of El Niño effects and also in routine weather forecasting. For astronomy there are demands on mass, and vibration from mechanical coolers can be an issue with sensitive detectors. The ESA Planck mission launched in 2009 brought together several technologies to produce temperatures down to 100 mK for measurements of the cosmic microwave background (Lamarre *et al.*, 2010). It is interesting to note that the 4 K cooler for this spacecraft has a helium inventory of a little over 1 gram and consumes around 100 W in operation. Small coolers based on flexure bearings have been extremely successful and have clocked up 52 years of in-flight operation (Gibson *et al.*, 2008). In addition, a laboratory-based version has run for over 18 years, which, at 40 Hz, is in excess of 10^{10} cycles. These coolers were originally developed for Earth observation at Oxford University and at the Rutherford Appleton Laboratory in the UK (Bradshaw and Orlowska, 1988).

Detector cooling

On the ground, cryocooler uses have increased due to increased demand for infrared night vision systems, surveillance and detection of bodies in emergency situations. For this application the challenges are size and mass, because of the need to produce truly portable systems. Requirements for lower noise in electronics, and the need to improve signal to noise ratio, have driven the use of cryocoolers in many types of detector.

Superconducting machinery

The use of superconducting machinery for both power generation and propulsion has taken a boost with the development of high-temperature superconductor (HTS) materials operating in the temperature range of 25–70 K (Campbell *et al.*, 2008a). These offer low-loss, lightweight and efficient solutions. Previous to their discovery, so-called low-temperature superconductors required temperatures in the region of 1.8–10 K. As discussed earlier, the efficiency of cryocoolers at low temperatures is quite poor; the development of HTS materials has made superconducting machinery a more practical possibility.

The use of superconducting generators in the nacelles of wind turbines has received much discussion as the solution offers lower mass and greater low-speed efficiency and may eliminate the need for complex gearboxes. However, the drive for this requirement is in the offshore environment, where larger turbines will be used for economic reasons. This is a challenging environment with access limited to periods of good weather, and five-year lifetimes without service must be seen as the goal.

Superconducting fault current limiters

Superconducting fault current limiters (FCLs) are being developed for use in power distribution systems, where they act to prevent the grid from unwanted surged current causing permanent fault (Campbell *et al.*, 2008b, and Noe and Streurer, 2007). The requirements for cooling are similar to those for wind power generation – long life and high reliability, but also fast and effective cooling of the FCL. This arises from the need to have maximum availability and minimum downtime for servicing, but also the fast cooling response, which cannot be provided by the closed-cycle refrigeration systems.

Summary

The trend over the past few years has been for helium costs to go up, and this is coupled with increasing problems of availability and security of supply. Added to this have been concerns surrounding the health and safety of working with liquid cryogens, the operational convenience of using cryocoolers and the potential cost benefits to cryostat design and use. These factors have driven the design of low-temperature equipment away from traditional wet cryostats to ones based on closed-cycle refrigerators and will continue to do so. In future the further development of high-temperature superconductors may also move some of the focus away from the operation of magnet systems, generators and machinery in liquid helium at 4 K to higher temperatures.

There is a hope that second-generation Gifford-McMahon coolers combined with the Joule-Thomson expansion valve may be a future answer to superconducting systems that demand more cooling power. Mechanical cryocoolers look set to continue to play an important role in these advances, for the practical, economical and design advantages that they bring.

Note

1 In estimating the relative CO_2 contribution of cryocoolers versus liquid helium production from natural gas there needs to be a starting point. It is assumed that a feed of 50 per cent helium (balance N_2 and CH_4) is available at an LNG plant. It is further assumed that the CO_2 impact of creating that mixture is accounted for elsewhere. To purify this up to Grade A quality will use about 10 per cent of the liquefaction energy, as it involves cryogenic processes. With regards to liquefaction, a 1990s liquefier uses about 300kW of electricity to create 450W of cold at 4K. In terms of liquid litres per hour (LL/h) that is 450•3600/(125gram/litre•20.6 J/gram)=620 LL/h. This ratio is better in large liquefiers, where the power ratio is more like 250:1 rather than 667:1 [=300,000/450]. Distribution costs, in terms of CO_2 impact, are estimated to be around 20 per cent of the liquefaction energy and given the average time "at sea" of 17 days, around half the liquid's "cold value" is lost (it is nearly all lost after 30 days). Thus, allowing for purification and distribution, we have: 300kWh•(1+0.1+0.2)=390kWh/620=0.63kWh/LL. Allowing a 100 per cent margin for other inefficiencies (e.g. decanting losses) the figure becomes 1.25kWh/LL. Given that 1 LL/h is worth (125•20.6)/3600=0.72 watt-hours, this gives a power ratio (=kWh electricity/kWh at 4K) of 1.25/0.00072=1730:1. For comparison, a typical GM machine has a power ratio of around 3700:1 (see text) i.e. its CO_2 impact is around 2.1 times worse than liquid helium. This assumes that the primary energy source used to drive the GM machine or liquefier is has the same impact in both cases and that the carbon impact of the manufacturing all the machinery involved is the same.... However, despite this, the CO_2 impact is not, of course, the whole story. Apart from the decanting losses mentioned above, topping up any superconducting machine with liquid helium causes thermal and therefore operational instabilities for a while afterwards. This can be a significant consideration with an MRI machine that needs to perform on demand and on schedule. Anything that causes a disruption or outage of a high value machine will have a cost impact much, much greater than the cost of the liquid helium used. For this reason there is a concerted trend towards cryocoolers in many applications, with the consequential energy impact identified in Table 1.1.

References

Bhatia, R. S. (2002) "Review of Spacecraft Cryogenic Coolers", *Journal of Spacecraft and Rockets*, Vol. 39: 329–46.

Bradshaw, T. W. and Orlowska, A. H. (1988) "Miniature Mechanical Refrigerators for Space Use", Proceedings of the Society of Photo-Optical Instrumentation Engineers, London: 39–43.

De Waele, A. T. A. M. (2000) "Pulse-Tube Refrigerators: Principle, Recent Developments, and Prospects", *Physica B*, Vol. 280: 479–82.

Gibson, A. S., Reed, J., Bradshaw, T. W. and Linder, M. (2008) "Heritage Overview: 20 Years Of Commercial Production Of Cryocoolers For Space, Advances in Cryogenic Engineering", *Transactions of the Cryogenic Engineering Conference – CEC*, Vol. 52, AIP Conference Proceedings, Vol. 985: 493–505.

Glaister, D. S., Donabedian, M., Curran, D. G. T. and Davis, T. (1998) "An Overview of the Performance and Maturity of Long Life Cryocoolers for Space Applications", *Proceedings of the Tenth International Cryocooler Conference*: 1–19.

Jewell, C. I. (1996) "The Thirtieth ESLAB Symposium on Submillimetre and Far-Infrared Space Instrumentation", ESA SP-388: 267.

Kirk, A. C. (1874) "Minutes of Proceedings – Institution of Civil Engineers", Vol. 37: 244–82.

Kittel, P. (2006) "Cryocooler Performance Estimator", Proceedings of the Fourteenth International Cryocooler Conference, Maryland: 563–72.

Lamarre, J.-L. *et al.* (2010) "Planck Pre-Launch Status: The HFI Instrument, from Specification to Actual Performance", *Astronomy & Astrophysics*, Vol. 520, Sept–Oct.: A9.

Campbell, A., Dew-Hughes, D., Leghissa, M., Middonen, R., Heinz-Wener, N., Sykulski, J. and Vajda, I. (2008) "Motors and Generators", Lectures on Superconductivity, ed. B. Glowacki, Applied Superconductivity and Cryoscience Group, University of Cambridge, 7 October, www.msm.cam.ac.uk/ascg/lectures/applications/motors.php.

Campbell, A., Antognazza, L., Gehring, R., Kinder, H., Kozak, J., Martini, L., Obradors, X. and Sicking, F. (2008) "Fault Current Limiters", Lectures on Superconductivity, ed. B. Glowacki, Applied Superconductivity and Cryoscience Group, University of Cambridge, 7 October, www.msm.cam.ac.uk/ascg/lectures/applications/fcl.php.

Campbell, A., Antognazza, L., Gehring, R., Kinder, H., Kozak, J., Martini, L., Obradors, X. and Sicking, F. (2008) "MRI", Lectures on Superconductivity, ed. B. Glowacki, Applied Superconductivity and Cryoscience Group, University of Cambridge, 7 October, www.msm.cam.ac.uk/ascg/lectures/applications/medicalmri.php.

NCNR-NIST (2010) "Closed Cycle Refrigerators", www.ncnr.nist.gov/equipment/displex.html.

Noe, M. and Steurer, M. (2007) "High-Temperature Superconductor Fault Current Limiters: Concepts, Applications, and Development Status", *Superconductor Science and Technology*, Vol. 20: R15–R29, doi:10.1088/0953-2048/20/3/R01.

Strobridge, T. R. (1974) "Cryogenic Refrigerators – An Updated Survey", National Bureau of Standards Technical Note 655, Washington, DC.

Walker, G. and Bingham, E. R. (1994) *Low-Capacity Cryogenic Refrigeration* (part of the "Monographs in Cryogenics" series, number 9, series editor R. G. Scurlock), Clarendon Press, Oxford.

12 Medical imaging

Why helium prevails

Adrian Thomas

The MRI industry

The current world market of the MRI business in 2011 was approximatelly 3,100 systems per year. From that, 85 per cent (~2,600 units) are made of NbTi low-temperature superconductor characterised by critical temperature $T_c = 11\,K$, and 15 per cent (~500 units) are permanent systems based on hard permanent ferromagnetic material NdFeB. Costs of the complete unit vary. An MRI system generating a field of 0.5 T may cost ~US$0.4 million, whereas the 3.0 T system may cost US$1.7 million. The market value of the overall MRI business is US$3.2 billion.

One of the largest commercial uses of helium is as a refrigerant for superconducting magnets used in MRI (magnetic resonance imaging) systems.

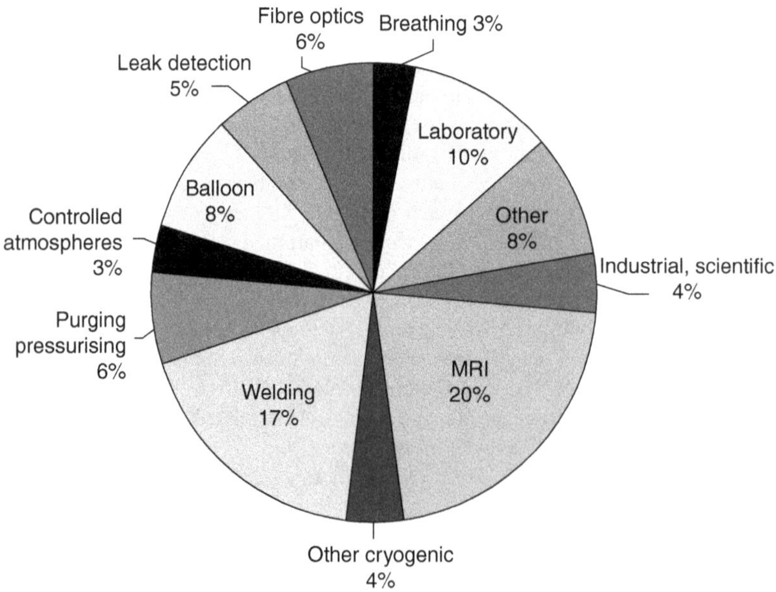

Figure 12.1 Global helium demand in 2006.

Helium has the lowest boiling point of all elements, at 4.2 K (under normal pressure) and is therefore ideal for keeping materials at the temperatures required to make them superconducting, a key enabler for the generation of the high magnetic field strengths required for MRI.

Magnetic resonance imaging – the need for field

Magnetic resonance imaging (MRI) is a powerful, non-invasive medical diagnostic technique. At its heart is the phenomenon of nuclear magnetic resonance (NMR), a process whereby the protons buried in atomic nuclei can be induced into a process of precession by the application of a magnetic field. While a full explanation of NMR is beyond the scope of this chapter, higher field strengths offer improved resolution and signal strength, giving faster images and greater 'information density' as well as enabling imaging of human functions such as brain activity and metabolic function, which, in addition to direct clinical benefit, also aid the development of pharmaceuticals through the ability to image their effectivity.

The typical improvement in 'information density' is illustrated in the schematic chart below, where magnetic flux density increases incrementally from left to right.

Lower field strengths can be achieved through the use of permanent magnets or resistive coils, usually combined with ferromagnetic materials (see Figure 12.2a). However, in order to achieve the high field strengths, which enable the higher diagnostic capabilities, it is necessary to use superconducting electromagnets with very high current density (Figure 12.2b).

Increasing the magnetic field strength generated by an electromagnet requires the current density in the windings to be increased. This can be achieved by increasing the current passing through the coil, or reducing the cross-section of the coil. For resistive materials, such as copper, both these approaches will increase the heat generated in the coil, which must be removed, usually by using high flow rates of air or water.

Table 12.1 Schematic improvement in 'information density'

Field (Tesla)	1.5	3.0	7.0		9.4	11.7
Resolution (μm)	1,000	200		70		50
Resolution (μg)	1,000	10		0.35		0.125
Measuring rate(s)	1	0.3		0.6 (^{13}C)		0.2 (^{13}C)
	← H (tissue) →			← ^{13}C (metabolic) →		
Diagnostic	Anatomical	Brain function		Early-stage cellular clusters and pharmaceutical effectivity		

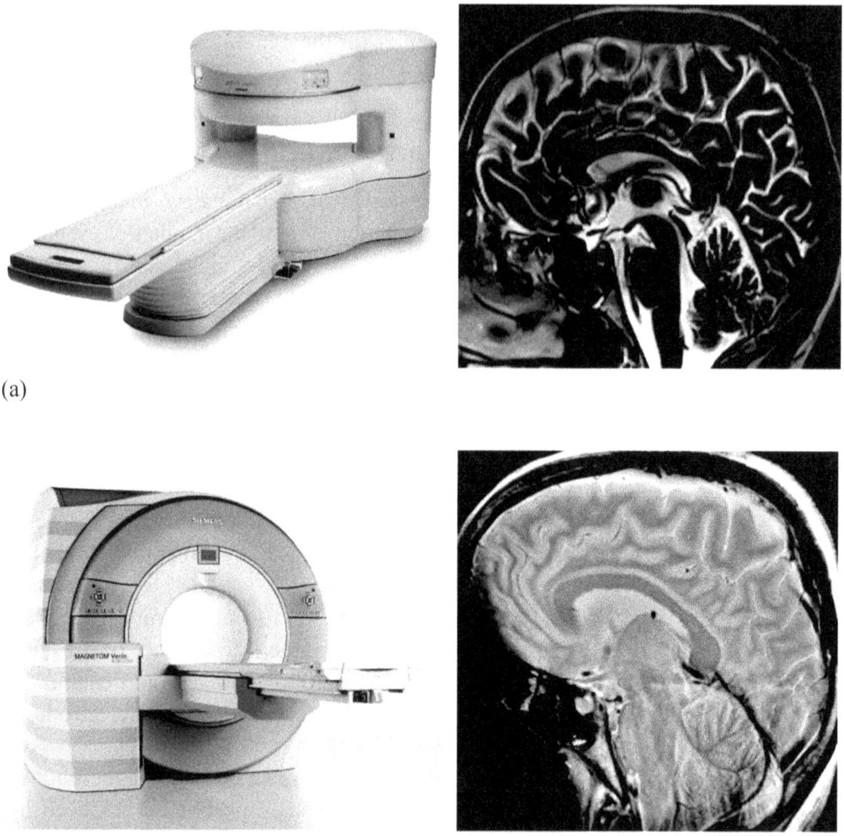

(a)

(b)

Figure 12.2 Examples of MRI systems: (a) 0.4 T permanent MRI magnet system; (b) 3.0 T superconducting magnet MRI system. The exemplary images provided show clearly the improved MRI processing with higher magnetic field density.

Superconductors and temperature

The successful liquefaction of helium in 1908 by Heike Kamerlingh Onnes in Leiden, and its associated achievement of lower temperatures than ever before led, two years later, to the discovery of the property of superconductivity in mercury. The complete absence of electrical resistivity led to the potential for achieving the high current densities required by magnet designers, without the subsequent generation of heat, although it took a further 50 years to produce conductors capable of tolerating high magnetic fields, such as NbTi and Nb_3Sn (T_c = 18.5 K). Since then, the major commercial application of superconductivity, and the one that has affected many people directly, is the MRI system with the superconducting magnet at its core.

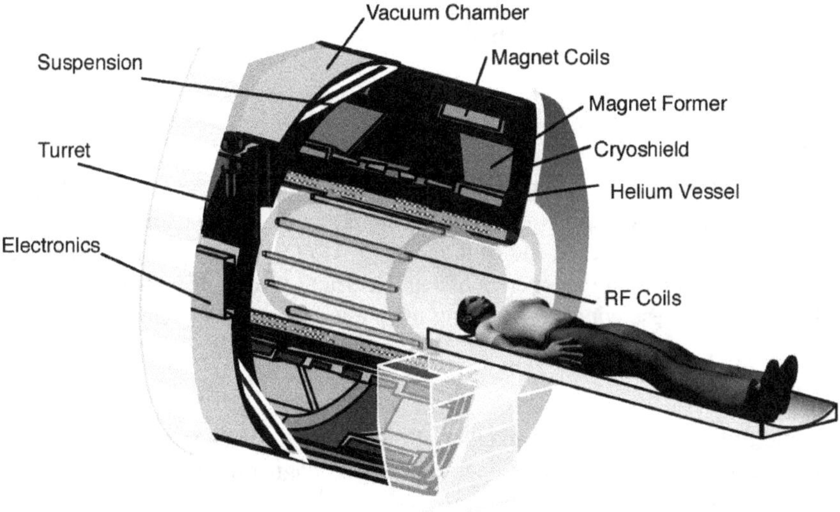

(a)

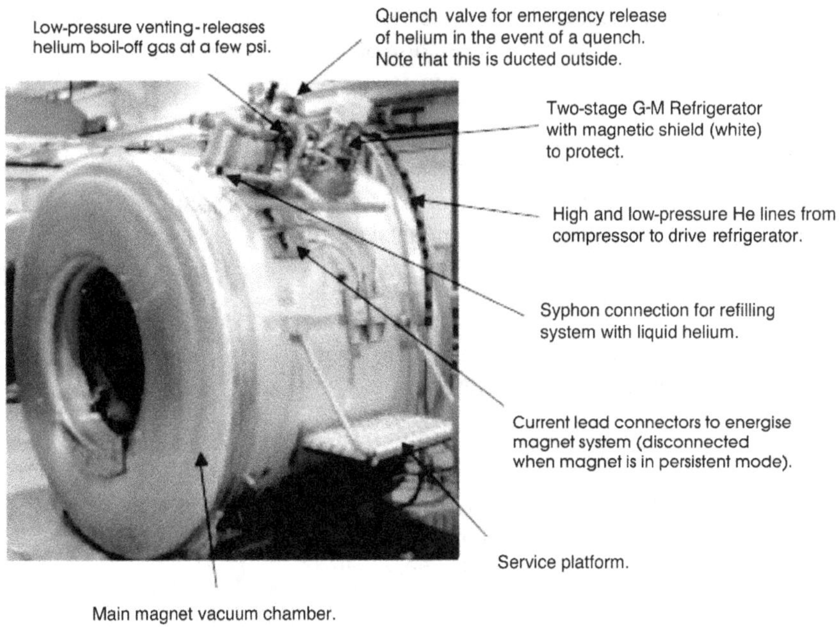

(b)

Figure 12.3 (a) Cross-sectional view of a modern MRI system showing details of the functional subsystems that constitute the working MRI unit; (b) Picture of the fully assembled MRI cryostat.

MRI system

A modern MRI magnet system is illustrated in Figure 12.3. Its major constituents are as follows.

Magnet coils

Windings of superconducting wire – usually containing the alloy niobium-titanium (NbTi). Discussion of the possible application of other superconducting materials is presented at the end of the chapter. Up to 50 km of wire is used in a clinical superconducting magnet. When an electrical current of up to 600 amps is passed through these windings, a strong magnetic field is generated.

Magnet former

The multiple coils within the magnet attract, or sometimes repel, one another with forces of up to 400 tonnes. Operation of the magnet requires the coils to be held in position by the magnet former.

Helium vessel

The helium vessel completely encloses the superconducting coils and former and is filled with liquid helium, at a temperature of 4.2 K (–269 °C). At this temperature, the coils take on the property of superconductivity.

Cryoshield

This is a thermal shield, which fully encloses the helium vessel to intercept heat from the room temperature environment outside.

Vacuum chamber

As with a conventional vacuum flask, heat conduction is prevented by fully enclosing the magnet in a vacuum vessel.

Suspension

The superconducting magnet and helium vessel must be supported within the vacuum chamber. These components must be strong enough to support the magnet weight, but have low thermal conductivity.

Turret

Access to the low-temperature region of the magnet is through a small access port. This port is where all cryogens (liquid helium), power and instrumentation wiring must pass. The turret also houses the refrigerator, which removes heat from the cryoshield and the helium vessel to prevent the loss of liquid helium (LHe).

Electronics

System monitoring of the magnet and control of internal systems is conducted automatically by dedicated electronics.

Many other components are required to convert the magnet into an MRI imaging system but are not illustrated here. These include superconducting switches, allowing the operation of the superconducting electromagnetic coils in a persistent mode (more like permanent magnet).

Superconducting materials

All superconductors have a critical temperature, T_c, above which the superconducting property vanishes. Below this temperature, their properties steadily improve – in particular, their ability to carry large currents, and their tolerance to magnetic field, both essential properties in the design of superconducting magnets.

Superconductors are generally high-cost materials, comprising 30–50 per cent of the cost of the whole superconducting magnet. Consequently, engineers are very keen to minimise the quantity of these materials used, resulting in a strong incentive to utilise them at the lowest available temperature to maximise their properties, as presented in Figure 12.4. An increase in operating temperature of a few degrees can reduce the performance of the electromagnet by a significant factor, requiring the use of more materials and therefore increasing the cost of the unit.

Using the superconductors close to their possible maximum critical parameters (see Figure 12.4a), however, carries the penalty that small increases in temperature can take the demands of the magnets beyond the properties of the superconductor.

This usually results in quench and loss of valuable helium. Replacing and refilling the MRI system in the hospital environment would be a costly and disruptive operation, as will be described more in detail later. This sensitivity is normally expressed as a 'temperature margin', and for a commercially competitive magnet design the margin must be kept down to a few degrees.

Boiling point, heat capacity

Most materials have very low heat capacities at cryogenic temperatures. For example, at a temperature of 4.2 K the most widely used superconductor in MRI applications, niobium titanium (NbTi), has a heat capacity of only $4.57 \cdot 10^{-4}$ $Jg^{-1}K^{-1}$, requiring less than half a joule to heat 1 kg by 1 K. This makes a superconducting magnet very sensitive to small inputs of energy created by mechanical movements, or electromagnetic heating. By immersing a magnet in a cryogenic liquid like helium at its boiling point, it is possible to take advantage of the comparatively large latent heat of 20.9 Jg^{-1} (45,000 times larger than the heat capacity, per degree, of NbTi) to maintain the superconductor at a constant temperature, and to absorb external sources of energy.

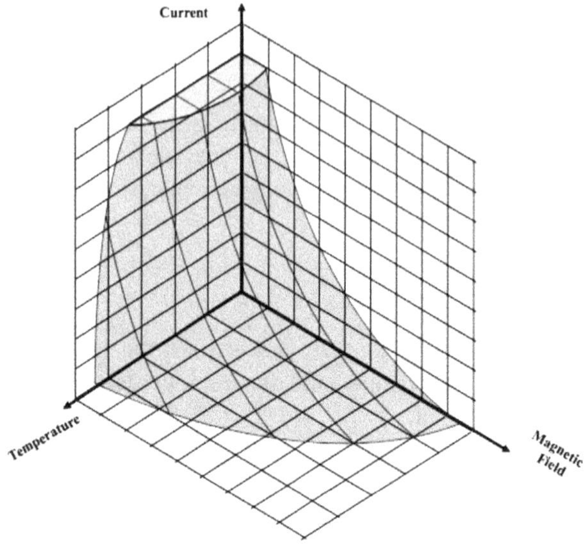

(a)

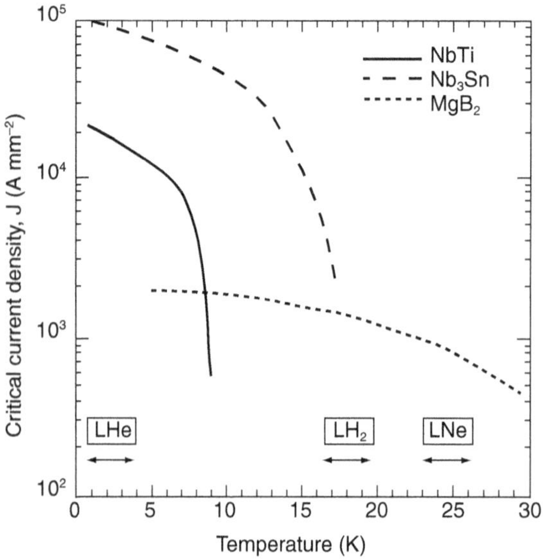

(b)

Figure 12.4 (a) Superconducting critical parameters phase diagram for NbTi (current – I, temperature – T, magnetic field – H); (b) Comparison of three main superconducting materials currently considered for MRI applications – NbTi, Nb_3Sn and MgB_2 – and their performance at different temperatures with three potential cryogens (despite the fact that there are other so-called high-temperature superconductors – Bi2223 and YBCO – their high production cost and limited reproducibility require further research if they are going to compete with current candidates).

Since the development of superconducting magnets, liquid helium has been used almost exclusively as the refrigerant. Despite the development of superconducting materials such as Bi-2212, Bi-2223 and YBCO with higher critical temperatures (see Chapter 16), their higher price and lower mechanical properties have led to limited applications using other refrigerants such as LN_2 or even LNe.

MRI system cryogen development

Until the mid-1980s a standard MRI system would have included not only a helium vessel for the superconducting magnet but also a thermal shield with a double-wall construction producing a vessel for liquid nitrogen.

Cryogen vessels must be maintained and handled with great care. An effect of the very low temperatures is that, if air, which inevitably contains moisture, is allowed to enter the vessel, the water will condense and freeze, creating an ice plug. In an extreme case, this ice plug can prevent the free venting of the cryogenic liquids. If left unchecked this will result in a catastrophic build-up of pressure. Fortunately, the nitrogen vessel is contained within the main insulating vacuum chamber of the system and any failure is contained. These nitrogen-cooled shields required frequent refilling and, although the cost of liquid nitrogen was not prohibitive, they required specialist maintenance, and carried risks associated with venting as well as the development of ice.

In the 1980s the development of the Gifford-McMahon cryocooler, delivering over 10 watts of cooling down to 10 K, permitted the replacement of the liquid nitrogen vessel with a single, self-contained cooling module, powered by a room-temperature compressor. The efficiency of such a system is not very high because for every 1 W of cooling power at the 4 K level, it requires 6 kW of power through the RT compressor. However, this development had the benefit that the cryostat became less complex to construct, and allowed a greater internal volume for the magnet.

This latter became more significant with the development of actively shielded magnets, which use opposite-polarity 'shield coils' to cancel the external field of the magnet and make siting in a hospital environment significantly easier. These opposing coils, however, need to be separated as far as possible within the cryostat to avoid excessive stresses and to maximise the efficiency of the magnet. These refrigerated systems, however, still required the replacement of the liquid helium on a periodic basis, since the cryocoolers were incapable of providing cooling power at 4.2 K. Although a few systems were developed with external 4 K refrigeration systems, much of the development for the next decade concentrated on minimising the heat leak into the helium vessel to maximise the interval between refills. Many of these developments were through the application of 'good engineering practice' and the development of new structural, functional and also superconducting materials.

One of the sources of heat load into the helium vessel is through the suspension system that supports the magnet in the middle of the vacuum chamber. This suspension needs to be connected to the cold magnet at one end and the main vacuum chamber at room temperature at the other end (see Figure 12.3), achieving the necessary strength while minimising the thermal conductivity. Composite materials are able to reduce the heat load into the system, while technology improvements in the magnet itself enabled the same performance to be achieved with less material, thereby reducing the suspended mass and the cross-section of the suspension. As a result of these and other developments, helium refill intervals have extended to over two years.

The Gifford-McMahon refrigerators were limited to a base temperature of about 10 K. The thermal cycle on which the refrigerators operate require a regenerator that is able to absorb heat at low temperature. The heat capacity of most materials reduces dramatically at low temperature, with lead offering one of the best performances but still having insufficient heat capacity for the refrigerators to operate below 10 K (see Figure 12.5).

A breakthrough came in the 1990s with the development of 'rare earth' compounds based on materials such as europium, holmium, dysprosium and gadolinium. These compounds undergo a magnetic phase change at very low temperatures, which, like physical phase changes such as melting or boiling, require energy to be absorbed and are able to act as low-temperature regenerators. The specific heat maximum is function of the external magnetic field presented in Figure 12.5. It important to realise that if the regenerator is to operate in proximity to high magnetic flux density this will reduce its efficiency.

With this development, refrigerators became capable of reaching temperatures below 4.2 K, although with low power, typically only 1 W of cooling power

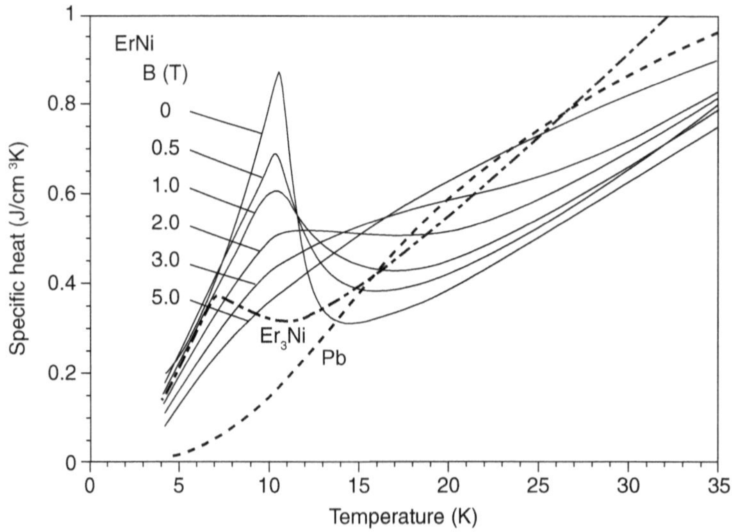

Figure 12.5 Specific heat versus temperature for lead and Er_3Ni at different magnetic flux densities (Yayama *et al.*, 1996).

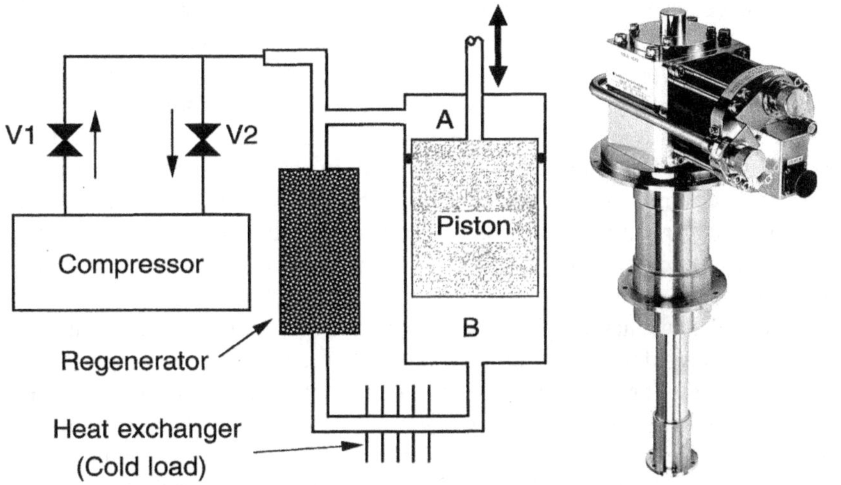

(a) (b)

Figure 12.6 (a) Gifford-McMahon single-stage closed-cycle refrigerator, basic working principium – V1 and V2 are valves, A and B are the working gas chambers; (b) Two-stage Gifford-McMahon cryocooler with 50 K and 4 K cooling stages (at the copper rims) (see also Figure 12.3b) (Courtesy: SHI Ltd.).

at 4.2 K. As was pointed out earlier, the lower the temperature the better the performance of the superconducting conductor and the less conductor is required, allowing the cost of the system to become more competitive.

Nevertheless, the development of cryostat design and technology had reached the point at which the heat load was below even this low power. MRI magnets were now able act as 'zero-loss' systems, requiring no cryogen refills once they were installed and working. Achieving a zero-loss system has represented a major development in terms of ease of operation within hospitals. MRI manufacturers, however, must also consider the logistics of transporting and installing systems from the point of manufacture to sites throughout the world.

Early systems were cooled and tested at the manufacturing base, but then warmed up and fitted with transport brackets to support the magnet during transit. The shock loading that magnet systems may be subjected to during road, air and sea transport were beyond those the suspension system could tolerate. The cooling down of a magnet system in a hospital, however, is a major logistical conundrum. Installation in a hospital can take several weeks, and when ordering cyrogenic liquids hospitals do not benefit from 'bulk buying' enjoyed by manufacturers. Superconducting magnets may sometimes change their behaviour as a result of the effect of changing temperature from 4 K to room temperature and back again.

The ability to ship magnets cold and to install them in a hospital more quickly and more cheaply, with their factory performance 'locked in', are significant factors in a commercially competitive environment. Consequently, MRI manufacturers have developed their systems to be capable of being shipped cold. It is generally not possible to keep the refrigerator running during transit, as about 6 kW of electrical power is required to drive the compressor on which the refrigerator depends. Many systems will be filled with LHe before shipment and allowed to gradually exhaust the helium gas during transport, arriving cold but often requiring refilling immediately on arrival. In order for a magnet system to arrive at its destination cold, the duration of the transport must be limited. For this reason longer journeys must be made by air, and the magnet manufacturers are required to comply with a variety of regulations and restrictions to ensure safe transit.

Many of the advantages of cold shipment have been enabled by the reduction in the cold mass for the majority of systems used for clinical applications. However, as described at the beginning of this chapter, there is always a demand for higher field-strength magnets to enable research developing the benefits of MRI. These systems are inevitably heavier and in many cases still require warm transportation, with the need for significant quantities of cryogenic liquids, primarily helium, to be available at the hospital site.

Limitations posed by the use of helium

The technical properties of helium make it a very useful material for operating superconducting magnets in a laboratory environment. However, when those magnets are installed into a clinical environment, some of helium's properties can become undesirable.

Availability of helium can be dependent on many factors outside the control of a clinical environment. It must be supplied as a liquid, and therefore geographical proximity to source or liquefaction centres, as well as worldwide supply and demand, can affect both the availability and price of LHe into the clinical environment. This can have a detrimental effect on the ability of developing countries to take advantage of this technology.

The operation of a superconducting magnet and the subsequent requirement to handle cryogenic fluids has been a triumph of the development of 'user-friendly' cryogenic equipment. Equipment that would normally be confined to laboratory environments for use by highly trained technicians has been made available in clinical environments around the world. Nevertheless, despite this success there are still limitations imposed by the use of cryogens within the clinical environment.

The expansion ratio of helium (increase in volume when changing from a liquid to a gas) of 757 dictates that a venting system must be installed to allow safe exhaust of the gas outside the building. This venting typically takes the form of a 100 mm diameter ducting and must be capable of withstanding the low-temperature exhaust gas, with welded steel being the usual solution. Depending upon the position of the imaging suite within the hospital building, this can represent a significant logistical and cost challenge in the installation of an MRI system.

To achieve expansion of the availability of MRI into developing countries, dependency on this logistically complex material (helium) must be reduced. The ability of modern MRI systems to operate in a 'zero-loss' configuration enables developments to reduce the dependency of the industry on helium, particularly in its liquid form. Early systems required large volumes of LHe in the magnet system to minimise the frequency with which the magnet needed to be refilled. Now that this requirement has been effectively removed, the potential exists to reduce significantly the dependency of the MRI industry, both in manufacture and clinical operation, on the availability of LHe.

However, the compensation of the thermal heat load with a refrigerator is not the only requirement for a low helium dependency. Transport logistics have been mentioned earlier in the chapter. If an MRI system is to lose its helium vessel and the 600 litres of liquid helium to keep it cold during transport, alternatives must be found. Operation of the refrigerator during transit may be possible, but changes in air and sea shipment would be required to make the necessary electrical power available.

A large helium vessel provides a significant buffer in the event of failure in the refrigerator, either through mechanical or power failure. The absence of the helium buffer would make the system very vulnerable to such breakdowns. It should be recalled that the temperature of the superconducting magnet need only rise by a single degree to lose the superconducting performance and for the magnet system to fail. Improvements in refrigerator reliability and the availability of back-up power are prerequisites to losing the helium buffer. A refrigerator provides a single point of cooling power, and this needs to be efficiently transported to all points of the superconducting magnet. Liquid helium is highly efficient at achieving this and would need to be replaced by a similarly efficient system.

Such challenges, however, are not insurmountable, and are an extension of the developments that have already brought superconductivity into all major hospitals in the world. A comparatively simple development is the reduction of the enclosed helium volume through the manufacture of a more complex vessel shape, which closely fits the superconducting coils but contains a fraction of the volume. An extension to this volume reduction is to contain the cryogen in a pipe system. This reduces the cooling contact area to the superconducting coil, and inevitably introduces a thermal obstacle in the additional material layers between the cryogen and the coil material. The final logical extension to this process is to replace the liquid cryogen as the heat transport system with a solid state system, using materials such as copper or aluminium to transmit cooling power from the refrigerator to the coils. Again, there is a loss in thermal efficiency with an associated increase in the risk of magnet failure.

Some of the difficulties associated with reducing helium dependency can be addressed through the use of alternative superconducting materials (such as those mentioned earlier in this chapter) that operate at higher temperatures, enabling the use of alternative cryogens or mitigating the thermal inefficiencies associated with dry systems. All of these technologies have been demonstrated on superconducting magnets in various forms. Dry magnet MRI systems using materials such as Nb_3Sn and, more recently, magnesium diboride (MgB_2) are available commercially (see Figure 12.4). However, these systems face commercial challenges. The superconducting material in a clinical MRI system based on niobium-titanium, operating at 4.2 K, represents 30–50 per cent of the cost of the magnet. Alternative superconducting materials, such as those mentioned above, are currently available for approximately five times the cost. This represents a major impact on the price of an MRI system. Future developments may result in these materials becoming more commercially attractive and lead to a large-scale movement away from cryogen-cooled MRI magnet systems.

Reference

Yayama, H., Hatta, Y., Tomokiyo, A., Okamura, M., Tokai, Y., Takahashi, A., Hashimoto, T. (1996) "Specific Heat of ErNi in Magnetic Fields and Its Performance as Regenerator Material in Gifford-McMahon Refrigerator at Cryogenic Temperatures", *Japanese Journal of Applied Physics*, Vol. 35, pp. 5545–9

13 Rising to the challenges of constrained helium supply in cryogenic systems for the research market

John W. Burgoyne and Michael N. Cuthbert

Introduction

While the application of superconducting magnets in commercial markets has been dominated to date by magnetic resonance imaging (MRI), the global market for low-temperature research instruments – encompassing superconducting magnets, ultra-low-temperature refrigerators and numerous cryostat types – remains a key enabler of physical science research. Oxford Instruments, a consistent leader and innovator across these low-temperature research tools since its inception in 1959, is one of several commercial companies serving this market. Fundamental to continued commercial success for any supplier into this market is a twofold approach to the increasing costs and reducing supply of liquid helium: first, responsiveness to changing customer needs for reduced or eliminated liquid helium usage; and second, proactive innovation and development of new products which offer new and extended capabilities to the researcher. This chapter discusses the drivers, enablers and realisation of these goals.

Estimates indicate that 28 per cent of worldwide helium gas use is for cryogenic applications, and although helium use for cryogenic applications has increased by 300 per cent in the past 20 years, the proportion has fallen from 44 per cent – in other words, there are increasing competing demands for non-cryogenic use of helium. The vast majority of cryogenic use is for MRI, NMR and major facilities such as CERN and national laboratories. The estimates put helium use for low-temperature physics at ~4 per cent of worldwide use.

In recent times significant price rises, shortages, supply disruption and termination of supply contracts have raised the profile of this non-renewable, strategic resource to new levels, and the low-temperature community has been working to leverage a strategic position within the funding organisations.

The research drivers for liquid helium use reduction and elimination

Liquid helium costs

The first and most obvious motivation for removing reliance on liquid helium is that of cost. There are clear differences across institutions in liquid helium costs, and particularly in their significance as a proportion of the operating budget; for large facilities, helium costs per litre are relatively low – both in absolute terms, driven by high usage volume, and because of on-site gas reclaim/reliquefier facilities; while for small, departmental-level low-temperature physics groups the cost per litre may be extremely high.

Liquid helium prices vary widely across the world, but for research laboratories prices range from as low as US$3 per litre up to US$20 per litre, or even more when import duties and transportation are included. In the former case, costs are frequently kept low by the laboratory having its own helium liquefier and helium gas recovery system to allow capture and reliquefaction of evaporated helium from a number of distributed cryogenic systems. However, large costs are incurred in sustaining such infrastructure, with attendant management overheads for the facility, and only the largest laboratories tend to have them. A number of laboratories have decommissioned liquefiers for this reason, particularly if liquid helium usage has dropped close to a cost break-even for the facility.

The rate of cost rise has further compounded the impact on researchers, making budgeting increasingly difficult. An example illustrates the situation well: one researcher reported that his supplier was going to enforce a 65 per cent increase in price per litre with short notice. Since the researcher's expenditure on liquid helium for the previous 12 months amounted to more than US$60,000 of direct costs, based on about US$5 per litre, or about US$100,000 with overhead, this meant a negative impact of approximately US$65,000 on their overall budget – about the equivalent of funding required to support one graduate student over the three-year period.

Overhead costs

A further issue in the use of liquid helium is that of staff health and safety, training and employment costs, particularly when set against the present background of typically reduced or frozen science budgets. The health and safety point is particularly relevant in beamline facilities, where the cryogenic system is situated on a neutron, synchrotron or other beam. Here the beamline safety restrictions and limited physical accessibility for liquid helium dewars mean that reduction in cryogen refill frequency (of both liquid helium and liquid nitrogen) is highly desirable to reduce in turn the frequency with which staff need to enter the beam hall. In such multi-user facilities, the scientists working on the experimental system are very often not expert, or even familiar, with cryogenic systems, and therefore a specialist technical team will service the equipment. With budgets under constant pressure, staff efficiency improvements are aided by the reduction of such routine tasks as cryogen refills.

Security and continuity of supply

Not only have rising costs limited research in many higher education departments and research establishments but security and continuity of supply has become of further concern. As one researcher put it, “Even if we can weather the cost increases it will be very frustrating to have to shut down good experiments due to lack of available helium.” Missing only one or two helium deliveries can mean losing experimental work for a month or more.

The choice of suppliers available within a geographic area is very limited, if indeed there is a choice at all. It is often found that alternative suppliers are not accepting new customers and cannot guarantee availability. In the US, which has previously been immune to liquid helium supply issues, the majority of users – over 80 per cent in small to medium-sized research facilities – have been put onto fixed allocations of liquid helium supply, typically two-thirds of their prior usage (see further below).

Thus the driver for many researchers to seek cryogen-free solutions which would be immune from these effects is not only one of cost but that of being able to guarantee the continuity of their experiments, which is of course fundamental to any research programme.

Results of a survey of US researchers

In 2007, NSF and Penn State University with US researchers compiled a survey of the impact of liquid helium costs and supply. In this survey:

- there were responses from 27 universites and national laboratories in the US;
- the respondents were mainly physics users;
- the geographic breakdown was dominated by East Coast and north-eastern users
- respondents were classified into small, medium and large users by liquid helium usage:
  - a small user was defined as consuming <100 litres per week
  - a medium user was defined as consuming <2,500 litres per month
  - a large user was defined as consuming more than >2,500 litres per month
- there were 12 large users, accounting for 86 per cent of the total He usage reported.

While the survey only covered the US, and the respondents were predominantly from within one region of the US (it should be stressed that this was a relatively simple survey within the community rather than a full, professional market research exercise), the outcomes shown in the table below are nevertheless instructive and illustrative of the general situation and the challenges facing researchers.

In these survey results we see researchers across the institutional scale being affected by both price increases and a lack of security of supply, in varying degrees.

- Price rises are similar for both small and medium users, while large users faced a smaller increase.

Table 13.1 Small user feedback containing data of helium usage, itemised prices and availability, and average of price increase

Small user feedback		
• Accounted for ~1% of liquid helium used		
• A constituency of 18.5% of respondents	Sum of annual He usage (litres)	12,000
• Average use 2,400 litres per year	Average of annual He usage (litres)	2,400
• Typical He budget US$10,000–15,000 per year	He cost per litre – min.	US$4.20
• 80% have seen recent price increases	He cost per litre – max.	US$7.76
• Average increase of 57%	He cost per litre – average	US$5.19
• Worst-case price rise of 117%	Most recent price – min.	US$5.13
• 20% have had an interruption to supply	Most Recent Price – max.	US$9.75
• 80% had been put on allocation	Most recent price – average	$7.79
• The average allocation was 60% of previous usage	Average price increase	57%
• None had liquefier on site		

Table 13.2 Medium user feedback containing data of helium usage, itemised prices and availability, and average price increase

Medium user feedback		
• Accounted for 13% of liquid helium used		
• A constituency of 37% of respondents	Sum of annual He usage (litres)	140,500
• Average use 17,500 litres per year	Average of annual He usage (litres)	17,563
• Typical He budget US$30,000–100,000 per year	He cost per litre – min.	US$2.40
• 90% have seen recent price increases	He cost per litre – max.	US$5.50
• Average increase of 60%	He cost per litre – average	US$4.11
• Worst-case price rise of 103%	Most recent price – min.	US$4.15
• 80% have had an interruption to supply	Most recent price – max.	US$9.08
• 88% had been put on allocation	Most recent price – average	US$6.40
• The average allocation was 67% of previous usage	Average price increase	60%
• 20% had liquefier on site		

Table 13.3 Large user feedback containing data of helium usage, itemised prices and availability, and average price increase

Large User Feedback		
• Accounted for 86% of liquid helium used		
• A constituency of 44% of respondents	Sum of annual He usage (litres)	919,000
• Average use 76,500 litres per year	Average of annual He usage (litres)	76,583
• Typical He budget >US$200,000 per year	He cost per litre – min.	US$2.50
• 92% have seen recent price increases	He cost per litre – max.	US$4.35
• Average increase of 36%	He cost per litre – average	US$3.43
• Worst-case price rise of 74%	Most recent price – min.	US$3.62
• 36% have had an interruption to supply	Most recent price – max.	US$5.65
• 55% had been put on allocation	Most recent price – average	US$4.51
• The average allocation was 63% of previous usage	Average price increase	36%
• 58% had liquefier on site		

- Average helium prices clearly scale inversely with usage, and particularly in the upper range of large users, where there is a liquefier on site.
- Medium users appear to be the most vulnerable to interruption of supply, being at the pinch point where the volume of liquid helium required is significant (greater than small users), yet they do not have the strategic buying power of large users.
- For those put onto an allocation of liquid helium, the reduction in allocation versus prior usage was similar across all user scales.

Assuming these figures give a strong indication of what is happening in most countries, the dual challenge of price and supply in the low-temperature research community is clear.

Research in developing economies

The expanding science base of the so-called BRICs (Brazil, Russia, India, China) and similar economies has significantly added to the call for liquid helium-free products. In these geographic areas liquid helium costs are typically at their highest, which, coupled with import duties, customs clearance times and high transportation costs, can form an insurmountable barrier to the use of low-temperature equipment. Indeed, we have heard anecdotally of one user whose liquid helium consignment was delayed in customs for a number of weeks, to the point that liquid helium completely boiled off and the researcher had paid thousands of dollars for an empty dewar.

Worldwide, as many new institutes and researchers seek to add low-temperature techniques to their research tools, taking cryogenic systems out of the sole realms of specialised laboratories, the need is growing continually for greater ease of use and less reliance on prior cryogenic knowledge and experience.

Enabling technologies

In the early 1990s, two key developments in technology worked to enable the development of cryogen-free superconducting magnets; cryocoolers and, later, full cryogenic systems.

4 K cryocoolers

Closed-cycle cryocoolers, using a closed loop of high-purity helium gas through a compression/expansion cycle, have been in use for many years on a variety of cryogenic equipment. The cryocooler system consists of a compressor and a "cold head", with interconnecting high-pressure gas lines and electric cable. These cryocoolers utilise the Gifford-McMahon (G-M) cycle, with either a single or double stage of cooling via a heat exchange medium at the cold end.

High-powered G-M cryocoolers, delivering typically 100 W of cooling at a ~77 K "first stage" and 5 W at 20 K, have commonly been used for radiation shield cooling on MRI magnet systems (removing the need for a liquid nitrogen-cooled shield), industrial magnetic separation systems (Figures 13.1 and 13.2), some semi-industrial research magnets (Figure 13.3) and in cryopumps. Such coolers can achieve a "base temperature" – i.e. the lowest temperature with no heat load – of around 10 K, and indeed were used to cool the very first cryogen-free superconducting magnets, built entirely from Nb_3Sn wire, which has a sufficiently high critical temperature (T_c) to realise a practical current and therefore field-generating capability (Figure 13.4). The conductor cost of Nb_3Sn superconductor, however, is prohibitive to fabrication of entire magnets from this material.

A development of the heat exchanging materials in G-M coolers enabled lower second-stage temperatures to be reached, down to a base temperature below 4 K – i.e. matching the atmospheric boiling point temperature of liquid helium, for which most low-temperature equipment is designed. The G-M cryocooler removes heat through a moving "displacer", which is filled with a high-heat-capacity "regenerator" material and reciprocates at the cold end with the helium gas compression/expansion cycle (see also Chapter 11). In 10 K cryocoolers the regenerator material is typically lead shot, but rare earth and similar compounds were developed (initially by Toshiba) which improved the heat capacity and thus offered lower base temperatures (see Chapter 12). Current G-M cryocoolers of this type offer second-stage cooling powers of up to 1.5 W at 4 K, with 30–40 W of first-stage cooling at ~50 K. This 4 K cooling capability allows the use of NbTi conductor for low-field magnets rather than the much more costly Nb_3Sn demanded by higher operating temperatures, and was a key step in the commercial realisation of cryogen-free superconducting magnets.

The role of pre-industrial applications

While we are focusing here on research systems, many of the drivers for cryogen-free operation and its supporting technology development originally came – and continue to come – from industrial applications (Burgoyne *et al.*, 2000), starting in an R&D or pilot-scale environment. The first main push for cryogen-free superconducting magnets came from Materials Processing researchers (mainly in Japan) in the 1990s, driven by (1) ease of use and system installation into materials science rather than physics labs, and in particular (2) the demonstration of applicability of these processes to an industrial setting, where cryogen-free operation was seen as a key enabler to the adoption of cryogenic technology. Applications have included magnetic separation, levitation for cold-crucible casting, electromagnetic braking in steel processing, silicon crystal pullers and many others. Clearly any of these would be attractive markets to the companies supplying superconducting magnet equipment, who have often been searching for the next big commercial application of superconductivity to follow NMR and MRI.

This industrial or at least pre-industrial drive has continued to change and develop. Some potential applications have fallen by the wayside, while others such as magnetic separation continue (see Figures 13.1 and 13.2).

Figure 13.1 A cryogen-free open-gradient magnetic separation (OGMS) magnet in a pilot-scale system for industrial magnetic separation. Material drops off a conveyor belt (right), falling through a magnetic field gradient projected outside of the vacuum casing. The superconducting magnet consists of a "racetrack" coil, cooled by two 4K G-M coolers (left). By making the magnet cryogen-free, not only is it more amenable to the industrial application, but the distance from the magnet coils to the outer vacuum casing surface is minimised, and hence the field gradient (or strictly, the field-gradient product) is maximised (courtesy of Carpco, now Outotec).

Pulse tube refrigerators and relative cryocooler merits

The disadvantages of the conventional G-M-type cryocooler are twofold, both relating to the moving parts (the reciprocating displacer) at the cold end. First, the displacer requires periodic maintenance due to saturation and the build-up of contamination, typically every 10,000 hours of operation, by exchange for a new part. This servicing at the cold end, i.e. on the cryogenic system itself, requires some or all of the system to be warmed up fully to room temperature and is a specialised operation since it involves high-pressure (16 bar), high-purity (99.999 per cent) helium gas. Secondly, the reciprocating displacer creates a significant mechanical impulse into the system and potentially unacceptable vibration into the experiment.

An evolution of these cryocoolers, based on the same G-M cycle, has been the pulse tube refrigerator (PTR). In this conception, the moving displacer is replaced by a "slug" of helium gas that travels up and down through the regenerator material, removing heat from the cold end. Very significant development effort has been put into PTR products over a period of approximately ten years, with the result that 4 K systems are now widely available. By elimination the

Figure 13.2 An industrial-scale high-gradient magnetic separation (HGMS) system, used to separate dark-coloured, para- or dia-magnetic impurities from kaolin in the production of a clean, bright white powder product. These superconducting magnet systems are typically 5 Tesla, with a 300–1,000 mm room-temperature bore. 20 K G-M coolers are used to cool radiation shields and minimise liquid helium consumption; some models have been built as recondensing systems (see later in this chapter). Working in geographically remote locations close to the mining source, immunity from temporary power outages is essential (courtesy of Carpco, now Outotec).

moving displacer, the PTR overcomes the above-mentioned disadvantages of the conventional G-M cooler, in that there are no moving cold parts to service, and vibration is very much reduced (although not entirely eliminated).

Nevertheless, the PTR does suffer some disadvantage compared to the standard 4 K G-M cryocooler. First, the orientation on which the cold head can be mounted is limited by the danger of significant loss of cooling power – typically 30° from vertical is the limit, whereas the 4 K G-M cold head will operate on any orientation, including horizontal and "upside down". This creates some limits for the designer, and if the system is required to operate in multiple

Figure 13.3 A superconducting gyrotron magnet system using a 20 K G-M cooler to cool radiation shields, reducing liquid helium consumption and removing the need for liquid nitrogen (courtesy of Oxford Instruments).

orientations – for example, to rotate a magnet room temperature bore from vertical to horizontal – then the 4 K G-M cryocooler is the one of choice. The PTR is also inherently less efficient than the G-M cryocooler system, requiring greater compressor power.

Important new applications extending the need for cryogen-free operation to ultra-low temperatures have emerged more recently, in particular with quantum computing, where dilution refrigerator temperatures are required to cool the quantum computer chips.

Even in this brief introduction to the topic, it can be seen that in a cryogenic system the appropriate cryocooler needs to be chosen carefully.

High-temperature superconductor current leads

The limited cooling power of the cryocooler (different models generally offering 0.5 W, 1.0 W or 1.5 W at 4 K) presents a dilemma as to how to introduce a necessary current into a cryogen-free superconducting magnet, where the finite power of the cryocooler must be able to disperse both the static losses (conduction and radiation) and dynamic losses (current lead heating and magnet AC losses). On one hand, high electrical conductivity is needed to minimise ohmic heating of the current leads during magnetic field ramping, while low thermal conductivity

Figure 13.4 An early 1990s 9 T cryogen-free magnet system using all Nb_3Sn magnet coils (courtesy of Oxford Instruments).

is necessary to reduce the static thermal load (Figure 13.5). With conventional brass or copper current leads, the heat loads would be unacceptable. These competing demands have been resolved by the use of high-temperature superconductor (HTS) current leads: since they are superconducting they present no ohmic heating and have low thermal conductivity, allowing a small cross-section (Ballarino and Carrillo, 2008).

Early HTS current leads were typically solid rods or cylinders of a suitable (generally $Bi_2Sr_2CaCu_2O_{8+x}$) material, but were not sufficiently robust for many applications. Therefore, the majority of HTS current leads use multiple strips of HTS tape conductor (the number according to the current capacity required) with a low-thermal conductivity alloy substrate/matrix. The $YBa_2Cu_3O_x$ and $Bi_2Sr_2CaCu_2O_{8+x}$ or $(Bi_{2-x}Pb_x)Sr_2Ca_2Cu_3O_y$ compounds are all used.

With critical temperatures of the HTS material in the 80–110 K range, the conventional arrangement is to anchor the upper end of the HTS current lead element to the first stage of the cryocooler (operating at 50–70 K), with the lower end having a transition to conventional "low" temperature superconductor – generally NbTi – into the magnet itself. It is an important feature of running a cryogen-free superconducting magnet that the HTS current lead temperature must be maintained sufficiently below its critical temperature, and diagnostic temperature sensors are often fitted for this purpose.

It should be noted that HTS current leads are now used not only in cryogen-free systems but often in the variants of them described below and in fully "wet"

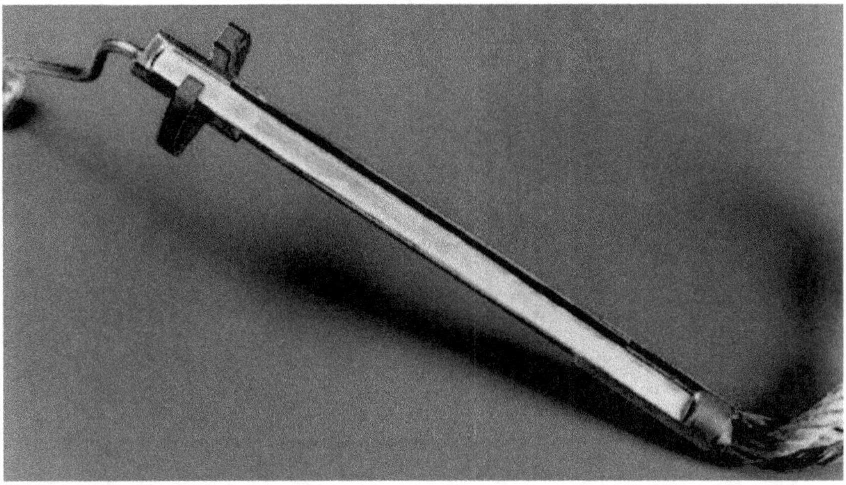

Figure 13.5 A typical HTS current lead using Bi-based HTS tape conductor (courtesy of Oxford Instruments).

systems, providing the same advantages of lower static and dynamic heat loads as in dry systems. The corollary to this is that suitable cooling and/or anchoring of the HTS current must be provided to avoid the risks of burnout.

Conduction-cooled cryogen-free magnet systems

Room temperature bore magnet systems

As noted above, the first drive from wet to cryogen-free research products occurred in stand-alone superconducting magnets – in the main with room temperature bores suitable for the materials processing and similar applications of interest at that time. These systems are relatively straightforward, with the superconducting magnet coils being conduction-cooled by the second stage of the 4 K cryocooler and a radiation shield cooled by the first stage (Figure 13.6).

Room-temperature bore (RTB) combinations now cover almost the full spectrum of magnet capability, with two 4 K cryocoolers being used to cool the largest magnets. Typical research magnet systems range from low field, wide bore (e.g. 5 Tesla, 300 mm RTB) up to high field with narrower bore (e.g. 16 T, 40 mm RTB) (see Figure 13.6). Many custom-engineered variants extend these ranges, with the most extreme offerings going up to high-field and wide-bore combinations such as 16 T, 200 mm RTB. A mid-range "workhorse" specification of 10 T, 100 mm RTB appears to offer an advantageous price-performance point. Split-pair magnets are also available in this format, although two 4 K cryocoolers are frequently required, one to cool each side of the split pair, due to the distributed geometry of the magnet.

Figure 13.6 A typical room-temperature bore cryogen-free magnet system, in this case a 10 T, 100 mm bore. Using a G-M rather than PTR cryocooler allows the magnet to be operated with the bore tube in either a vertical or horizontal orientation. Note the compact size compared to a conventional "wet" system (courtesy of Oxford Instruments).

Specific applications of RTB magnets have been addressed through these designs, in particular for magnetic separation with longer uniform field regions, and gyrotons for high-power microwave generation systems, in which multiple coils are used to shape and trim a very specific magnetic field profile (Figure 13.7).

Magnet systems with integral low-temperature inserts

Although cryogen-free magnet systems have been a significant step forward, the more significant one for the research community is the integration of low-temperature experimental inserts to provide a complete cryogen-free "B-T

Figure 13.7 A cryogen-free gyrotron magnet, specifically designed for rugged, semi-industrial use (courtesy of Oxford Instruments).

environment" (Figure 13.8). We will come to discuss ultra-low-temperature systems later in the chapter, and here focus on those magnet systems with an integrated variable temperature insert (VTI), providing typical cooling ranges from 1.5 K up to 300–400 K.

The challenge in such systems is to subdivide the finite cooling power of the 4 K cryocooler between the magnet and VTI cooling, particularly in the dynamic states of magnet ramping and high or changing VTI temperature – all steps must be taken to provide highly efficient heat transfer throughout the system, and to limit the heating generated in the magnet coils. Some type of heat switch is typically employed between the cryocooler second stage and the VTI, to enable isolation during VTI heating while allowing good thermal transfer for cooling.

Cryogen-free VTI operation is achieved in a number of detailed ways, each being proprietary to the manufacturer but generally using the principle of a pumping system drawing cold helium gas through an expansion valve. The helium gas is generally held in a separate supply tank.

Integrated cryogen-free magnet/VTI systems are now widely available in the market from a number of vendors, including solenoid magnets and more recently optical split-pair systems (Figure 13.8).

The design challenges in conduction-cooled systems

Cryocooler-based systems present some particular challenges to the designer, some of which have already been touched upon, but are worth highlighting as they are key points to consider for users in specifying cryogen-free systems.

Figure 13.8 A cryogen-free 7 T split-pair magnet system (right) for optical applications, with integral VTI (i.e. the magnet and VTI are cooled by the same, single 4 K PTR). This product takes the Oxford Instruments Spectromag® product from "wet" (left) to "dry" technology (courtesy of Oxford Instruments).

The first challenge is that of *thermal budget*. In a "wet" system, both static (i.e. the fixed conductive, radiative and convective heat loads on the system) and dynamic (ohmic heating, superconductor losses, experimental heating) thermal inputs are absorbed by the specific heat capacity of the liquid helium and result in boil-off of the liquid helium (and, where applicable, liquid nitrogen). The system designer will take steps to limit and reduce these heat loads so as to create an efficient system, but there is no hard limit. In a cryogen-free system the thermal budget at 4 K (and indeed, at 40–50 K for shield, etc. cooling) is strictly fixed by the available cooling power of the cryocooler second stage. Thus, severe steps must be taken to limit heat inputs to the system and a rigorous thermal budget kept.

Practical outcomes of this limited thermal budget typically appear in lower rates of change of insert temperature and magnetic field than available in a "wet" system, to minimise losses. In the latter case, low-loss conductors are frequently used to keep magnet energisation times to a minimum, but with a requisite cost impact.

Initial cooldown time of the system is naturally determined by the cryocooler power relative to static heat loads and the mass to be cooled. This typically results in cooldown from room temperature of the order of 24–48 hours, depending on the type of system, and it can be much longer for very large systems. Some manufacturers have introduced ways to accelerate the cooldown by using precool loops or heat switches between the cryocooler stages.

A significant factor resulting from limited cooling power is that of system recovery time following a magnet quench. Whereas in a liquid helium system, heat generated in the magnet by a quench is absorbed by the liquid and for a large quench typically results in a high volume of helium boil-off, the heat generated in a cryogen-free system has no such way to be absorbed and therefore heats the magnet coils themselves. Specific design and protection of the magnet is necessary to account for this, but the coils can still reach temperatures of several tens of kelvin, and will typically take a number of hours to recool to the operating temperature.

A second important factor is that of *vibration*, due to the cryocooler cold head being mechanically linked to the magnet and insert system. For certain experiments, particularly optical or scanning probe experiments, this can present a significant problem. Certainly, the development of pulse tube refrigerators has been a significant step in overcoming this, by removing mechanical reciprocating displacer. Nevertheless, it would be quite wrong to see the PTR as vibration-free, as anyone with experience of using one knows. There are a number of solutions to vibration, implemented in varying combinations by different system manufacturers.

The cold head itself can often be supplied with a "remote motor" (i.e. one physically detached from the system), which reduces some mechanical vibration and also removes sources of potential electrical noise. However, it seems that the majority of low-frequency vibration arises from the movement of the helium gas through the gas lines from the compressor and in the cold head. Therefore, secondary methods have been employed whereby the mechanical coupling between the main system and the experiment is reduced by mounting the insert or sample stick from a separate anti-vibration mount to the main system, and completing the vacuum path with a thin-walled bellows arrangement. While efficient in reducing vibration, this inevitably adds cost.

Options and variants on cryogen-free operation

It would be wrong to see the choice of cryogenic system cooling as a polarised one between fully "wet" liquid helium and totally "dry" cryogen-free operation. As we have seen above, the strictly limited 4 K thermal budget places constraints upon system design that may not be acceptable for some applications. In fact, many systems, particularly complex magnets, do not lend themselves well to full cryogen-free operation, and the use of multiple cryocoolers would introduce undesirable costs. Further, not all applications themselves are inherently appropriate for completely dry systems. Intermediate options exist; their nature and advantages/disadvantages are explored below.

Recondensing systems

In a liquid helium recondensing system, a 4 K cryocooler is used to reliquefy helium with a cryogen bath similar to a conventional "wet" system, rather than cooling the superconducting magnet or low-temperature insert by direct cooling. Recondensing systems can have helium refill intervals up to several years, depending on the duty cycle of the magnet, and therefore can – depending on circumstances and use – be considered to be effectively cryogen-free, in that no or very infrequent cryogen addition is required. The majority of modern MRI magnet systems operate on this principle, and since the magnet is set in persistent mode the liquid helium refill interval is typically five to ten years (the cryogen consumption is generated by the remaining static heat loads, plus induced boil-off from the gradient coil pulsing). The first stage of the cryocooler is connected to a radiation shield in the same way as the fully "dry" system, removing the need for liquid nitrogen. In some magnet systems, this in itself is a significant advantage (e.g. for beamline applications, as described above).

In the specification and design of superconducting magnets, the benefits of a liquid helium reservoir should not be forgotten: liquid helium offers an excellent temperature-stabilising environment and allows full use of the 4.2 K critical current of the superconductor, critical current enhancement by reducing the liquid helium partial pressure to provide 2.2 K cooling and the widest range of magnet geometries and added features (for example, independent shims, modulation coils, etc.). Where the cooling power of a single 4 K cryocooler may be insufficient to cool a magnet and its associated low-temperature inserts fully, it may be preferable to adopt the recondensing approach rather than add significant costs through multiple cryocoolers.

Figure 13.9 A liquid helium recondensing magnet system for neutron scattering. These systems can have enough cooling power to recondense not only the static boil-off from the cryostat but also the helium gas flow from the VTI (courtesy of Oxford Instruments).

With efficient design of static heat loads, a recondensing system will have an excess of 4 K cooling power to maintain zero static liquid helium loss, and a control loop can be used to prevent the helium bath pressure dropping to sub-atmospheric levels, which would run the risk of sucking in air. Alternatively, with appropriate cryostat construction, the liquid helium can be deliberately allowed to cool to below 4.2 K to provide a greater thermal margin in the liquid. Since the enthalpy of cryogenic liquids can be used, the cooldown of even a large recondensing magnet system can be accomplished quite quickly, at least within 24 hours, where a comparable fully dry system would take five days or more. Depending on application and the usage cycle of the system, this can be a significant consideration. A further advantage of the recondensing systems is that this cooldown can be performed away from the experiment if desired (particularly, for example, on a beamline), and the system then moved fully cold to the experimental location and the cryocooler connected.

Many magnet systems have been supplied using this technology, with applications ranging from magnetic separation to neutron scattering (Lelièvre-Berna *et al.*, 2009; Kirichek *et al.*, 2009). In the latter case recondensing technology has found wide acceptance, as it allows high-field, complex geometry, split-pair magnets (for example, up to 14 T) in combination with a low-temperature insert.

The latest systems have demonstrated that it is possible to realise zero liquid helium boil-off with the magnet operating in persistent mode, including the recondensation of the gas flow (drawn from the main liquid helium bath) from a variable temperature insert (Kirichek *et al.*, 2010), since there is excess cooling power in the cooler over that required for recondensation of the static boil-off.

Vibration in recondensing systems is generally of less concern than that in a conduction-cooled system, as the mechanical linkage between the cryocooler cold head and the experiment is much weaker.

Retrofittable recondensing cryostats

In addition to a priori recondensing system design, cryostats are also available which offer a retrofit solution to existing equipment. In Oxford Instruments' installed base alone there are over 4,000 cryostats/dewars installed worldwide and not all of them can be replaced with new cryogen-free systems in the near term. Specifically designed to be a direct replacement for a traditional "bucket" dewar, with 100 per cent compatibility to existing magnets and low-temperature inserts, these recondensing dewars offer researchers a way to reduce liquid helium costs and reliance, while preserving their existing core experimental equipment and protecting both the financial and intellectual investments made in it – experiments need not be redesigned, and expensive "wet" dilution refrigerators do not need to be replaced (see Figure13.10). Systems of this type retrofitted to an existing magnet and dilution refrigerator have been demonstrated to reduce liquid helium usage by up to 90 per cent.

Figure 13.10 A recondensing cryostat designed to be retrofitted to existing standard magnet/low-temperature insert systems (courtesy of Oxford Instruments).

A further advantage of recondensing technology is the ability to continue to operate the system with the cryocooler switched off. This is of particular significance to those researchers performing very vibration-sensitive measurements such as SPM (scanning probe microscopy) or AFM (atomic force microscopy); in these it is typical to perform the initial sample-screening experiments with the system fully recondensing, and then to switch off the cryocooler for the duration of the most sensitive experiments on "good" samples. The hold time of the liquid helium in the recondensing cryostat is entirely sufficient to enable this for several hours.

Minimum-volume condensing systems

A further development of the recondensing principle has been to cool the superconducting magnet with only a very small "skin" volume of helium condensate around. Such a system is designed to obtain condensation of liquid helium directly in the system from a helium gas reservoir, and thus to the user the system is truly cryogen-free, never requiring liquid helium filling, despite liquid being in the system. By using the liquid helium condensate to distribute "cold" around the system, more complex geometries of magnet are enabled than might be possible with a purely dry system. The disadvantage of this technology lies in the cooldown time to condense the liquid and cool the system, and in the necessary footprint to accommodate the helium gas reservoir.

Compact helium reliquefiers

In addition to "built-in" cooling, a number of manufacturers offer compact mini-liquefiers, which can reliquefy the helium gas exhaust from one or two experimental cryostats. These have the advantage of maintaining the whole existing experimental set-up, but require additional space in the laboratory compared to retrofitting a recondensing dewar, as described above. Nevertheless, if such a cryostat is not available for the equipment, such mini-liquefiers can offer a practical solution.

System autonomy considerations

In selecting or designing any cryocooled magnet system, due consideration must be given to the effect of plant failure on both the magnet and process in which it is operating. A cryogen-free magnet will quench in a few minutes following power failure to the cryocooler compressor, while a magnet with minimal helium condensate will quench in tens of minutes to hours, depending on the exact conditions. However, a recondensing system may last for days or weeks before quench. These relative merits should be balanced against the impact of delays to research, and the costs of supporting devices such as an uninterruptible power supply (UPS).

Small cryogenic systems

While the above has focused on cryogen-free technology as applied to the larger scale of research systems, with superconducting magnet and low-temperature or dilution refrigerator combinations, there is a significant market for smaller cryostats, predominantly but far from exclusively for cold optical measurement techniques spanning wavelengths across ultraviolet and visible (UV, UV-Vis), infrared (NIR, IR, FIR) and terahertz (THz) applications. These tend to be compact cryostats as dictated by the need to integrate onto an optical bench, and to maintain a short working distance from the outer vacuum windows to the sample position. Many cryostats of this type operate on a continuous flow of liquid helium, to avoid the need for a large cryogen bath. Cryogen-free systems are particularly suited to offering a short working distance, as the outer vacuum chamber can be brought close into the experimental stage since there are no liquid baths to accommodate.

The drivers towards cryogen-free operation are the same as with all systems, with perhaps even more emphasis on the need for simplicity of operation (as these products are frequently used outside the traditional physics department environment for cryogenic equipment), and the need for compactness since the access for liquid helium storage dewars in optical laboratories tends to be particularly limited.

A very wide range of cryocooled products is available, with base temperatures ranging from 77 K down to 4 K (Figure 13.11). At their simplest these offer a sample-in-vacuum mounting within the outer vacuum chamber space, with the sample mounted on to the second stage (or in some higher temperature options, the single stage) of the cryocooler and a heater allowing control of sample temperature from cryocooler base up to room temperature or above. More complex systems offer exchange gas environments or the full capability of a variable temperature insert, typically ranging from 1.5 K to 300–400 K.

Figure 13.11 Small cryogen-free optical spectroscopy cryostats, with options of sample in vacuum (left) or sample in exchange gas (right).

In these products vibration is of course a concern in the same way as highlighted for conduction-cooled systems above. Again, various methods are employed to overcome this, in some products to the extent of wholly separating the cryocooler from the experimental stage and using cold helium gas as a heat-transfer medium between the two units.

A significant cost challenge lies in these products, as the cryocooler cost becomes wholly dominant in the product cost, in an otherwise inexpensive cryostat. These cryostats are seen as the ideal research tool by many scientists in developing economies, enabling them to enter areas of research to which liquid helium would have previously raised an insurmountable barrier. There is therefore significant cost-reduction effort to be made on such products.

Cryogen-free ^{3}He and dilution refrigerator integrated systems

Arguably the most significant advance in cryogen-free technology in recent years has been its extension to sub-1 K cooling through cryogen-free ^{3}He and ^{3}He/^{4}He dilution refrigerators. In these, rather than using liquid helium to provide cooling for the inner vacuum chamber (IVC), the second stage of a PTR is used. The principles of operation are broadly identical to those of traditional "wet" ^{3}He and dilution refrigerators (DRs), with the key components of 1 K pot, still and mixing chamber present in the dilution unit. Standard commercial DRs provide cooling to <10 mK and cooling powers typically up to 400 μW at 100 mK, with custom-designed DRs extending this range (Figure 13.12).

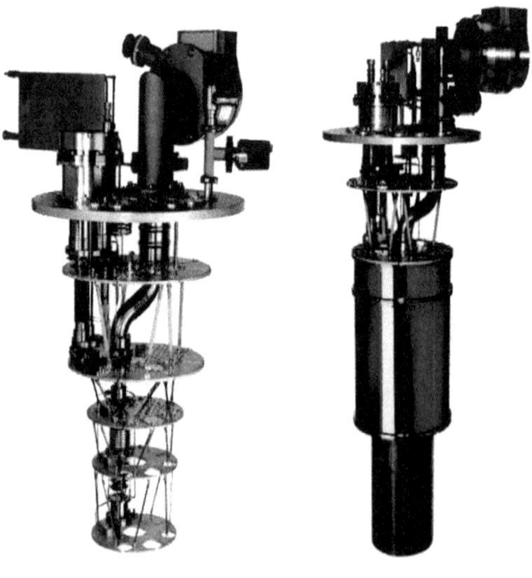

Figure 13.12 A cryogen-free dilution refrigerator, offering base temperatures below 10 mK, and cooling power of 200 μW at 100 mK, shown (right) with a fully integrated 12 T superconducting magnet (courtesy of Oxford Instruments).

Yet more recently, fully integrated cryogen-free DR and magnet systems have been made available (Batey *et al.*, 2009), replicating all but the most extreme field-temperature combinations previously only available "wet". This capability is being continuously extended by manufacturers, not only in cooling power and available magnetic field but in configurations including vector rotate magnets (allowing the tilt and/or rotation of the magnetic field vector), complex wiring options and UHV compatibility. Quantum computing research is a very strong driver of the capability of these instruments, with increasing processor complexity and multiplicity requiring ever more complex wiring looms, with the requisite implications for thermal budget as discussed above.

In common with other cryogen-free products, these ^{3}He (Figure 13.13) and dilution refrigerator systems are making new areas of research available, both to established and emerging scientific communities, by reducing the knowledge barriers associated with setting up and running such a complex piece of equipment and by offering increasing automation of the operation of the experiment. By removing the need for liquid cryogens, these systems have become more compact – particularly in relation to the ceiling height required, making them easier to locate in laboratories without the need for high ceilings or pits in which to sink the cryostat. The adoption of cryogen-free dilution refrigerator technology has been extremely rapid, moving in only two to three years from a predominantly "wet" market to one that seems likely become completely dry in the near future.

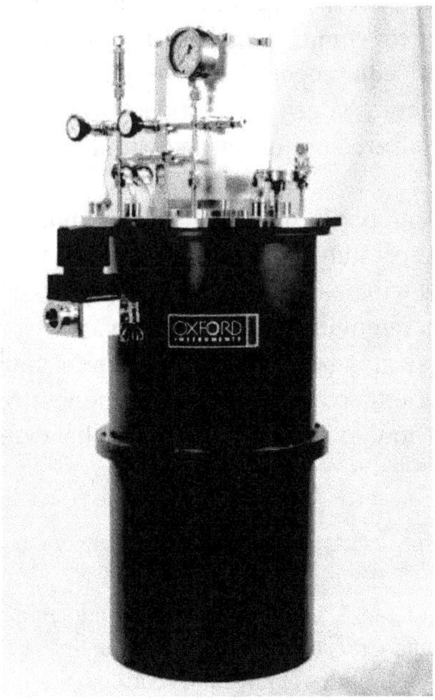

Figure 13.13 A compact ^{3}He cryogen-free system, providing sample cooling to below 300 mK (courtesy of Oxford Instruments).

^{3}He supply issues

As a final note on in our discussion of how the research community and cryogenic equipment manufacturers have responded to the pricing and supply pressures of liquid helium, it is essential to highlight the simultaneous and even greater pressure now being experienced by the subset of researchers reliant upon the ^{3}He isotope for the operation of sub-1 K and dilution refrigerators. With growing demand from the security and homeland defence sectors as a neutron detector for "dirty bomb" material, as well as increasing production costs (^{3}He being produced as a by-product from nuclear reactors), the price of ^{3}He has risen by at least an order of magnitude in the past two to three years, and scarcity of supply is affecting many researchers and even some manufacturers. The scarcity of supply versus demand has been exacerbated as the US has moved to privatise its ^{3}He reserves. Even at prices now in excess of US$3,000 per litre, researchers frequently find themselves unable to source sufficient quantities for existing or new equipment.

Much lobbying continues by the scientific community users – mainly from neutron beamline facilities and the low-temperature community – to ensure that sufficient supply is made available for research. At the same time, manufacturers are working to reduce the volumes of ^{3}He required by their products.

Summary

In the above, we have seen the immense pressures being experienced by those members of the low-temperature research community who are reliant upon liquid helium to run their experimental equipment. However, we have also observed the opportunities offered by new cryogen-free products, which have been developed by manufacturers to overcome or at least mitigate those pressures.

Driven partly by economic necessity and partly by active product innovation, a new generation of cryogen-free low-temperature equipment has grown to a significant level of maturity. These products will enable new science and wider participation in low-temperature research by removing barriers to entry, particularly in economies where liquid helium has always been prohibitive. The supply of liquid helium seems unlikely to become easier or cheaper, but – supported by appropriate funding – the opportunities now exist to remove our reliance upon this finite commodity.

References

Ballarino, A and Carrillo, A. (2008) "Current Leads", Lectures on Superconductivity, ed. B. Glowacki, Applied Superconductivity and Cryoscience Group, University of Cambridge, 7 October, www.msm.cam.ac.uk/ascg/lectures/applications/currentleads.php

Batey, G., Buehler, M., Cuthbert, M., Foster, T., Matthews, A. J., Teleberg, G. and Twin, A. (2009) "Integration of Superconducting Magnets with Cryogen-Free Dilution Refrigerator Systems", *Cryogenics*, Vol. 49(12): 727–34

Burgoyne, J. W., Daniels, P. D., Timms, K. T. and Vale, S. H. (2000) "Advances in Superconducting Magnets for Commercial and Industrial Applications", *IEEE Transactions on Applied Superconductivity*, Vol. 10: 703–9

Kirichek, O., Evans, B. E., Down, R. B. E. and Bowden, Z. A. (2009) "Cryogen Free Low Temperature Sample Environment for Neutron Scattering Experiments", *Journal of Physics: Conference Series*, Vol. 150: 012022

Kirichek, O., Down, R. B. E., Kouzmenko, G., Keeping, J., Bunce, D., Wotherspoon, R. and Bowden, Z. A. (2010) "Operation of Superconducting Magnet with Dilution Refrigerator Insert in Zero Boil-Off Regime", *Cryogenics*, Vol. 50: 666–9

Lelièvre-Berna, E., Brown, J., Jones, H., Kouzmenko, G., Losserand, O., Pickering, P., Tonon, X. and Turc, S. (2009) "Zero-Boil-Off Cryomagnets for Neutron Scattering", *Journal of Physics: Conference Series*, in preparation

14 Helium and nuclear fission energy

Richard Stainsby

Gases have been used as reactor coolants since the earliest days of nuclear power. Early reactors, such as the Windscale piles in the UK, intended for production of weapons-grade plutonium, used atmospheric air for the coolant in an open cycle. These reactors used air at atmospheric pressure and the flow rate was sufficiently large to limit the temperature rise and thus limit the rate of oxidation of the core graphite. This meant that the main blowers had to pump a large volume of low-density air, with the consequent high power demands that this imposed. For a power reactor the coolant temperature rise has to be greater to achieve a reasonable thermal efficiency, and the pumping power has to be minimised. The pumping power is minimised by increasing the density of the gas by containing it in a pressurised closed circuit. However, there are two main problems with using high-pressure, high-temperature air as a (thermal) reactor coolant. First, nitrogen is a strong neutron absorber, which in the process of absorbing neutrons produces radioactive carbon-14. Second, a gas-cooled thermal reactor contains a graphite moderator, and this will oxidise rapidly at high temperatures. Similarly, metallic structures within the reactor will be oxidised more rapidly as the coolant temperature is increased.

For the first generation of gas-cooled power reactors, the gaseous coolant of choice was carbon dioxide. CO_2 is heavier than air and its density is increased further by holding it under pressure within a closed circuit. Many CO_2-cooled thermal reactors have been built and many are still operating – over 1,000 reactor-years of operating experience has been gained in the UK alone with CO_2-cooled thermal reactors. In the first generation of CO_2-cooled reactors, for example the Magnox reactors in the UK, the coolant temperatures were sufficiently low for the CO_2 to be chemically stable. In the second generation, the advanced gas-cooled reactor (AGR), the coolant outlet temperature was raised to 650 °C. At these temperatures CO_2 dissociates into carbon monoxide and free oxygen, in the presence of gamma and neutron radiation (radiolytic dissociation). The free oxygen oxidises the graphite and reactor structures aggressively, so coolant chemistry must be controlled to limit this oxidation. In AGRs, methane (CH_4) is added to the coolant, which itself dissociates into carbon and hydrogen – the hydrogen then combines with the free oxygen to produce water vapour and the carbon deposits back onto the graphite to replace the carbon lost by oxidation. At higher temperatures still, the rate of dissociation of CO_2 in a reactor core becomes unacceptable.

To alleviate the problem of radiolytic dissociation associated with carbon dioxide, high-temperature thermal reactors (HTRs) use helium as their gaseous coolant. A number of HTRs have operated, such as Dragon (UK), AVR and THTR-300 (Germany), Peach Bottom and Fort St Vrain (US), HTR-10 (China) and HTTR (Japan), so a significant amount of operating experience with helium-cooled reactors has been accrued. HTRs have coolant outlet temperatures in the range of 750–950 °C, which allows both high-efficiency electricity generation and the production of high-quality process heat.

All of the power reactors in use today are "thermal" reactors, in which heat is generated from the fission of uranium-235 by means of low-energy or "thermal" neutrons. Uranium-235 is a very finite resource in that it constitutes less than 1 per cent of natural uranium. For nuclear fission to be considered sustainable, the utilisation of natural uranium has to be improved considerably. A practical improvement by about a factor of 60 is achievable through the breeding of plutonium from the more abundant isotope uranium-238, and burning of this plutonium in a class of fission reactors known as "fast" reactors. Fast reactors make use of fission resulting from high-energy, or fast, neutrons. As such, fast reactor cores are very compact to preserve the neutron energies and by minimising the amount of structural and coolant materials with which the neutrons can interact. Such compact cores require a very effective coolant to transport the heat – traditionally this has been a liquid metal, such as sodium. However, despite its excellent heat-transfer performance, handling and neutronic problems associated with sodium serve as drivers to develop an alternative system. One such system is the gas-cooled fast reactor.

Gas-cooled fast reactor concepts, in which both carbon dioxide and helium have been proposed as coolants (as well as steam and dinitrogen tetroxide), have been explored since the 1960s. Generally for moderate temperatures CO_2 was considered to be an acceptable coolant, whilst the majority of the concepts, and specifically those with high outlet temperatures, were helium cooled. Interestingly, the CO_2 dissociation problem is less of an issue in gas-cooled fast reactors because there is no graphite to oxidise; however, there would still be a problem with metallic components.

Current high-temperature reactor projects

Historically, commercial-scale helium-cooled HTR projects were pursued in Germany and the USA. Whilst the German programme ended in the 1990s, activity continued in the USA, which in collaboration with Russia led to the development of the gas turbine modular high-temperature reactor (GT-MHR) concept. This reactor developed the prismatic fuel element core of Fort St Vrain into a small modular design that generates electricity using a direct-cycle helium gas turbine. The German reactors were based on an alternative fuel concept: the pebble bed. In such a reactor, the core is filled with a large number of randomly packed fuel spheres, with each sphere being slightly smaller than a tennis ball.

After pebble bed reactor development ceased in Germany the technology was developed further by South Africa, in the pebble bed modular reactor (PBMR) project, and by China, first as the HTR-10 experimental plant, followed by the commercial-sized HTR-PM (high-temperature reactor – pebble modular). The original intention was for the PBMR to drive a direct-cycle helium gas turbine; however, recent changes to the design have downrated the unit and now propose coupling with a steam cycle. The intention for HTR-PM from the outset was to generate electricity using a steam cycle.

Within Europe HTR development has continued in the form of the ANTARES reactor from AREVA. The ANTARES reactor is based upon the GT-MHR, but in this instance is coupled with an indirect-cycle gas turbine using a helium-nitrogen mixture as the working fluid, with a steam turbine making use of the waste heat from the gas turbine exhaust.

Japan has developed a variant of the block-fuelled prismatic core concept in which pin-in-block fuel elements are used. An experimental reactor, the high-temperature test reactor (HTTR) has been constructed and is currently in operation. The GTHTR-300 concept extends the technology up to a 300 MWth plant driving a direct-cycle helium gas turbine.

In all of the above, the reactor cores are cooled by high-purity helium. The main difference between the direct and indirect cycle plants is that the same high-purity helium is used to drive the turbine in the former, while in the latter there is freedom to use helium, other gases or gas mixtures or water/steam. Further, in the direct-cycle plants, the need to maintain a high level of purity in the helium precludes the use of submerged oil-lubricated bearings in the turbomachinery, resulting in the need for either submerged electromagnetic bearings, or external oil-lubricated bearings together with suitable dry gas seals. A further complication is the transport of graphite dust from the core to the turbomachinery and, conversely, the transport of wear and erosion products from the turbine through the reactor core. The indirect-cycle plants avoid many of these problems, at the expense of requiring large intermediate heat exchangers and gas circulators for the primary, reactor core, cooling circuit. The lubrication of the gas circulator bearings should, ideally, be oil-free, but because the gas-circulators are much smaller than power conversion turbomachinery, the bearings are much lighter, simpler and achievable using existing technology.

Generation IV advanced reactors

The Generation IV initiative was launched by the United States Department of Energy (USDOE) in the early 2000s as the basis for a global collaboration on the development of the next generation of nuclear fission power reactors. Generation IV is so named because all the power reactors in operation in the world and the advanced light water reactors, both proposed and currently under construction, have been deployed in three waves, or three generations, with the technology of each generation being a significant step forward from its predecessor. The high-level goals for Generation IV systems were laid down ahead of the selection of the most promising systems that were most likely to fulfil those goals, and are summarised as:

- improved safety;
- improved economics;
- minimisation of wastes;
- proliferation resistant;
- sustainable.

Six systems were proposed for further development and these systems were selected on the basis of their potential for contributing to the fulfilment of the above goals. Of these systems, three are fast reactors and the remaining three are thermal or epithermal spectrum reactors, although, more recently, fast-neutron spectrum variants for two of these have been proposed.

With regard to systems that require helium, two of the six systems – one fast reactor and one thermal reactor – require helium as the coolant for the reactor core (primary coolant). Of the others, for those that are capable of operating at high temperatures, options are proposed in which helium is used in a secondary circuit to drive a gas turbine. Here, consideration will be given only to the systems that rely on the availability of helium to provide their primary coolant, and it is simply assumed that alternative working fluids can be deployed in the secondary circuits if the availability of helium becomes an issue. Specifically, the systems in which the reactor core is cooled by helium are the very high-temperature reactor (VHTR) and the gas-cooled fast reactor (GFR).

The VHTR (Figure 14.1) is an extrapolation of existing high-temperature reactor technology, aiming at core outlet temperatures approaching 1,000 °C. Outlet temperatures in this range enable electricity to be generated with high efficiency using closed-cycle gas turbines (either direct or indirect) and the production of high-quality process heat, suitable, say, for the thermochemical splitting of water for the carbon-free production of hydrogen on an industrial scale.

The Generation IV GFR system (Figure 14.2) aims to capitalise on fast-reactor and high-temperature experience. As such, the aim is to derive a system that avoids the problems associated with liquid metal coolants, offers the same performance and range of applications as a high-temperature reactor, but is suitable for deployment long into the future as it is based on a plutonium fuel cycle. A further, and significant, advantage of fast reactors in general is their ability to fission the long-lived radionuclides (specifically the minor actinides – neptunium, americium and curium) that exist in nuclear waste as a result of fast neutrons – both in their own wastes and in legacy wastes arising from earlier generations of thermal reactors. In this respect, the gas-cooled fast reactor is particularly effective owing to the larger proportion of high-energy neutrons present in its neutron spectrum compared with a sodium-cooled reactor. Stated simply, this is because the number of coolant atoms per unit volume is smaller in a gas-cooled core compared with a liquid sodium-cooled core; therefore, the amount of moderation, or the slowing down of neutrons, is smaller in the former.

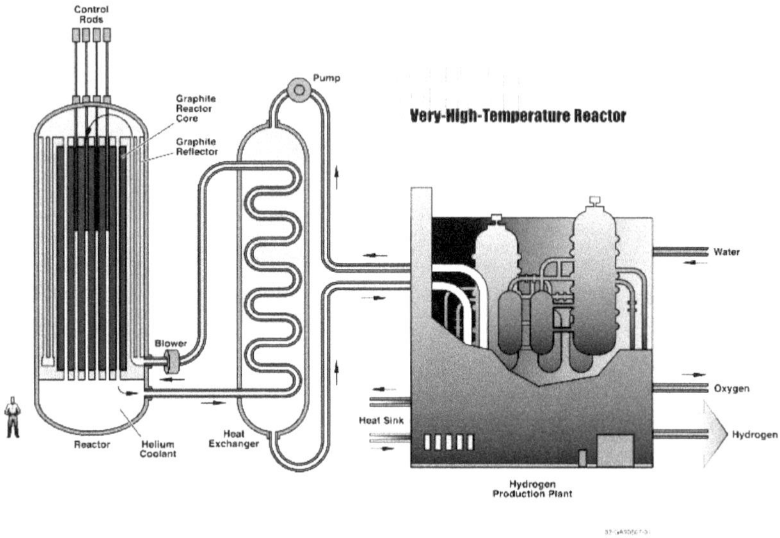

Figure 14.1 Generation IV International Consortium: very high temperature reactor (for process heat applications) (source: Idaho National Laboratory).

Of the available gaseous coolants, helium offers the best performance in terms of high-temperature chemical stability and a high level of nuclear stability. The gas atoms have only a small influence on the neutronic behaviour of the core and, as a coolant, it does not give rise to the production of radioactive nuclides during neutron irradiation. Other gases have advantages in terms of offering higher density for a given pressure and temperature, resulting in lower pumping power and better natural convection performance, but either their chemical and/or nuclear stability are much reduced.

Helium demand

The precise helium demands for Generation IV systems and nearer-term high-temperature reactors are not known with any certainty. The main factors affecting the helium demand will be the helium inventory of a single reactor unit, the rate at which new systems are rolled out and the rate of helium leakage expected.

With regard to the helium inventory of a single reactor, we can base an estimate on current proposed designs for HTRs. The direct-cycle PBMR version in South Africa is a 400 MWth driving a direct-cycle gas turbine with a helium inventory of about 4 tonnes. In China, HTR-PM is a 250 MWth reactor driving an indirect steam cycle and has a helium inventory of 2.44 tonnes. From these values, it is observed that both of these reactors require approximately 1 tonne per 100 megawatts of thermal power, or 25 tonnes per gigawatt of electricity, assuming an overall efficiency of about 40 per cent.

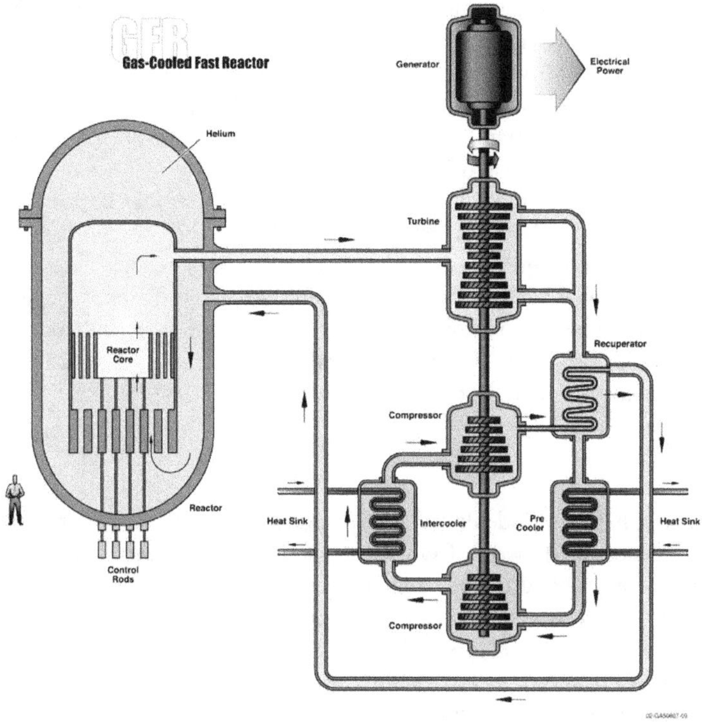

Figure 14.2 Generation IV International Consortium: gas-cooled fast reactor (source: Idaho National Laboratory).

The rate of roll-out of HTR systems is not certain, but an order-of-magnitude estimate can be made by looking at the size of the nuclear element in the UK electricity supply. The UK produces 20 per cent of its electricity through nuclear power (about 10 GW) – if this was replaced by HTRs we would need 250 tonnes of helium for the initial charge of the reactors. If all of the UK's electricity came from HTRs, the UK would need 1,250 tonnes of helium for the initial charge. However, HTRs are being developed for their non-electricity uses, such as the synthesis of hydrogen, synthesis of methane and liquid fuels; therefore, the ultimate demand for HTRs should exceed the numbers required just for electricity production. This raises an interesting question: when we live in a world in which we have to synthesise methane, suggesting that natural gas is not available or uneconomical to extract, where will the helium come from?

There is a certain amount of historical data regarding helium leakage rates from HTRs. The target leakage rate for Dragon was 0.1 per cent of inventory per day, although this was not measured directly. Indications from individual component tests that the actual rate was possibly very much lower at 3×10^{-7} per cent per day. Commercial plants will have more leak paths than an experimental reactor – e.g. more shaft seals, and more cable and pipework penetrations, plus more fuel charging and discharging operations. The Dragon limit of 0.1 per cent per day suggests that the inventory is replaced every 1,000 days (approximately every three years). For a 60-year reactor life, the helium inventory would be changed 20 times during the plant's lifetime. Hypothetically speaking, to satisfy the *current* UK electricity demand over 60 years could require up to 25,000 tonnes of He (say, between 2020 and 2080).

The electricity demand in the UK is dwarfed by that in rapidly expanding China. Development of HTR-PM is proceeding well, with construction work having started in 2009, and deployment is expected to be widespread.

At the moment the gas-cooled fast reactor concepts are not sufficiently developed to estimate their helium requirements rigorously. However, the higher power density of large gas-cooled fast reactors should require only about 25–50 per cent of the inventory of HTRs for the same power output.

Conclusions

Gas-cooled reactors have been deployed on an industrial scale for over half a century. However, most of these are CO_2-cooled thermal reactors with moderate operating temperatures. The drive towards higher operating temperatures, for improved efficiency and to widen the range of applications beyond electricity generation into process heat and synthesis of hydrogen and hydrocarbons, requires a coolant with good chemical and nuclear stability. Helium fits this requirement well and is the coolant of choice for high-temperature thermal reactors (HTRs)

Gas-cooled fast reactors are a route by which HTR technology can be made more sustainable by utilising natural uranium reserves more effectively, whilst minimising the volumes and radiotoxicity of wastes. Again, the chemical and nuclear stability of helium makes it a good candidate for the coolant in these systems.

The Generation IV initiative was established to formalise international collaboration on the development of advanced reactor systems. The very high-temperature reactor and the gas-cooled fast reactor are two out of six systems being developed within Generation IV and the two systems in which helium has been chosen as the reference gaseous coolant.

The helium demand of Generation IV systems will be large and the questions to be asked are:

1 Can the helium supply industry cope?
2 One of the main drivers for Generation IV systems is to lessen dependence on hydrocarbons or to provide energy at a time when hydrocarbons are not available (and before fusion is available commercially) – where will the helium come from in this future?

15 Helium and fusion energy

Richard H. Clarke and Zhiming Cai

An aim of this book is to identify potentially significant users of helium in the medium and long term. What distinguishes this helium application from the others is both its potential scale and its timing. As we shall see, despite the large research efforts going on around the world, the first fusion power-generating reactor is not expected before 2035 and more probably 2040. This places a potentially growing demand for helium on the *downside* of the production curve described in Chapter 10. What this demand will be, of course, is conjecture, but this chapter describes three fusion helium demand scenarios, of which two would appear to be unworkable. We argue that this challenge can be met – but only if the problem is continuously and logically addressed from now on.

In the first part of the chapter we examine what fusion is, the scale of what is being done in the research arena, and when and how likely it is that a fusion power plant will be built. In the later sections we consider the potential rate at which further reactors might be built, how the helium demand of individual reactors arises and how successive generations of machine might become more efficient.

What is the aim of fusion energy research?

The attraction of fusion is that, in one sense, we know it works. It is the process that powers the stars, and ever since this was first understood in the early twentieth century scientists have been fascinated with the idea of harnessing the tremendous power source that transmuting, or fusing, light elements might bring (Eddington, 1920). The "Atoms for Peace" movement initiated by the United Nations was instrumental in bringing together researchers – physicists, engineers and scientists – from many nations (Herman, 1990; Braams and Stott, 2002).

Today, this work continues in many nations and through international, collaborative programmes. The Culham Science Centre near Oxford (England) is the home of the world's largest fusion experiment "JET" (Joint European Torus). It began operating in 1983. As the name implies, it is a joint project of the European Community. JET is a fusion device of the "tokamak" type. The word tokamak derives from a Russian acronym, kamera s aksial'nym magnitnym polem, meaning *toriodal chamber with an axial magnetic field.*

Fusion researchers' efforts are focused on certain, known, thermonuclear reactions, which – it has been demonstrated – make much more energy than is put in to get the reaction going. There are two major approaches: magnetically confined fusion (MCF) in tokamaks, and inertially confined fusion (ICF). ICF is considered later. A variant of MCF is the stellarator (Hirsch *et al*., 2008).

MCF involves heating two isotopes of hydrogen to around *200 million ℃* and holding enough atoms together in a tokamak chamber for long enough for a reaction to take place. For physics reasons this is best achieved under high vacuum conditions, not least to ensure that air (a contaminant) is excluded from the chamber. At the temperatures involved, the atoms are fully ionised and in the form of a plasma. To make MCF fusion work, such high temperatures are needed to give the plasma ions enough energy to overcome the coulomb force that makes the hydrogen nuclei repel one another. Only when the ions approach closely will the fusion reaction occur. John Lawson identified the required conditions early on (Lawson, 1957) – the so-called "triple product". The progress towards it has been, relatively, as fast as Moore's law, which has been central to the microelectronics revolution. Figure 15.1 illustrates the most promising (lowest-temperature) reaction, that between deuterium (a stable form of hydrogen, with one neutron, that has a natural abundance in seawater of one part in 6,700) and tritium (a radioactive form of hydrogen, with two neutrons, that has a half-life of 12.3 years and consequently has a very low abundance in nature).

The neutrons produced escape the tokamak and bombard a "lithium blanket", designed but not yet tested, that surrounds the reactor vacuum vessel to produce more tritium. Figure 15.2 shows a schematic representation of a fusion power plant (Llewellyn-Smith, 2008).

The fissionable and fusionable energy resources of the Earth are potentially many times greater than those of coal, oil and natural gas combined (Hubbert, 1971). If fusion were to be developed, then there could be an abundant energy

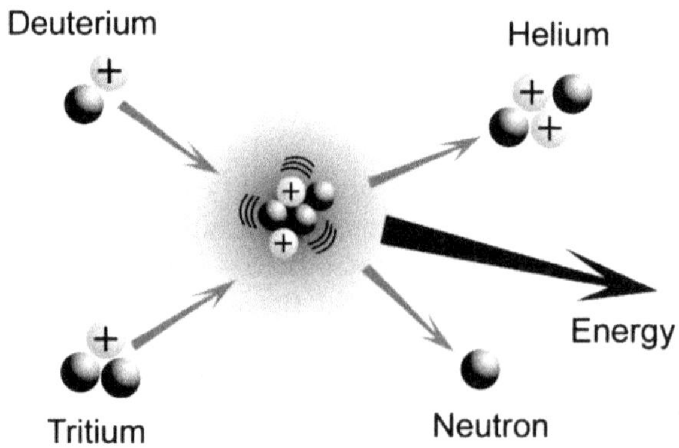

Figure 15.1 Deuterium-tritium fusion reaction; although helium-4 is produced, the quantities are negligible compared to a fusion reactor's needs (credit: Lancaster University).

supply for millions of years. Beyond electricity and transporation supplies (hydrogen?), there might be sufficient energy to recover the precious minerals and fresh water resources that seem to be central to a technological society.

These ideas excite serious commentators:

> Even if ITER [the US$10bn next generation fusion machine after JET] runs well over budget, its spending level is unlikely to exceed $1bn a year, a small price to pay for a reasonable chance to give the world another energy option for a time when it will no longer be possible to burn fossil fuels on the profligate scale of the early 21st century.
>
> (*Financial Times*, 2004)

To get to where we are today has taken a huge amount of research effort. This can be partly attributed to the realities of confining plasma. There have been many challenges, such as instabilities, magnetohydrodynamics and turbulence. Most of the issues have been addressed and control methods have been developed, at JET (Keen, 1987) and elsewhere, to obtain the desired conditions – principally to avoid the plasma touching the wall of the tokamak. A sense of the scale of the work that has been required can be gleaned by comparing the early JET with what it is today (Figure 15.3), a US$100 million per year research programme.

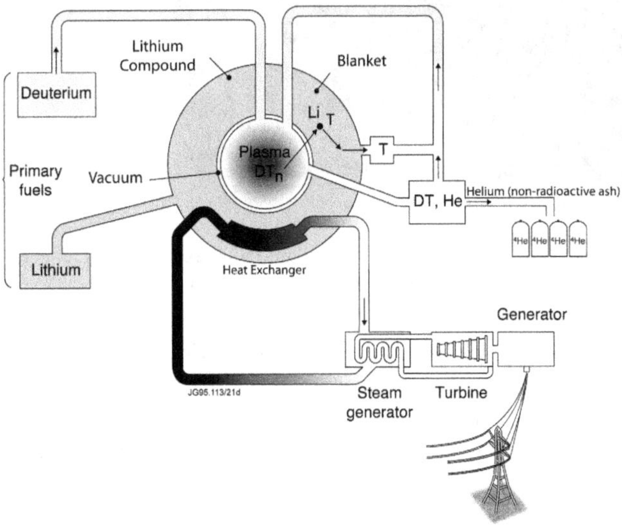

Figure 15.2 A fusion power station is conceptually similar to an existing thermal power station but with a different furnace and a different fuel. The figure is not to scale. In reality, the fusion core would be a very much smaller part of the whole power station, and the "blanket" would be about 1 m thick, while the plasma (which would be contained in a tokamak chamber) would occupy 1,000–3,000 m^3 (Llewellyn-Smith, 2008).

(a) (b) (c)

Figure 15.3 Progress at JET. (a), (b) The machine in 1983; (c) The machine in 2010 (credit: United Kingdom Atomic Energy Authority).

However, even if the physics is now within grasp, serious engineering issues remain. What materials should face the plasma, inside the tokamak? They must not poison the plasma with impurities, nor must they disintegrate under the extreme conditions of temperature (heat flux) or integrated neutron flux (fluence). How, at full scale, will the plasma be heated to the extreme temperatures required? Should the reactor be pulsed or continuous? What can be done to prevent or manage fuel retention inside the tokamak structure, not least to ensure that the delicate supply–demand balance for tritium is not upset? Current thinking is to build the plasma facing wall – the first wall (see Figure 15.4) – out of beryllium/tungsten tiles, and so the inside of JET was thus equipped in 2011 in preparation for ITER, which will become the world's largest fusion machine by 2020. As of 2012, the new wall's performance has been very good. ITER is being built

Figure 15.4 Inside JET, 2011: the newly installed beryllium/tungsten plasma facing tiles are designed to be resilient against plasma-induced erosion. The present chamber is about 5 m high × 9 m diameter. A robotic arm (left) was used to install tiles (credit: United Kingdom Atomic Energy Authority).

in Southern France under a global collaboration and it is designed to demonstrate a sustained fusion power output of 500 MW (thermal), ten times the power that will need to be input. In parallel with ITER, a whole raft of materials testing and development work will be conducted, ultimately paving the way to DEMO – the first fusion power station which, it is anticipated, will be connected to a grid around 2035–40.

So far we have focused mostly on technological issues, yet there is no doubt that the rate of funding for fusion is a fundamental determinant of progress. Llewellyn-Smith (2008) notes that fusion is not alone in having seen a stalling of funding growth; the level of energy R&D funding worldwide, in real terms, is half what it was in 1980. Interestingly, in 1978, the US helium administration was tasked under a US Department of the Interior contract to explore the interrelationship between funding, fusion R&D (both MCF and ICF) and a significant future demand for helium. Figure 15.5, which was used in the US DoI report (Ferrer, 1978), suggests we are currently operating somewhat below "Logic II", given the current 2035–40 estimate for DEMO's start up. The risk here is that if the overall fusion budget is trimmed still further the timescales would tend to flatline, as in "Logic I".

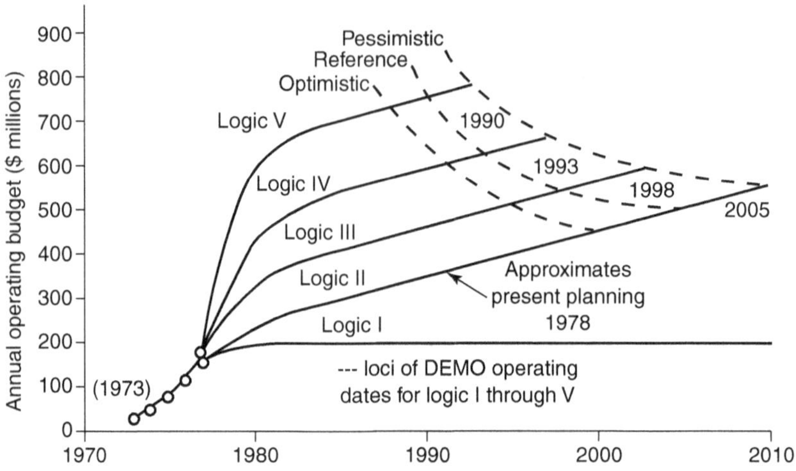

Figure 15.5 1976 US Energy Research and Development Administration (now part of the US Department of Energy) estimates for the start up of DEMO indicated that a critical level of funding is needed to achieve the project's delivery (Dean, 1998). Logic levels: I (physics interest only) … V (fast track).

Helium use in fusion reactors: the potential scale of demand

With the system dynamics model of helium resources and market developed by Zhiming Cai (Cai *et al.*, 2010) in place, we saw that it could be utilised to assess the potential impact of rolling out a generation of fusion power plants. It soon became clear that fusion's needs are neither trivial nor are they (yet) fixed in stone. If fusion researchers deem helium to be one of their "must-have" materials then the fusion community needs to take this issue seriously – it is both long term and high impact.

While this same community could, or perhaps should, take an active interest in helium resources, the latest supply developments and market trends (prices, resources), it is unlikely – even if it were possible – they would want to intervene. Conservation may save money or enable hedging against short-term supply disturbances. That is economically rational and achieves technological learning ("leak learning", as we call it). A rational, well-managed transition is needed so that plant and processes are designed to use a *needs-based quota* of what is predicted to be available. It is here that this chapter may be of value to policy-makers and technologists in the fusion and energy communities.

The components of helium use in an MCF reactor

By exploring the components of a fusion reactor's helium requirements it is possible to build several scenarios to investigate the possible impacts of future helium demand from fusion on helium market dynamics. In doing so, we suggest steps that might be adopted by the fusion community to overcome the potential constraints to the future development of fusion power technology imposed by limited, long-term helium resources. The prospect of a backstop resource – helium from air, as examined in Chapter 8 – turns out to be rather important.

First, let us consider helium production in a fusion reactor. The reaction shown in Figure 15.1 might produce around 80 Mcf (~400 kg) helium "ash" per gigawatt-year. This includes the side reactions that occur when neutrons hit the lithium blanket. Even in the best-case scenario we shall set out, this production rate is negligible. The helium ash would be extracted in a specialised gas processing plant that separates deuterium, tritium, helium and other impurities and recycles the deuterium and tritium for further use. Both the separation process and the re-injection process (e.g. via frozen pellets of deuterium and tritium) involve the use of cryogenic systems requiring helium gas as a refrigerant. There is no currently proven alternative if helium ash is to be successfully removed from the reactor, as it must be.

Equally important, the plasma needs to be kept pure to avoid the cooling effect of thermal radiation by contaminants. A very high vacuum (~10^{-7} mbar) must be achievable in the tokamak vessel. Large cryopumps (essentially bare or carbon-coated cold plates onto which gases freeze or adsorb) use cold helium (at about –269 °C) to remove gaseous impurities. It may be possible, as shown in Chapter 16, to mimimise helium requirements using an *indirect hydrogen* (i-LH_2) concept but in any case some helium is required. Fortunately, low-temperature helium circuits are normally very leak-tight; they have to be for the refrigeration system to work properly.

Apart from the main (divertor) cryopump, which captures the helium ash, there may be other large cryopumps in the so-called additional heating (or "current drive") systems. These subsystems enable the plasma to be heated to the ~200 million degrees Celsius needed to sustain fusion reactions. There may also be high-pressure, –196 °C helium-filled shields, which are externally cooled with liquid nitrogen, to reduce the thermal load on the lower-temperature helium circuits (see Table 15.1 for an explanation).

In ITER (Nuclear Fusion, 2007), large quantities of cryogenic helium will be used in the niobium poloidal and toriodal magnet systems that confine and control the plasma shape. While it is expected that new, high-transition-temperature (high T_c) superconducting magnets will be developed by the time fusion becomes commercialised, so far this has been found to be a challenge. The magnets have to achieve a high field (in the range of 6–13 Tesla or more), high current density (J) and be resistant to radiation (neutrons) and the huge dynamic forces exerted on the large magnets. An achievable goal may be to operate high-T_c magnets at –253 °C (where the i-LH_2 loop, described above, might be employed) and then perhaps –246 °C (liquid neon, inert but recoverable from air, as described in Chapter 8). It is doubtful that higher temperatures will be used, as none of the known superconducting materials meet all the above criteria at, say, the boiling point of liquid nitrogen, –196 °C.

In the fusion reaction, the *extractable* energy appears mainly in the form of neutrons with high kinetic energy which convert into heat in the tritium-breeding blanket. This has to be transported away by a heat-transfer medium. High-pressure helium would be suitable, as Richard Stainsby shows in Chapter 14, since the gas is unaffected by neutrons and is unreactive. Indeed, the leading blanket concepts in Europe use helium cooling (Poitevin *et al.*, 2010). A helium-water heat exchanger could be used to produce steam, which would be used to drive a conventional turbine generator set to produce electricity. Alternatively, an advanced Brayton cycle (Figure 14.2) could be employed to raise the plant efficiency from 33 per cent to perhaps 40 per cent or even 50 per cent – although, as the pebble bed modular reactor (PBMR) experience shows, the practice is much harder than the theory. Moreover, keeping these systems leak-tight is a significant challenge in itself.

In future, helium requirements from fusion reactors could escalate strongly – how much depends on the technologies chosen for helium ash removal, additional heating systems and the choice of high-temperature heat-transfer medium. Today, helium is considered an ideal fluid for all these applications, without much regard to helium's future availability. With the exception of ash removal, which is likely to remain a critical cryogenic helium demand, the other areas are possible candidates for substitution.

In terms of independence, fusion is intended to be a sustainable energy source for the long term. At present it appears some raw materials essential to the operation of fusion reactors will have to come from non-renewable resources. This is reasonable provided the non-renewable routes are used as "stepping stones" to future sustainability. Currently, helium is one such material. Table 15.1 summarises the applications of helium, and alternatives that might be expected in current and future fusion reactors. In this table, question marks mean the choice is not yet known.

Helium usage in a fusion research reactor – a case study

In this section, a case study of helium usage in the experimental fusion reactor JET is used as the basis for estimating future helium demand per GWe of power plant. The JET cryogenic plant is nominally a sealed system. It has an inventory of approximately 20,000 litres (2.5 tonnes) liquid helium equivalent, though at any one time the helium can be found in liquid, low-pressure pure gas or high-pressure impure gas forms around the plant system. Although over 98 per cent of the helium is recovered, because the plant is so big (it liquefies and reliquefies about 1 tonne of helium per day), the losses are typically about 800 litres liquid helium equivalent per week. Thus losses are twice the plant's inventory per year, although that does include gas lost through operational incidents such as power cuts and losses from small dewars used elsewhere at the site. Most of the plant's emissions are from the gas recovery system, which consists of large helium balloons, high-pressure (200 bar) recovery compressors and a gas store. Figure 15.6 shows the part of the plant where helium distribution takes place. The black, superinsulated transfer lines send cold helium to the JET machine's cryopumps and deuterium pellet system about 80 metres away.

Table 15.1 Helium applications in MCF fusion research reactors

Application	*JET, currently the world's largest tokamak, achieved 16 MW of D-T fusion power in 1997*	*ITER, the 500 MW (thermal) proof-of-concept experimental reactor*	*DEMO, the generic name for the first fusion power plant demonstration reactor*
Cryopumping 4.5 K[1] (divertor cryopump)	Supercritical[2] helium	Supercritical helium	Supercritical helium
Cryopumping 80 K (shielding)	Liquid nitrogen	Gaseous helium, cooled by LN_2	Liquid argon[4]?
Diagnostic	Liquid helium	"Dryogenics"[3]?	"Dryogenics"?
Magnets	Not used in JET (copper conductor magnets used)	4.5 K helium, needed for low-T_c magnets	Indirect liquid hydrogen? Liquid neon[5]? (If magnets become medium T_c)
Heat transfer (heat removal circuits)	Not used in JET (water)	Gaseous helium+Li pellets *or* Pb-Li?	As ITER, or gaseous neon?[6]

Notes

1 Degrees Kelvin (K) = degrees Celsius +273.15.

2 Supercritical helium is used for long horizontal loops in the tokamak to prevent boiling heat transfer problems that might otherwise occur.

3 "Dryogenics" means obtaining 4.5 K by means other than liquid helium; e.g with a Gifford-McMahon cooler or a Stirling engine, although some helium is still needed as the refrigerant (Chapter 11).

4 Nitrogen is rejected as a choice for the 80 K circuits because of the formation of ^{14}C by neutrons; the alternative – boiling argon – forms a short-lived radioisotope, ^{41}Ar, and that would need to be assessed.

5 Neon production (from air) would increase if helium were extracted from air, as they are jointly collected, as described in Chapter 8.

6 In 1960 (OSTI, 1960), when helium supply was scarce, research showed that neon is the second-best reactor coolant after helium (good conductivity, low pumping power and low neutron absorption cross-section). However, even at high pressure, with helium the pumping power alone could exceed 10 per cent of the reactor's output. With neon it would be worse.

Figure 15.6 JET's cryogenics system, 2005 (credit: J. Polverini and L. Antalova).

If we accept, as a starting point, that future cryogenics plants will be modelled on the JET design, then it might be possible to estimate the inventory of ITER, DEMO and so on by scaling the JET inventory. How should this scaling be done? Using the refrigerator capacity at –269 °C (JET 1.2 kW, ITER 65 kW) is one possibility, although following discussions with David Ward (2008), who leads power plant conceptual studies at the Culham Centre for Fusion Energy (CCFE), the nominal reactor thermal power seems a better choice. Most of the cryogenic equipment in a power plant would be associated with crypoumps, cryogenic pellet fuelling, fuel gas recovery and magnets, and they, in turn, are sized to handle the loading from the plasma and other sources. In normal deuterium-deuterium operation JET makes less than 100 kW of fusion power,[1] but the machine has an injected (additional heating) power of 25 MW. Conversely, ITER should *make* 500 MWth (thermal). To obtain the ITER inventory, I_{ITER}, from JET's, I_{JET}, it is necessary to scale up as follows: $I_{ITER} = I_{JET} (500/25)^s$, where s is a scaling exponent. It turns out that this scaling exponent has a strong impact on the results: the scale-up is huge, from JET to a rolled-out fusion power industry. Process engineering provides some insight about choosing s. For example, if there is a section of cryogenic line in which the pressure drop is optimised for thermodynamic reasons, the design engineer might still want the same pressure drop in a plant twice as big. By adjusting the pipe diameter to achieve this, it is found that the inventory scales with s ~0.75. If we apply this to ITER the calculated inventory is 23.6 tonnes, in good agreement with the 24 tonnes cited in the current ITER design (Serio, 2010). On top of the inventory, the annual leakage loss needs to be calculated. This is more difficult, since today there is no operational experience with ITER. The data from JET suggests that (annual loss rate)/(inventory) ~2. The Tore Supra experiment in France also calculates this value at around 2. In their case, the main reason is quenches (sudden loss of superconductivity) in the magnets.

Nonetheless, it seems that the JET cryogenic plant design is unusual in having a balloon-based helium recovery system with a relatively high leakage rate. A study conducted at Culham Science Centre (Mathis, 2004) revealed that, without its recovery system, JET would have had a loss/inventory ratio of ~0.7. Luigi Serio, Head of the Plant Engineering Division at ITER, estimates (Serio, 2011) that there will be less than 30 per cent annual losses during commissioning and they expect to approach 10 per cent during operation, based on the experience of other large machines like the LHC at CERN. Accordingly, we have used a loss ratio of 0.7 as being representative of a 1983 cryogenic design (albeit for a fusion device having magnets made from ambient-temperature copper coils).

Next, the heat transfer loops must be considered. The demand for helium as a heat transfer medium is based on its use in the high-temperature gas-cooled reactor. South African pebble bed reactor data (a 165 MWe HTGR) shows the initial charge is about 7.3 tonnes of helium, with a loss of 2.6 tonnes per annum or ~0.5 std cc/s (Van Aswegen, 2008). Much of it is lost through the gas circulator drive shaft: 36.5 per cent per year seems quite high, but it is adopted in the modelling anyway. Again, this inventory is scaled as (power ratio)$^{0.75}$. This gives 24.4 tonnes of inventory and 8.9 tonnes per year loss for DEMO's heat transfer loops, assuming DEMO has a 2500 MW_{th} output.

Table 15.2 summarises the helium requirements of JET, ITER and DEMO to establish the helium usage requirements of fusion per GWe installed. Three fusion helium use scenarios are developed: the "base case", the "intermediate case" and the "best case". Their details will be described later in the chapter. For DEMO, assumed to be the first 1 GWe plant, the "best-case" engineering is different from the other two cases, hence the two sections in the table. In the "best" case the effect of two design improvements is explored:

1 generation efficiency is increased to 40 per cent from 33 per cent;
2 in both the heat transfer system and the cryogenics system, the annual loss rate/inventory is reduced to 50 per cent of the "base" case.

To achieve the heat transfer target, very low-loss circulators would be required for the high-pressure coolant loops. Similarly, very low-leakage main compressors would be needed to drive the helium refrigerators. Improving generation efficiency is harder and it would be a challenge to reach even 40 per cent by the time DEMO is built: higher thermal efficiency requires higher reactor wall

Table 15.2 Helium requirements of JET, ITER and DEMO used to estimate consumption of MCF fusion reactors per GWe produced

Fusion reactor	*Subsystem*	*Comments*	*Fusion power MWth*	*Electric power MWe*	*Inventory tonnes He*	*Loss tonnes/y*	*Loss Inventory ratio*
JET 1983→	He cryogenics (80K loop is LN_2)	4.5K systems[+]	25*	0	2.50	1.75	0.7[a]
	Magnets	Cu magnets					
	Heat transfer	No HT coolant loops (for power generation)					
ITER 2018→	He cryogenics	4.5K systems[+] / 80K / magnets	500*	0	23.64[o]	5.67[1]	0.241[a]
	Magnets	Nb-alloy (4.5K) magnets					
	Heat transfer	No HT coolant loops (for power generation)					
DEMO "Base" case 2035→	He cryogenics	4.5K systems[+]/ 80K / magnets	2,500	825	79.06	11.38	0.144[a]
	Magnets	Nb-alloy (4.5K) or HTS					
	Heat transfer	Based on PBMR = 165 MWe PBMR efficiency = 33 per cent PBMR He inventory = 7.29 tonnes			24.38	8.90	0.365

Helium use in 2035 per GWe (initial fill + operate) = **28.0** MMScf/y, and in other years ("topping up") = **4.6** MMScf/y

DEMO "Best" case 2035→	He cryogenics	4.5K systems+/ 80K/magnets		2,500	1,000	79.06	5.69	0.072
	Magnets	Nb-alloy (4.5K) or HTS						
	Heat transfer	Based on PBMR	= 165 MWe			24.38	4.46	0.183
		PBMR efficiency	= 33 per cent					
		PBMR He inventory	= 7.29 tonnes					

Helium use in 2035 per GWe (initial fill + operate) = **22.3** MMScf/y, and in other years ("topping up") = **2.0** MMScf/y

Notes

"Base" case DEMO: efficiency 33 per cent; heat transfer inventory = $7.29 \cdot (0.33 \cdot 2500/165)^{0.75}$; cryogenic inventory = $2.5 \cdot (2500/25)^{0.75}$

"Intermediate" case = "Base" case, except both HT and cryogenic losses reduce by 2 per cent per year on new plants after 2035

"Best" case DEMO: efficiency 40 per cent; heat transfer and cryogenic inventories same as for "Base" case and annual loss/inventory is half of "Base" case; it is assumed that the heat transfer and cryogenic loop losses reduce by 3 per cent/year on new plants after 2035

+ 4.5K (non-magnet) systems consist of cryopumps, pellet (fuelling), and current drive (Neutral Beam cryopumps)

* No electricity produced; in the case of JET this is current drive / additional heating power (but see Chapter 15 end notes)

a Cryogenic annual loss/inventory ~0.144 but no further reductions after 2035 [$0.7 \cdot (1-LLR)^{(2035-1983)} = 0.144$; LLR = 3 per cent /year]

o Compares well with 24 tonnes calculated in proposed ITER cryogenics design (Serio, 2010)

1 Towards mid-range of 2.4–7.2 tonnes/year as estimated by Luigi Serio, Head of Plant Engineering Division, ITER project.

temperatures (Maisonnier *et al.*, 2006). Achieving higher efficiency is, however, of generic interest across the nuclear industry – particularly if helium is the working fluid for a Brayton cycle.

For DEMO, in the "base" and "intermediate" cases, the start-up year requirement for helium (all inventories plus all leakages) is estimated to be about 28 MMscf (143 tonnes per GWe), with demand in subsequent years being 4.6 MMscf/year for "topping up". In the "best" case the figures are 22.3 MMscf and 2 MMscf/year respectively. It is assumed that the life cycle for a fusion reactor is 40 years, and once a fusion reactor is installed, its helium leak rate remains constant. However, a principle assumption is that year-by-year design experience will lead to reduced leakage rates. The inventory, however, is probably constrained by thermal-hydraulic factors unless there is technological substitution (e.g. helium-free HTS magnets replace helium-cooled LTS magnets). This leak learning rate (LLR) is one of the critical factors in determining the consumption path of the fusion industry over the next century. Coupled to this is the leak learning that is, or is assumed to be, taking place now. For example, as Luigi Serio indicates, the ITER cryogenic system should have an annual loss/inventory of better than 0.24 (see Table 15.2) and that equates to an LLR, from JET to ITER, of 3 per cent per year. In the "best" case we have also assumed a similar rate of improvement in the heat transfer loops, from the helium-cooled PBMR (estimated to be a 2012 design) to DEMO, where 2034 is the target date. Whether such improvements will happen and would continue beyond DEMO is debatable.

It was found that doing nothing about leak rate between now and DEMO has a significant negative impact on fusion's helium sustainability in a post-abundant world. This is an important policy message for the fusion research community: set a target and continuously strive to achieve it, starting now. For example, according to Howden UK Ltd (Meiklejohn, 2009), it is possible to design extremely leak-tight high-pressure gas circulators having a leak rate of $\sim 10^{-4}$ std cc/s. Further integration work is needed to bring these into play. This requires coordinated, medium-term R&D. The leak learning rates chosen for the "intermediate" and "best" cases (2 per cent and 3 per cent, respectively) are believed to be achievable, as industrial history shows.

Consider the learning curve for steam engines. In 1727, when Newcomen's pumping engine was well established, 45 lb (pounds) of coal were required per horsepower-hour (Figure 15.7). After a century of improvements and developments, notably by John Smeaton and James Watt and others, the specific power had dropped to under 2 lb per horsepower-hour. What is interesting here is that most progress was made when there was competition, from 1770 onwards. Large deviations from the learning curve would require energetic catch-up later.

Roll-out of fusion and the associated helium demand

Having established, to some extent, specific helium demand scenarios for each 1 GWe of fusion power plant, we now consider the potential cumulative capacity of fusion power plants.

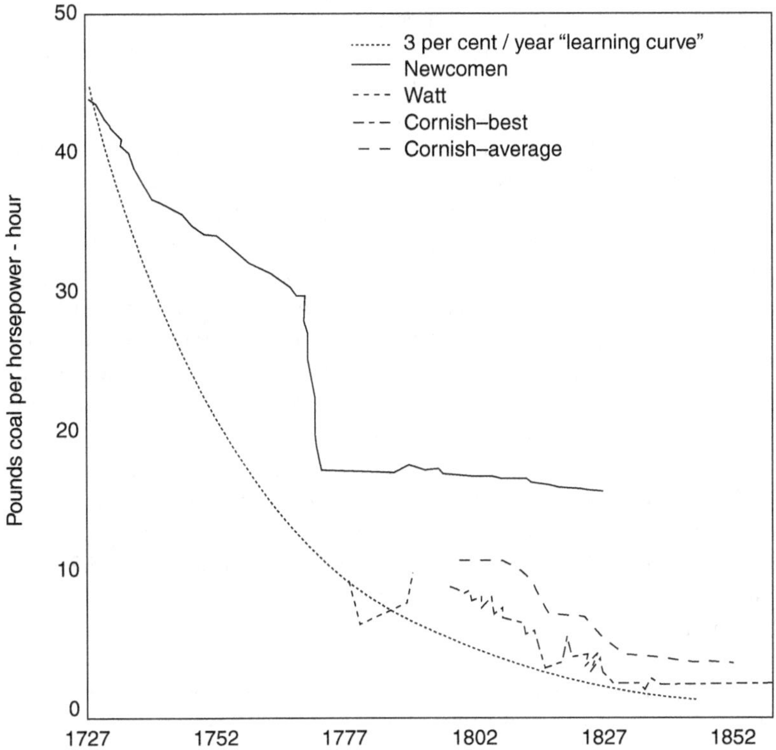

Figure 15.7 Coal consumption in pumping engines: pounds coal per horsepower-hour (Allen, 2009) shows a 3 per cent per year learning curve between 1727 and 1837.

The demand for helium from fusion reactors is certainly hard to predict due to uncertainties associated with macroeconomic conditions, geopolitics, population growth, competition from nuclear fission alternatives – such as Generation IV reactors – and not least the progress of fusion research itself. We suppose that fusion reactors will be commercialised by 2035, and that by 2075 the total installed capacity will be 350 GWe. Fission also had penetrated 15 per cent of the electricity market in 40 years (MacKay, 2008). Ultimately fusion capacity could reach 3,500 GWe, or perhaps 30 per cent of the electricity market in 2120. Some might argue this is too ambitious. We believe it is necessary to understand how big the industry *could* be in order to determine whether there will be sufficient helium in the market to permit fusion to reach a sustainable equilibrium.

It is our opinion that, in the early decades, fusion capacity will grow in a compound manner (at around 16 per cent per year) rather than linearly (at, say, ~18 GWe per year). There are two reasons for this. First, the fusion community and its users will want to gain confidence in the reliability of the new plants. That takes time and, if there are any setbacks, further capacity may be delayed. Second, expertise has to be gained and *a generation of engineers educated* – so building 18 GWe of capacity in 2035 would be extremely challenging, but perhaps not so in 2067 as the compound growth requires. Nonetheless, such curves do not continue indefinitely. It is a given that at some point the trend will break. This has been a characteristic of most forms of industrial energy; e.g. fission's growth in the US was stalled by the Three Mile Island (TMI) accident involving loss of coolant, even though the external consequences were relatively minor. Therefore, we estimate for the sake of a smooth curve that in 2080 there will be a breakpoint. Assuming it is not a major problem we suggest that, afterwards, capacity will be added linearly at 87.5 GWe/year globally. Allowing for old reactors being retired, an equilibrium online (i.e. connected to the grid) capacity of 3,500 GWe would be achieved by ~2120. MacKay (2008) observes that in 1984 nuclear plants were being built at the rate of 30 GWe per year, mostly in the West. With the BRIC economies added, 87.5 GWe per year of fusion *might* be possible in 2080 although, as yet, it is not known what is entailed in building even the first power plant.

The nominal (most likely) helium supply prediction from Zhiming Cai's (2010) helium market model was coupled with the fusion roll-out model described here to determine whether fusion might face helium supply issues sometime in the future. The three scenarios already outlined in Table 15.2 were tested. The parameters used are set out in Table 15.3.

Predictions from the fusion roll-out model are shown in Figures 15.8, 15.9 and 15.10 (the "base", "intermediate" and "best" cases respectively). Note that in each figure the vertical axis is logarithmic and the units are as given in the key. The x-axis is in years. In the "base" and "intermediate" cases there is no contribution from air separation units (ASUs) because the proportion of a fusion plant's power that would be needed to operate the ASUs would be too high. The rationale for using helium from air and the associated power demands are discussed in Chapter 8. If the fusion helium demand line crosses the helium production line then we anticipate that shortly afterwards the further operation of fusion plants will be jeopardised, along with the supply to other helium users.

In the "base" case, as with the other cases, in the first few years of fusion's roll-out there are some years when no reactors would be built. After 2120 the number of reactors that are retired starts to be significant, so the cumulative fusion capacity online separates from the capacity online curve. In the earlier study by Ferrer (1978), he predicted that by 2030, 20 years after the first reactor, the proportion of electricity from fusion reactors would comprise 11–17 per cent of the market; i.e. ~530 GWe using current US Energy Information Administration projections for 2030. Our model assumes a slower initial roll-out (16 GWe in 2054), although it picks up later.

In the "base" case there is no leak learning and at a helium consumption rate of 4.6 MMscf per year (Table 15.2) it is estimated that around 24 per cent of fusion's electrical output would be needed to drive ASUs of sufficient capacity to meet its helium needs. Clearly then, helium from air cannot make a useful contribution to the "base" case scenario (even if enough air separation demand were to exist).

Table 15.3 Three main scenarios tested in the fusion roll-out model

Parameter used in model (for case descriptions, see table footnote)	*"Base" case*	*"Int." case*	*"Best" case*
Compound growth rate (per cent/year)	15.8	15.8	15.8
Scenario start year	2034	2034	2034
Scenario end year	2150	2150	2150
Compound growth breakpoint year	2080	2080	2080
Breakpoint generation capacity (GWe)	714	714	714
Post breakpoint capacity acquisition rate (GWe/year)	87.5	87.5	87.5
Fusion long-term goal (GWe on-line, i.e. on the grid)	3,500	3,500	3,500
Leak learning rate (per cent reduction per year)	0	2.0	3.0
Reactor life (years)	40	40	40
Electrical generation efficiency (GWe/GWth) (per cent)	33	33	40
Inventory scaling exponent, s, based on thermal power	0.75	0.75	0.75
Cryogenics losses/inventory (per cent/year) in 2034	14.4	14.4	7.2
Heat transfer loop losses/inventory (per cent/year) in 2034	36.5	36.5	18.3
Retired reactor helium recycling factor (per cent)	0	0	90
X_{ASU}, constant in formula[a] describing switch to helium from air (in brackets, uptake percentage if A = 1 per cent)	0.001 (~0)	0.001 (~0)	2.1 (~55)
Helium *from* fusion (kg-helium per GWe-year output)	400	400	400

Notes
"Base case" – no leak learning, "standard" technology, no helium from air.
"Intermediate case (int.)" – some leak learning, "standard" technology, no helium from air.
"Best case" – full leak learning, "advanced" technology, helium from air, much helium recycling.
[a] ASU uptake fraction algorithm = $1 - \tanh(A/X_{ASU})$ A = per cent of fusion GWe used in ASU (see p. 131). 'A' will fall as leak-learning proceeds.

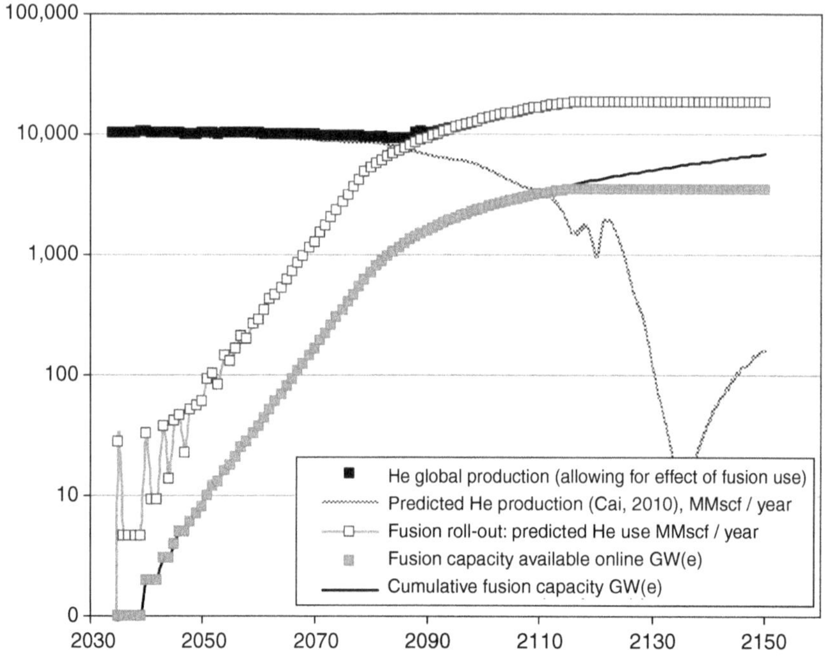

Figure 15.8 MCF "base" case: MCF fusion helium demand exceeds supply in 2087.

Fusion's predicted helium use zigzags in the early years, as the amount of helium needed to charge up the new reactors is higher than the next year's "top-up" requirements. It is not surprising then, that in this scenario without leak learning, fusion demand for helium exceeds world supply from natural gas sources around 2087. Moreover, another factor is predicted to emerge. The rate of helium's global demand growth would exceed 6 per cent per year, a figure BOC-Linde (Ward, 2007) suggests is the maximum rate at which helium refining capacity could be brought to market.

In the "intermediate" case some steps are taken to reduce helium consumption. A leak learning rate of 2 per cent is applied each year, whether or not new reactors get built. New reactors use less "top-up" helium than older ones, so the helium demand levels off even though capacity steadily increases. However, once again, fusion's demand would exceed the market's capacity (around 2098, although the market would attempt to respond, as we shall see). Eventually, even though the fusion capacity online stabilises at 3,500 GWe, helium demand starts to fall as old reactors are replaced with more leak-tight ones. "Leak learning" will have reduced the losses of both the cryogenic and heat transfer loops. Nonetheless, the minimum demand is set by a reactor's helium inventory and, unless recycled partially or wholly, this creates a significant loss each time a reactor is retired. In 2150, around half the predicted use is associated with inventory replacement.

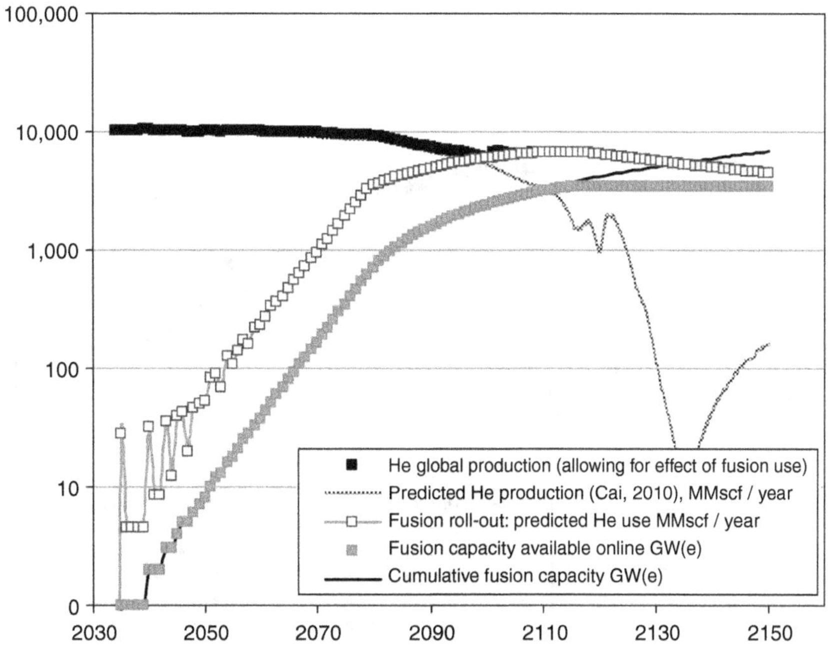

Figure 15.9 MCF "intermediate" case: MCF fusion's demand exceeds supply in 2098.

In the "best" case scenario (Figure 15.10) there are many kinds of interactions and that results in a complex helium demand pattern from fusion, but eventually, after 2100, fusion could head towards being self-sustaining in helium. The biggest factors are the higher leak learning rate (3 per cent) and the recycling of used helium. It is also very significant that both the cryogenics and heat transfer systems had started leak learning long before DEMO, as Table 15.2 shows, emphasising the importance of reducing helium leaks in fusion research today. Leak learning need not be solely about reducing leak rates: it can be about substitution, particularly in the heat transfer loops, as Table 15.1 illustrates. Although, in Figure 15.10, the fusion helium demand curve exceeds supply in 2128, we expect the market would make up the excess (see Figure 15.13). The role of ASUs (helium from cryogenic air separation) in the "best" case is also significant. The uptake rises over 100 years, as shown. Without ASUs (and without recycling) there would be an excess demand of perhaps 500 MMscf in 2138. By then, we expect ASUs to be cost-effective enough to meet modest helium demands (up to ~100 MMScf/year or around 2 per cent of today's helium market, see Chapter 8). Beyond that, an expansion of the air separation industry would be required. The concern here is that much of the world's oxygen demand is associated with economic growth and large-scale steel or chemicals manufacture. We also need to note that in the twenty-second century other users will want helium from ASUs as well, unless novel ground sources develop.

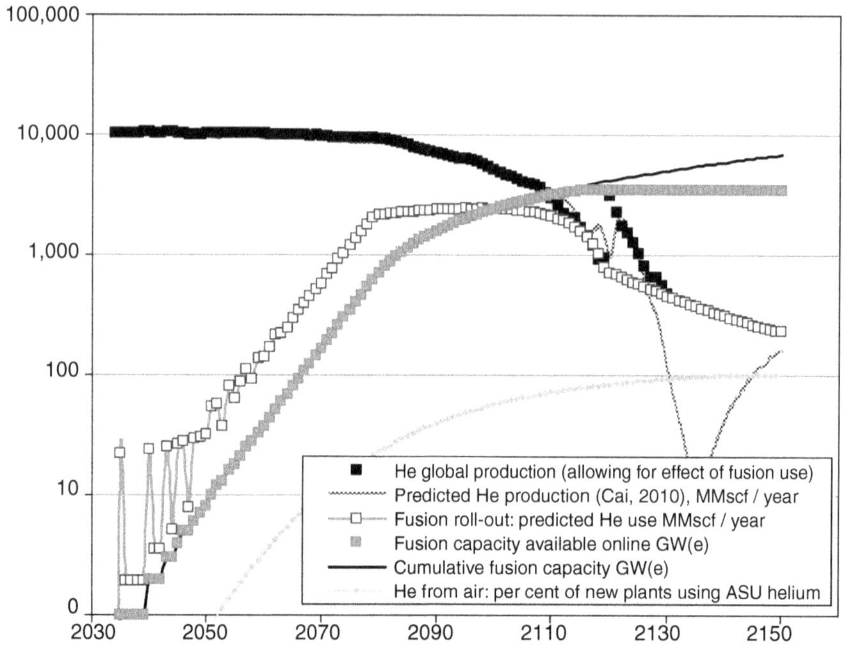

Figure 15.10 MCF "best" case: supply capacity could closely match demand (note: the 'predicted He production' curve (Cai, 2010) only refers to natural-gas associated sources of helium).

The main, perhaps the most important, contribution to fusion's needs would be helium recycling. If 90 per cent of the helium in old reactors were recovered and stored then the supply–demand gap identified above could be closed. Yet, however simple this sounds, it is fraught with difficulties. Who would own the stored helium? Who would be its custodian? What would stop the helium being inadvertently sold or vented? We only have to look at the US Department of Energy's handling of helium-3 to see that even within one government department there can be such mistakes (Wald, 2011). The result is severe rationing and, unsurprisingly, the cost of helium-3 has soared. The hope for helium-4 would be that rising prices encourage novel exploration and development and the management of earthbound resources, as Ralph Scurlock argues in Chapter 17. Ultimately, this would ease the transition to air-sourced helium.

Our sense is that the substantially reduced scale of the helium market in 2150 will warrant any large-scale users building secondary containment systems. For example, by hermetically sealing the compressor house of a cryogenics plant, where flanges could leak or relief valves lift, it may be possible to recover the building air and reprocess it in an ASU to purify the helium. To rebuild a whole inventory could take *years* if the only source were the ASU associated with a fusion plant. If we were discussing tritium instead of helium, none of these precautions or subsystems would be unusual. What will be required is a different way of thinking: helium will no longer be a "commodity", though for technical reasons the gas never made it to the futures market (Cai *et al.*, 2010). Price and availability may lead to its becoming a special gas like neon. In just 200 years we will have gone full circle with helium: rare – abundant – rare.

Inertial (ICF) fusion helium demand

Chatain and colleagues (2008) describe the cryogenic systems associated with ICF's deuterium-tritium fuel production where, to achieve the "triple product", the target (called a hohlraum) needs to contain the frozen hydrogen isotope mixture at ~1,500 bar and –255 °C (18 K). It may be possible to minimise helium requirements using the indirect hydrogen (i-LH_2) concept that works at this temperature. Currently, it is not clear what the refrigeration capacity for a full-sized reactor will be. We estimate that the demands of pellet (or target) fuelling, in-vessel cryopumping and deuterium-tritium fuel recycling and helium removal might be similar for ICF and MCF. We anticipate that the helium leak rate/inventory would be comparable to that of an MCF cryogenics plant. There will be no need for helium-cooled magnets and "current drive" in ICF, although we believe it is reasonable to assume that by the time DEMO is built, high-temperature superconductors (e.g. liquid neon-cooled YBCO, see Stautner (2011)) magnets would have been developed. This is described in Chapter 16.

We would expect the heat transfer loops of MCF and ICF to have similar characteristics. Within the laser systems of ICF, there is apparently no need for helium. The long-term helium demand of ICF could be consistent with the projected helium production rates shown in Figure 15.13, provided a "best" case-type strategy is developed.

Like MCF, ICF has its own particular set of challenges, not least the cryogenic hohlraum production, which has to be extremely precise. Rayleigh-Taylor thermal-hydraulic instabilities and laser radiation uniformity are also subjects of close attention. Meyerhoffer and colleagues (2011) stated that ICF, in the form of the OMEGA laser at the University of Rochester, New York State has achieved a Lawson "triple product" similar to that of the JET tokamak.

Fusion's interaction with the helium market

Figure 15.11 summarises the fusion helium demand for the three scenarios. In the MCF "base" case, although the fusion roll-out target is challenging, it is clear that the helium demand is completely unsustainable and this would be a serious technological roadblock for fusion. The demands of the "intermediate" scenario are unsustainable too, although if fusion had significantly rolled out in *2008* instead of *2108*, as was assumed by Ferrer (1978), such a demand curve might have been met by the helium industry. As Richard Stainsby indicates in Chapter 14, this is what the Generation IV fission community tacitly assumes for their reactors. This leaves only the "best" case or something similar, a challenging situation that can be resolved only if a range of measures are taken to reduce, recycle and produce helium from ASUs or, alternatively, to substitute helium in the relevant subsystems. This is just one of many compelling arguments showing that if fusion is to play an important role, the pace of its research developments must be accelerated. A delay of ten years (i.e. DEMO in 2044) could make even the MCF "best" case supply–demand balance unattainable. The situation would be similar for ICF.

Figures 15.12 and 15.13 show the impact of the fusion helium demand scenarios on the global helium market's prices and global consumption using the Vensim® system dynamics model developed by Zhiming Cai (Cai *et al.*, 2010). The extra helium demand accelerates the price rise still further, because the fusion demand is modelled as an exogenous variable. Although fusion's total demand for helium increases in a prescribed way (i.e. has an elasticity of zero), the additional increase of helium supply to the whole market is not as high as that shown in Figure 15.11. This is due to the interrelatedness of helium users. An extra increase in price may drive some other helium users out of the helium market altogether. Note that in the "base" and "intermediate" cases, the market collapses (at least, the model itself collapses) about 10–12 years later than Figure 15.8 and 15.9 show. The market will respond, and will most likely fail, to meet the sudden demand. For the "best" case, the fusion demand is much more in tune with projected availability and even in the supposed "end game" of the traditional helium market there would be suppliers willing to fill the modest requirements that fusion might then present.

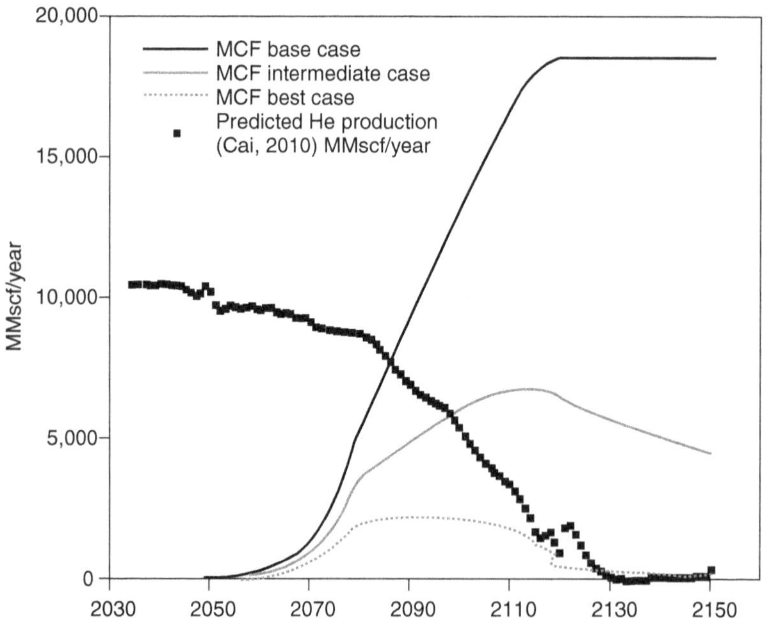

Figure 15.11 Helium demand projections for the scenarios specified in Table 15.3.

Sensitivity analysis of the "best" case

Another way of considering the many facets of the "best" case analysis is to carry out a sensitivity analysis. Figure 15.14 shows the results of this, in relative terms, for MCF fusion in the year 2150 (assuming the use of HTS magnets). Note that some changes, based on a 1 per cent perturbation, are positive and some are negative (*negative is defined as parameter reduction results in increased helium use*). The Appendix shows how the data for Figure 15.14 was calculated.

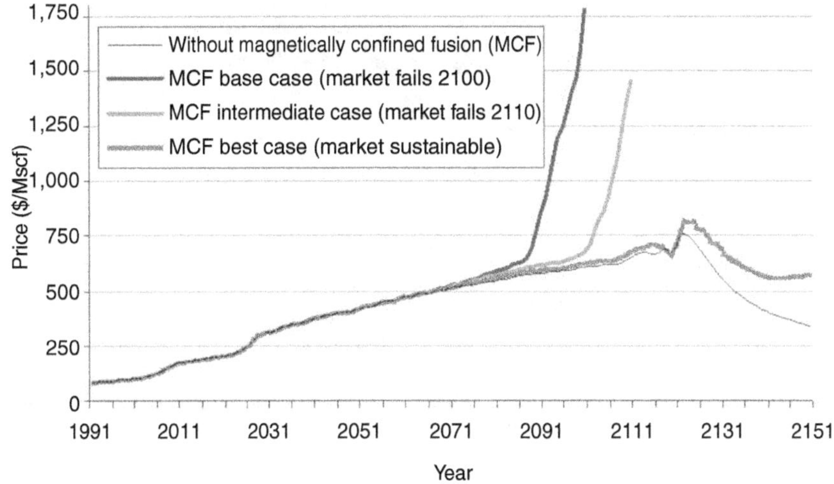

Figure 15.12 Projected Grade-A helium wholesale market price for the three scenarios of anticipated helium demand from fusion.

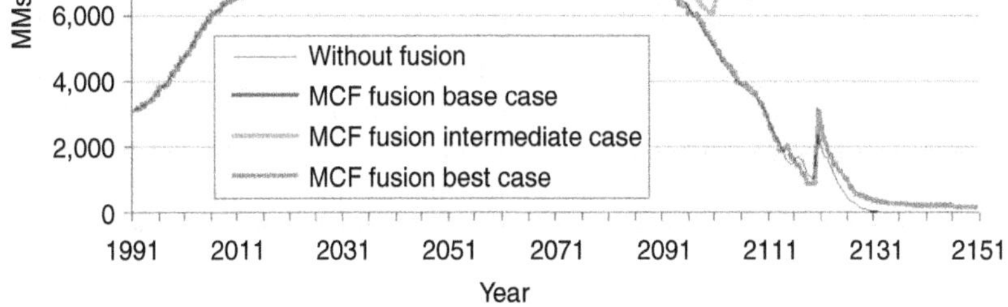

Figure 15.13 Future helium demand (whole market) for the three scenarios of anticipated helium demand from fusion; some users would be displaced by fusion's use.

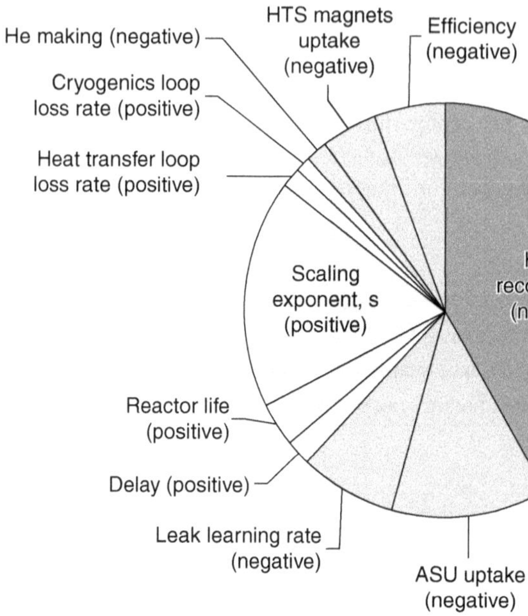

Figure 15.14 Effect of recycling, helium from air, etc. on the MCF "best" case in 2150: slices represent the relative importance of each parameter.

As the earlier discussion indicated, recovering helium from retired reactors may need to be a dominant activity in the twenty-second-century fusion industry. This might be ameliorated slightly by reducing the helium *inventory* through technological substitution (e.g. by using neon-cooled YBCO magnets). The effect of this is shown in Figure 15.14, although the inventory issue also applies to neon (it would have to be economically extracted from air). ASU uptake offsets the need for natural gas-sourced helium. A 2150 fusion plant could be self-sufficient in helium provided the following assumptions are made: the coils use HTS superconductors; make-up helium comes from air separation units (ASUs); and 90 per cent of the helium from retired reactors is recycled. The effect of fusion's early delivery is positive (i.e. delay increases helium demand). Delay would make it harder for fusion to become self-sustaining in helium, or the extent of fusion's roll-out may be constrained. The importance of leak learning rate has already been emphasised: by 2150 most of the learning should have been done, hence its smaller significance here. Shortening reactor life is helium efficient; it retires the gas-guzzlers. In the case study the scaling exponent, s, was determined to be 0.75. If $s > 0.9$ there would be difficulties even in the "best" case. The initial loss rates in the two helium loops are not highly significant, as they have effectively been learned out of the system by 2150. Helium making (the helium produced by fusion's reactions and blankets) is and always will be a small factor. Finally, as for most aspects of a power plant, the generation efficiency has a useful effect in reducing capital and operating costs (in this case to the 0.75 power). However, switching to a helium-filled Brayton cycle may, at best, be helium neutral (assuming the fusion blankets will one day permit the higher temperatures required). The potential leakiness of a Brayton loop may offset the reductions in helium consumed by fusion plant subsystems elsewhere.

Discussion and recommendations

Consideration has been given to fusion, a potential key player with needs in the future helium market. An assessment has been made of the prospective demands from magnetically confined fusion, MCF, and inertially confined fusion, ICF. We have taken into account the potential scale of fusion's roll-out over the next 100 years or so and the many roles that helium plays, or could play, in future reactors. The targets are deliberately ambitious. Why would the fusion community set their sights on anything less? An overview is given of the history and likelihood of fusion playing such a significant energy role, the technologies involved and the helium needs of major subsystems.

A numerical model was developed to investigate the impact that fusion could have on the helium market. As fusion is likely to emerge as a major energy source (whether for electricity, hydrogen, desalination, CO_2 mitigation or other applications) only towards the end of the twenty-first century, it is crucial that the fusion community understands the importance of assessing and planning helium demand.

For MCF, three scenarios were developed: the "base", "intermediate" and "best" cases. If nothing is done to mitigate helium use ("base" case), then it is clearly predicted that helium demand will outstrip market supply early on during fusion's roll-out. That would be a major technological roadblock for fusion. A modest attempt to reduce leakages (the "intermediate" case) reduces peak demand by 63 per cent, yet already it is too late for that. Fusion would need to have been significantly rolled out by now for that scale of helium demand to track the projected global supply.

Only in the "best" case with its many different approaches, including leak learning rate, substitution, recycling and obtaining helium from air using air separation units (ASUs), is there a reasonably assured path to self-sustainable helium use. The message is that, starting now, the leak learning and substitution process should begin (or in some cases continue) so that when the first reactor, DEMO, is built around 2034 its helium demand will be perhaps half that of the equivalent today. Given the projections for ITER this, at least, seems feasible.

Recommended helium use per electrical gigawatt of fusion capacity

One of the outcomes of this study, assuming the helium production (market model) and fusion use analyses are essentially correct, is a specific recommendation about helium use per electrical gigawatt (GWe). We presume that fusion would become the predominant demand. First, the generation efficiency needs to be calculated so that the reactor thermal power can be determined.

For a given reactor thermal power (Q_{TH} in megawatts),[2] we can define a *needs-based quota* of helium which would be the target for the cryogenics inventory (cryopumps, pellets, magnets), I_{CRYO}, and the high-temperature heat transfer (blankets and divertor) loops, I_{HT}, namely:

$$I_{CRYO} = A \cdot Q_{TH}^{0.75} = 0.048 \cdot Q_{TH}^{0.75}\,\text{MMscf}$$

$$I_{HT} = B \cdot Q_{TH}^{0.75} = 0.015 \cdot Q_{TH}^{0.75}\,\text{MMscf}$$

where (from Table 15.2) $A = K \cdot 2.5$ tonnes $\cdot$ $(25\,\text{MW})^{-0.75}$
and $B = K \cdot 7.29$ tonnes $\cdot$ $(165\,\text{MW}/0.33)^{-0.75}$
and $K = 0.2133$ MMscf/tonne of helium

The cryogenics loop loss rate, L_{CRYO}, and heat transfer loop loss rate, L_{HT}, are:

$$L_{CRYO} = 0.072 \cdot LLF \cdot I_{CRYO}\,\text{MMscf/year}$$

$$L_{HT} = 0.183 \cdot LLF \cdot I_{HT}\,\text{MMscf/year}$$

where leak learning factor, $LLF = (1\text{–}0.03)^{(PSY-2034)}$ and PSY is the plant start-up year, with 2034 being the base year.

Thus, in the first year (the plant start-up year) the total demand is:

$$I_{CRYO} + I_{HT} + L_{CRYO} + L_{HT}$$

And in subsequent years:

$$L_{CRYO} + L_{HT}$$

As an example, consider a 2,500 MWth MCF plant, starting up in 2050. The first year's demand would be 23.7 MMscf. This subsequently drops to 1.35 MMscf/year. The formula might be adapted for use by other large-scale helium users. (Note: it is assumed that, once built, a plant retains its plant start-up year LLF. That does not, of course, preclude efforts to reduce leakages on an existing plant).

Most of the discussion in this chapter has been concerned with MCF rather than ICF. It appears that less is known about the scaled-up helium demands of ICF, but we suggest that, if commercialised, the helium requirements of the two fusion strategies would be similar.

Overall, leak learning and substitution are by far the most useful areas to focus on in terms of fusion technology development. Looking at the terms L_{CRYO} and L_{HT} we see that these become $0.0035 \cdot LLF \cdot Q_{TH}^{0.75}$ and $0.0028 \cdot LLF \cdot Q_{TH}^{0.75}$ respectively, i.e. of similar significance.

Helium use probably cannot be avoided in certain cryogenic systems. The question is whether any part of the refrigeration system could work with a more abundant fluid, such as hydrogen. This could reduce, markedly, the size of the helium refrigeration plant, as is explained in Chapter 16. Coupled with the adoption of HTS magnet technologies, the pace towards helium self-sufficiency could be accelerated.

At the high-temperature end, it is the compressor and high-pressure circulator designs that will be pivotal in reducing leak rates. One possibility, in the context of fusion, is to work with KIT (Karlsruhe Institute of Technology) (Ghidersa *et al.*, 2006) and others, whose high-pressure helium test loops may be suitable platforms for proof-testing advanced turbo-machinery. This would complement their work on high-temperature helium-cooled components, such as the blanket test modules, etc.

Other factors that stand out from the modelling are the effects of delayed roll-out and recycling retired reactor helium. These are more policy issues than engineering ones. The first is clearly related to R&D spend (as Figure 15.5 highlights). We should note, however, that even a 15-year delay could constrain fusion's long-term ambitions to <3,000 GWe, possibly favouring less helium-intense Fifth-Generation power plants, such as thorium molten salt or accelerator-driven reactors. Also, if the leak learning curve follows a Newcomen-like path (Figure 15.7) rather than the 3 per cent per year curve, that too could limit future capacity. Recycling either requires a robust administration or, more likely, that old helium is monetised in the expectation that helium could again be purchased at a later date for a future reactor. However, in the traditional helium market's "end-game" prices and availability may be highly volatile.

If helium is still to play a role in large-scale applications such as fusion, then great care must be taken to protect against sudden massive loss of helium inventory. Leak-tight secondary containment systems may need to be employed to permit the recapture of lost helium.

One of the most obvious, yet difficult, requirements for fusion and the wider nuclear industry is to increase the overall efficiency of their electricity generation. At around 33 per cent, many baseload power plants are no more efficient than 1960s coal plants. Natural gas plants with combined-cycle gas turbine (CCGT) technology have raised the bar to 55–60 per cent. In fusion there will need to be ongoing efforts to address the apparent temperature constraints of fusion materials. All possible coolant and breeder options need to be examined, so that the higher temperatures needed for increased efficiency may be accessed. We note that helium could be a key component of advanced (Brayton) power plant cycle designs.

Finally, there is a balance to be struck: be sparing but not too sparing with helium. We recall that helium is a by-product of other, dominant, industrial gas processes (natural gas, CO_2, etc.). If demand slackens, even more helium will be vented. That might be a sign that major helium users have taken a less than optimal path on their way to helium self-sufficiency.

Appendix Sensitivity analysis for MCF "best" case in 2150 (with and without HTS) – relates to Figure 15.14

<u>LTS case assumptions</u>: Nb-alloy magnets; 90 per cent retired reactor helium recovery rate (RRRF); 40 per cent generation efficiency (Effy); 3 per cent per year leak learning rate (LLR); 2034 DEMO plant start-up year (PSY); 40-year reactor life (Rlife); 0.75 scaling exponent (s); ASU uptake X_{ASU}=2.1 (Table 15.3); 0.183 HT loop loss/inventory (HT l/i); 0.072 Cryoloop loss/inventory (Cryo l/i); 400 kg/MWe-year helium making (HeFus); 3,500 GWe of installed plant

MCF with LTS 2150 GSHeD=230.69 MMscf

<u>HTS case assumptions</u>: YBCO magnets; 90 per cent retired reactor helium recovery rate (RRRF); 40 per cent generation efficiency (Effy); 3 per cent per year leak learning rate (LLR); 2034 DEMO plant start-up year (PSY); 40-year reactor life (Rlife); 0.75 scaling exponent (s); ASU uptake X_{ASU}=2.1 (Table 15.3); 0.183 HT loop loss/inventory (HT l/i); 0.072 Cryoloop loss/inventory (Cryo l/i); 400 kg/MWe-year helium making (HeFus); 3,500 GWe of installed plant

MCF with HTS 2150 GSHeD=125.24 MMscf

Variation	*GSHeD*♦	*SF**	*Delta*	*Per cent*	*Variation*	*GSHeD*♦	*SF**	*Delta*	*Per cent*
RRRF=89.1	246.69	1	–16.0	36.8	RRRF=89.1	135.10	1	–9.86	41.8
Effy=39.6	233.35	1	–2.66	6.1	Effy=39.6	126.59	1	–1.35	5.7
LLR=2.97	235.43	1	–4.74	10.9	LLR=2.97	127.07	1	–1.83	7.8
PSY=2033	225.63	See notes[×]	+1.16	2.7	PSY=2033	123.55	See notes[×]	+0.46	2.0
Rlife=39	225.63	100/40	+2.02	4.6	Rlife=39	123.25	100/40	+0.80	3.4
s=0.7425	221.47	1	+9.22	21.2	s=0.7425	121.06	1	+4.18	17.7
X_{ASU}=0.001	527.75	100	–5.28	12.1	X_{ASU}=0.001	297.94	100	–2.98	12.6
HT l/i=0.18117	229.93	1	+0.76	1.7	HT l/i=0.18117	124.82	1	+0.42	1.8
Cryo l/i = 0.07128	229.72	1	+0.97	2.2	Cryo l/i=0.07128	124.97	1	+0.27	1.1
HeFus=396	231.37	1	–0.68	1.6	HeFus=396	125.61	1	–0.37	1.6
					No HTS°	230.69	100	–1.05	4.5

Notes

Vensim® SD model indicates that conventional market could supply 210 MMscf/year in 2150. Chapter 8 shows that air-sourced (ASU) He supply could be ~100 MMscf/year (in 2150 <5 per cent of this is needed to cover leaks ("best" case+HTS); the rest could be used to build "inventory" stocks).

* SF=Scaling factor used for integers (years) and options (ASU, HTS) to obtain 1 per cent effect.

× For one year earlier PSY, SF=100/(2034–2011) where 2034 is DEMO target start-up year.

♦ GSHeD=ground-sourced helium demand (MMscf) (e.g. residual methane-associated reserves).

° Including HTS magnets is assumed to reduce cryogenics inventory by half.

Notes

1 JET has operated with deuterium-tritium fuel (Keilhacker *et al.*, 1999) and produced 16 MW of thermal fusion power. In normal operation the less reactive, pure deuterium plasma mode is used.

2 For consistency, Q_{TH} should include the additional heating (or current drive) power as well as the power produced during fusion reactions.

References

Allen, R. C. (2009) *The British Industrial Revolution in Global Perspective*, Cambridge University Press, Cambridge, UK

Braams, C. M. and Stott, P. E., (2002) *Nuclear Fusion: Half a Century of Magnetic Confinement Research*, Institute of Physics Publishing, London

Cai, Z., Clarke, R. H., Glowacki, B. A., Nuttall, W. J and Ward, N. (2010) "Ongoing Ascent to the Helium Production Plateau – Insights from System Dynamics", *Resources Policy*, Vol. 35(2), June 2010: 77–89

Chatain, D. Perin J. P., Bonnay P., Bouleau E., Chichoux M., Communal D., Manzagol J., Viargues F., Brisset D., Lamaison V., and Paquignon G. (2008) "Cryogenic Systems for Inertial Fusion Energy", *Laser and Particle Beams*, Vol. 26: 517–23

Dean, S. O. (1998) "Fusion Power by Magnetic Confinement Program Plan", *Fusion Energy*, Vol. 17(4): 263–87

Eddington, A. S. (1920) "The Internal Constitution of the Stars", *Nature*, Vol. 106: 14–20

Ferrer, J. C. (1978) "Commercialization and Helium Requirements of Nuclear Fusion Power Reactors between the Years 2000–2030", Fusion Energy Corporation, Princeton, New Jersey, US, prepared under contract PO 480371 for the Amarillo Helium Operations Office – US Department of the Interior – Bureau of Mines, Fusion Energy Project JN-77–1201 (21 December 21 1977, revised 16 January 1978)

Financial Times (2004) "Global Fusion", leader, *Financial Times*, 25 November, www.ft.com

Ghidersa, B. E., Ionescu-Bujora, M. and Janeschitz, G. (2006) "Helium Loop Karlsruhe (HELOKA): A Valuable Tool for Testing and Qualifying ITER Components and their He Cooling Circuits", *Fusion Engineering and Design*, Vol. 81(8–14): 1471–6

Herman, R. (1990) *Fusion – the Search for Endless Energy*, Cambridge University Press, Cambridge, UK

Hirsch M., Baldzuhn J., Beidler C., Brakel R., Burhenn R., Dinklage A., Ehmler H., Endler M., Erckmann V., Feng Y., Geiger J., Giannone L., Grieger G., Grigull P., Hartfuß H-J., Hartmann D., Jaenicke R., König R., Laqua H.P., Maaßberg H., McCormick K., Sardei F., Speth E., Stroth U., Wagner F., Weller A., Werner A., Wobig H., Zoletnik for the W7-AS Team (2008) "Major Results from the Stellarator Wendlestein 7-AS", *Plasma Physics and Controlled Fusion* Vol. 50 (5) pp. 1–204

Hubbert, M. K. (1971) "The Energy Resources of the Earth", *Scientific American*, Vol. 225(3), September 1971: 60–70

Keilhacker, M., Gibson, A., Gormezano, C., Lomas, P. J., Thomas, P. R., Watkins, M. L., Andrew, P., Balet, B., Borba, D., Challis, C. D., Coffey, I., Cottrell, G. A., De Esch, H. P. L., Deliyanakis, N., Fasoli, A., Gowers, C. W., Guo, H. Y., Huysmans, G. T. A., Jones, T. T. C., Kerner, W., König, R. W. T., Loughlin, M. J., Maas, A., Marcus, F. B., Nave, M. F. F., Rimini, F. G., Sadler, J., Sharapov, S. E., Sips, G., Smeulders, P., Söldner, F. X., Taroni, A., Tubbing, B. J. D., von Hellermann, M. G., Ward, D. J. and

JET Team (1999) "High Fusion Performance from deuterium-Tritium Plasmas in JET", *Nuclear Fusion*, Vol. 39(2): 209–21

Keen, B. (ed.) (1987) *Fusion Technology*, Vol. 11(1): whole volume

Lawson, J. D. (1957) "Some Criteria for a Power Producing Thermonuclear Reactor", *Proceedings of the Physical Society B*, Vol. 70: 6

Llewellyn-Smith, C. and Ward, D. (2008) "Fusion", *Energy Policy*, Vol. 36: 4331–4

MacKay, D. J. C. (2008) "Sustainable Energy – Without the Hot Air", UIT (and online)

Maisonnier, D., Cook, I., Pierre, S., Lorenzo, B., Luigi, D. P., Luciano, G., Prachai, N., Aldo, P. and PPCS Team (2006) "DEMO and Fusion Power Plant Conceptual Studies in Europe", *Fusion Engineering and Design*, Vol. 81(8–14): 1123–30

Mathis, N. (2004) "2004 Helium Project at JET", Fourth Year Internship Report, Institute National des Sciences Appliquées (INSA), Lyon, France

Meiklejohn, P. (2009) "Development of Circulators for Gas Cooled High Temperature Reactors", Future for Helium Conference, Hughes Hall, University of Cambridge, March

Meyerhoffer, D. D., McCrory, R. L., Betti, R., Boehly, T., Casey, D. T., Collins, D. T. J., Craxton, R. S., Delettrez, J. A., Edgell, D. H., Epstein, R., Fletcher, K. A., Frenje, J. A., Glebov, Y. Y., Goncharov, V. N., Harding, D. R., Hu, S. X., Igumenshchev, I. V., Knauer, J. P., Li, C. K., Marozas, J. A., Marshall, F. J., McKenty, P. W., Nilson, P. M., Padalino, S. P., Petrasso, R. D., Radha, P. B., Regan, S. P., Sangster, T. C., Séguin, F. H., Seka, W., Short, R. W., Shvarts, D., Skupsky, S., Soures, J. M., Stoeckl, C., Theobald, W. and Yaakobi, B. (2011) "High-Performance Inertial Confinement Fusion Target Implosions on OMEGA", *Nuclear Fusion*, Vol. 51: 1–5

Nuclear Fusion (2007) Vol. 47(6): whole volume

OSTI (1960) "Neon as Coolant for Graphite-Moderated High Temperature Reactors", *Atompraxis* (West Germany), Vol. 6, www.osti.gov/energycitations/product.biblio.jsp?osti_id=4176614 (accessed May 2011, author unknown)

Poitevin, Y., Boccaccini, L. V., Zmitko, M., Ricapito, I., Salavy, J.-F., Diegele, E., Gabriel, F., Magnani, E., Neuberger, H., Lässer, R. and Guerrini, L. (2010) "Tritium Breeder Blanket Design Technologies in Europe: Development Status of ITER Test Blanket Modules, Test and Qualification Strategy and Roadmap towards DEMO", *Fusion Engineering and Design*, Vol. 85(10–12), December: 2340–7

Serio, L. (2010) "Challenges for Cryogenics at ITER", *AIP Conference Proceedings*, Vol. 1218, 9 April: 651–62

Serio, L. (2011) Private Communication

Stautner W., Xu M., Laskaris E. T., Conte G., Thompson P. S., van Epps C., Amm K. (2011) "The Cryogenics of a Thermosiphon-cooled HTS MRI Magnet – Assembly and Component Testing", *Applied Superconductivity*, Vol. 21(3): 2096–2099

Van Aswegen, B. (2008) "Sustainability of the Use of Helium in the Pebble Bed Modular Reactor", Arcus Gibb Sustainability Report, 16 January, WSP Environmental, Johannesburg, South Africa

Wald, M. L. (2011) "Agencies' Lack of Coordination Hindered Supply of Crucial Gas, Report Says", *New York Times* (US edition), 28 May, www.nytimes.com/2011/05/29/us/29helium.html (accessed May 2011)

Ward, D. (2008) Private communication

Ward, N. (2007) Private communication

16 Substituting hydrogen for helium in cryogenic applications

Bartek A. Glowacki

In 2011 we celebrated the centenary of the discovery by two Dutch physicists, Gilles Holst and Heike Kamerlingh Onnes, of superconductivity in mercury, cooled by liquid helium, which was liquefied for the first time in 1908. We should do more than just reminisce about how it happened and who discovered superconductivity. Today could be the turning point in applied superconductivity and cryoscience R&D, where electric materials engineering and cryogenic technology combine with construction engineering to pave the way for providing a sustainable future energy and cryogenic infrastructure. It is expected that the hydrogen economy and related technological advances may reduce political tensions driven by the need for access to fossil fuel resources.

Global energy needs are expected to double by 2030, increasing by a factor of five by 2100 due to human population growth and accelerated global industrialisation. Considering the projected demand for the future of energy supply, transfer and storage, and also noting the rapidly increasing demand by transportation, one can envisage that superconductivity interlocked with a liquid hydrogen economy could be the solution to most of our energy and transport problems.

Introduction

Cryogenic applications involving superconducting and non-superconducting materials require operation at a wide range of temperatures, from 1.8 K to 164 K. Such a wide temperature spectrum includes all cryogenic liquids ranging from LHe (0.95–5.19 K), LH_2 (13.9–33 K), LNe (24.6–44.5 K), LO_2 (54.4–154.6 K), LN_2 (64–126 K), LAr (83.8–150.9 K), LAir (57–132 K), CH_4 (90.7–190 K). Due to fact that liquid hydrogen under normal pressure has a temperature of 20 K and is the only element which is both a cryogen and an energy carrier, it may become the natural future choice of cooling medium for indirect or even direct cooling of superconducting devices.

Practical superconducting materials can be divided into three main categories, as represented in Figure 16.1:

- low-temperature metallic superconductors – NbTi (T_c = 11 K), Nb_3Sn (T_c = 18 K), Nb_3Al (T_c = 17.6 K);

- medium-temperature superconductors – eg. MgB_2 ($T_c = 39\,K$);
- high-temperature oxide ceramic superconductors – $YBa_2Cu_3O_7$ ($T_c = 91\,K$), $(Bi,Pb)_2Sr_2Ca_2Cu_3O_x$, ($T_c = 117\,K$), $(Tl,Hg)Ba_2Ca_2Cu_3O_x$ ($T_c = 134\,K$).

The critical temperature, T_c, defines the minimum cooling condition required to induce superconductivity but it does not automatically define the applicability of a given superconducting material for electromagnetic applications. This is because superconductors are also characterised by upper critical magnetic induction, B_{c2}, irreversibility induction, B_{irr}, as well as critical current density, J_c. B_{irr} defines the boundary conditions on the $B_{irr}(T)$ graph above which a superconductor will not be able to carry the superconducting current, as presented in Figure 16.2.

From Figure 16.2 it is evident that $(Bi,Pb)_2Sr_2Ca_2Cu_3O_x$ conductors, despite a high T_c value of 117 K, can reach their full potential only at temperatures close to that of liquid hydrogen, 20 K. It is also evident that there is a crossover between MgB_2 and Bi-based conductors at 20 K, which makes the low-cost MgB_2 conductor the favourable candidate for some applications at the LH_2 temperature.

Before reaching more general conclusions on the possible transition from helium to hydrogen as a cooling medium, let us analyse a few representative examples that may indicate the necessity of such a transition. After the discovery of a variety of so-called high-temperature superconducting materials

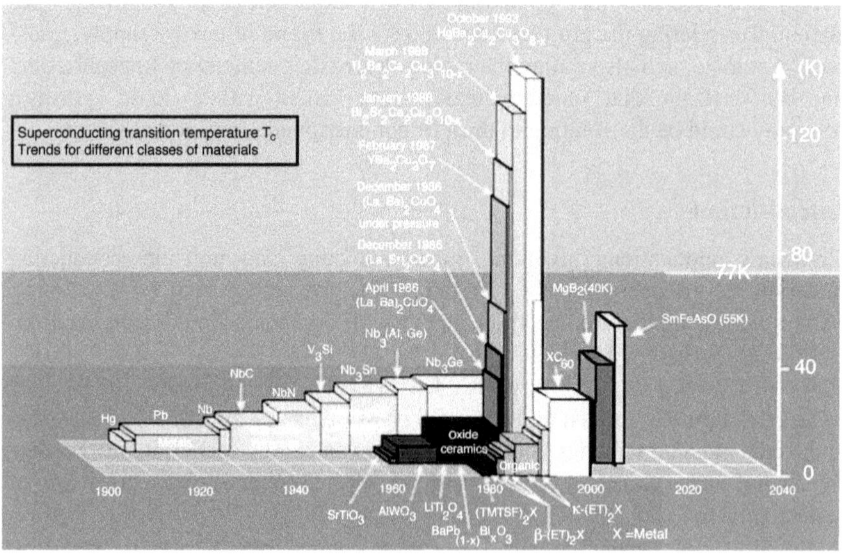

Figure 16.1 Superconducting transition temperature, T_c, for different classes of materials: metals, oxide ceramics and metal-organic. It is evident that the progress with new materials in the medium-temperature range will facilitate use of liquid hydrogen as a cooling medium.

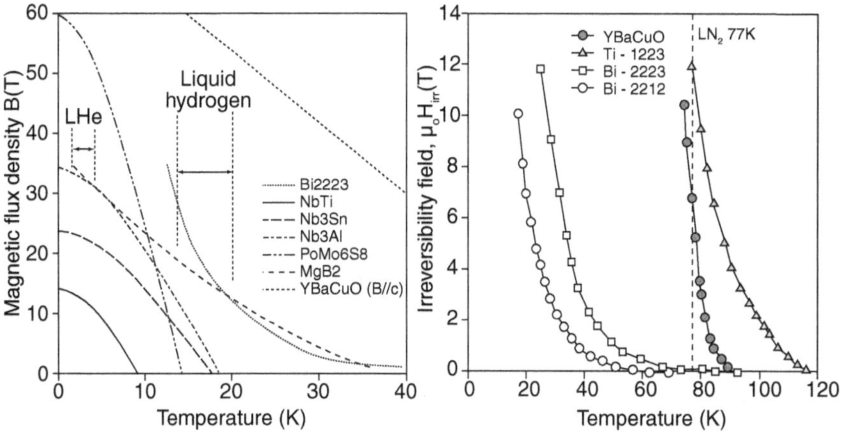

Figure 16.2 Plot of magnetic flux density irreversibility, B_{irr}, versus temperature, T, for: (a) low-critical-temperature superconductors, medium critical temperature superconductor MgB_2 ($T_c = 39\,K$) and high-critical-temperature superconductor $YBa_2Cu_3O_7$ – magnetic flux density, B, is parallel to crystallographic c-direction of the superconducting $YBa_2Cu_3O_7$ layer (Glowacki and Nuttall, 2010); (b) $(Bi,Pb)_2Sr_2Ca_2Cu_3O_x$, $Bi_2Sr_2CaCu_2O_x$, $(Tl,Hg)Ba_2Ca_2Cu_3O_x$ and $YBa_2Cu_3O_7$.

more that 20 years ago, it was expected that shortly afterwards most of the electrotechnology applications would be fully developed. Some technological and economical issues like granularity of the YBCO conductors, $B_{irr}(T)$, and the manufacturing cost of Bi-based conductors, prevented this from happening quickly. Despite the fact that YBCO-coated conductors have been under development for the last 25 years using physical and chemical deposition techniques (Chen *et al.*, 2009), long pieces of such conductors for applications are rarely available even now. Until the performance of such conductors reaches the desired level, YBCO electromagnets generating very high magnetic induction at 15–20 T will need to operate at a temperature of 20–30 K (see Figure 16.3).

For this purpose there are only two cryogenic liquids that can be considered between nitrogen (77 K) and helium (4.2 K). These are: very expensive neon (25 K) or cheap but potentially hazardous hydrogen (20 K). The R&D community is making steady progress improving the superconducting properties of conductors to apply them in everyday life (see Figure 16.4).

Recently conducted simulation of the global resources of helium presented in Chapter 10 predicts that there will be a continuous price increase (Cai *et al.*, 2010), which will cause a reduction in demand. Figure 16.5 indicates that helium's wholesale price tends to increase with time, although in principle the price may drop if oversupply occurs. This may happen if the industry overestimates future demand, resulting in overinvestment.

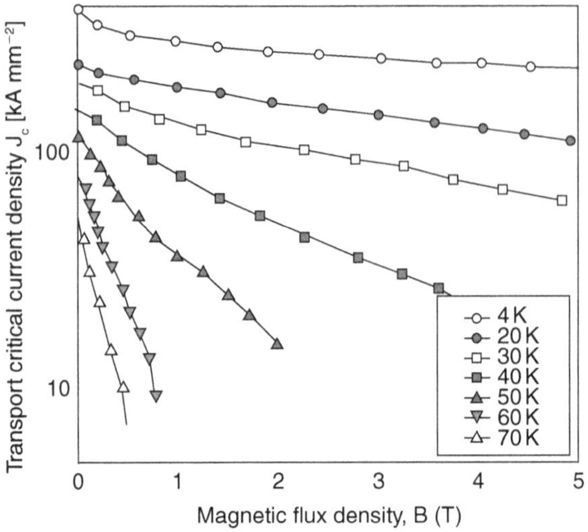

Figure 16.3 Critical current density versus magnetic induction $J_c(B,T)$ for $YBa_2Cu_3O_7$ conductor.

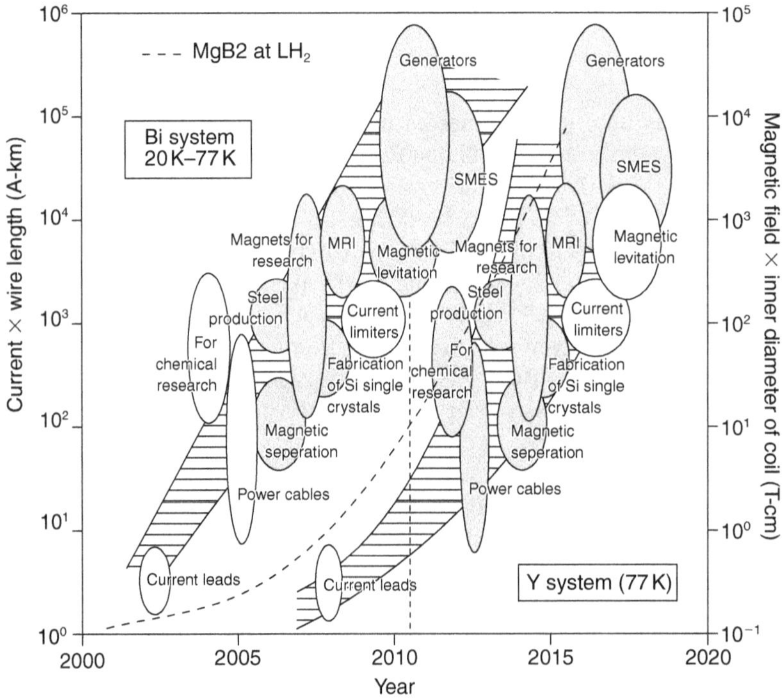

Figure 16.4 $(Bi,Pb)_2Sr_2Ca_2Cu_3O_x$, MgB_2 and $YBa_2Cu_3O_{7-x}$ conductors market penetration prediction (modified in 2010). One can notice that Bi-based conductors and MgB_2 conductors are envisaged to operate at a temperature of 20 K.

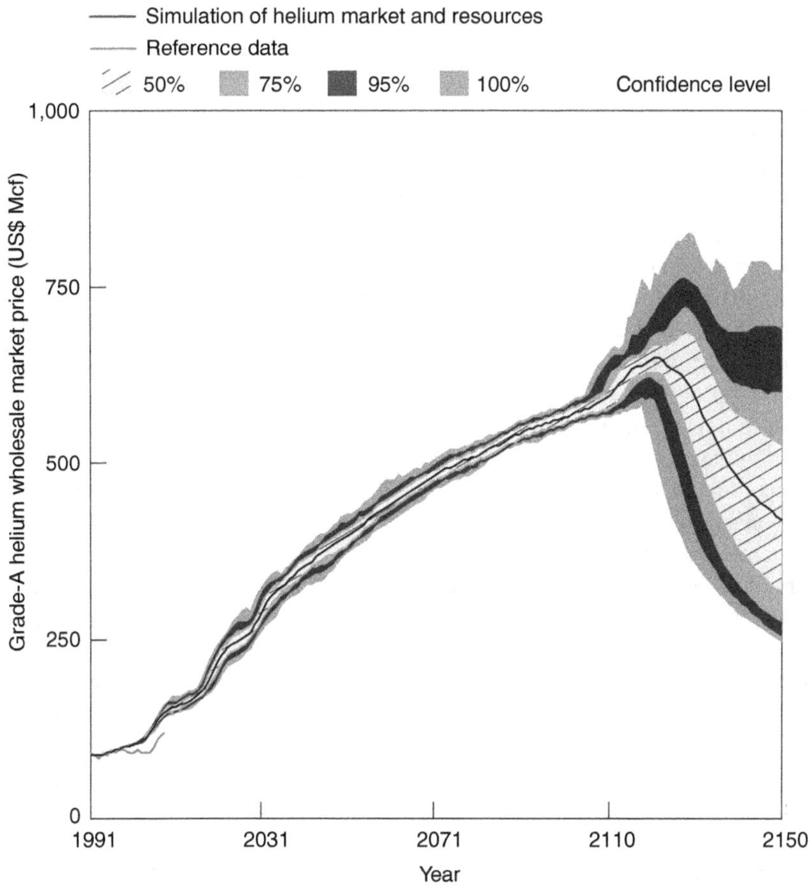

Figure 16.5 Helium wholesale price US$/MCF (2008) from Monte Carlo simulation, (Cai *et al.*, 2010).

It was noted from the recent economic slowdown that prices stabilised rather than fell and that, more specifically, the USBLM price effectively became a floor price for the market. Then, in the long-term contracts under which wholesalers must supply helium, under supply is likely. Price variation is mainly caused by two factors: one is the production cost; another is the imbalance between demand and supply, which is called excess demand in the model. This constant rise in price will put a downward pressure on demand, eventually curtailing market growth as consumers seek alternative technologies to reduce their helium dependence. Because helium is a by-product of natural gas production, its value is very small compared with the value of natural gas (even though it costs more per Mcf, the raw helium concentration is nearly always very small). Therefore, any effort to reduce the end demand of helium via recycling or alternatives technologies does not significantly change the depletion path of helium resources. It is happening already in some parts of the world, where low-temperature research has had to stop due to the very high cost of the liquid helium that in 2010 had reached ~£50 (US$75) per litre of LHe.

Long-term operation of efficient cryomagnetic applications requires reliable Gifford-McMahon close-cycle or pulse tube cryorefrigerators (see Chapter 11) to provide shielding and condensation of the helium vapour at 4.2 K. Such systems have been perfected by Siemens Magnet Technology. Combined with superconducting switches, electromagnets trap large magnetic fields and have so-called zero helium boil-off (see Chapter 13). The drawback is that low-power cryorefrigerators operating with helium gas in a closed-cycle generate 1 W of cooling capacity at 4.2 K, but they require an input power of 3 kW at room temperature. In the current climate of energy-efficiency awareness, this may create commercial problems. There are other ways to minimise helium from the superconducting device, such as thermal conductance gaskets for zero-boil-off superconducting magnets.

The continuously increasing price of helium has inspired the development of precautionary measures for distributed cryogenics, such as decentralised LHe-microplants that control liquid helium consumption, generation and conservation. There are companies such as Cryomech Inc., who manufacture LHe-microplants that allow the user to liquefy helium locally from any room-temperature helium source (see Figure 16.6b).

Such devices are ideal for laboratories, to help researchers to manage their own liquid helium consumption and conservation. Despite the positive outcomes of such measures, a high energy consumption is required to run such systems. Also, the system itself is relatively expensive and does not provide a long-term answer to the more fundamental question about the economical provision of helium gas.

As was predicted by M. A. Green (Green, 2001), the cost of niobium titanium coils cooled by a two-stage cryocooler at 4 K can be substantially lower than coils made with an HTS conductor such as Bi2223 cooled to 20–30 K by a cryocooler. The development of cryocoolers that reliably operate below 4 K will allow magnets made from LTS conductors to remain competitive for many years to come. A key enabling technology for the use of low-temperature cryocoolers on LTS magnets has been the development of HTS leads (see Figure 16.6b) (Ballarino and Carrillo, 2008). Companies are routinely producing superconducting cryogen-free magnet systems capable of generating an induction of 18 T, but tests show that cooling them from room temperature to 10 K takes about 60 hours. There are also limits involved in the use of 4 K cryocoolers. They do not work well when the magnet cold mass is over a few tonnes or when the lead current is over 500 A.

Low-energy-consumption devices can rely solely on closed-cycle cooling systems. However, in some cases – such as with fault current limiters, transformers, fusion electromagnetic coils, etc. – intensive direct cooling is required rather than just conduction cooling. Also, many low-temperature physics oriented research facilities require active cooling by liquid helium.

There is often a necessity for direct cooling by helium, such as in superconducting devices like MRI or NMR electromagnets, and the powerful toroidal and poloidal electromagnets of the ITER fusion reactor (ITER) currently being constructed from metallic superconductors such as Nb_3Sn and NbTi, as shown in Figure 16.2.

It is important to realise that there are other devices, such as astrophysics detectors and sensors (Withington and Goldie, 2008), that need to operate at ultra-low temperatures (in the milli Kelvin (mK) range) where dilution refrigerators or sophisticated Lambda plates using liquid helium are required. In dilution refrigerators a mixture of two isotopes of helium, ^3He and ^4He, are used. Below 870 mK the mixture undergoes a spontaneous phase separation to form a

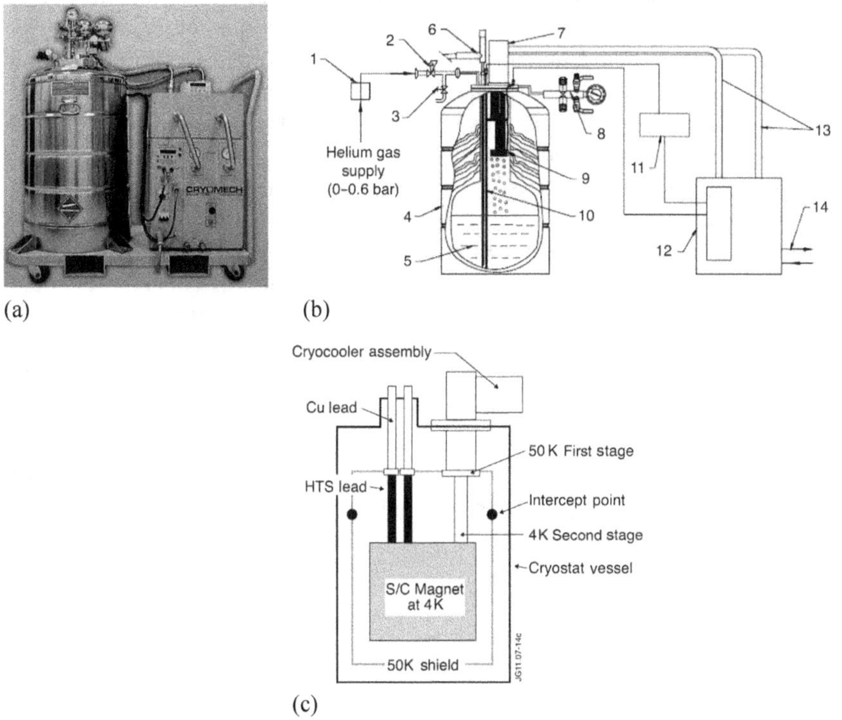

Figure 16.6 (a) LHe-microPlant, that allows the user to liquefy helium locally from any room-temperature helium source (courtesy: Peter E. Gifford from Cryomech); (b) Schematic of helium liquefaction system. 1. Mass flow meter; 2. Pressure regulator; 3. Vacuum valve for dewar pumping; 4. Liquid helium dewar; 5. Liquid helium; 6. Liquid withdrawal line; 7. Cold head of pulse tube cryocooler; 8. Safety unit; 9. Temperature sensor; 10. Liquid helium level sensor; 11. Liquid helium level indicator/controller; 12. Helium compressor of cryocooler; 13. High and low SS flexible lines of cryocooler; 14. Cooling water (after Wang, 2009); (c) A schematic diagram of a superconducting magnet with superconducting current leads characterised by very low thermal conductivity and radiation shield cooled with a two-stage 50 K and 4 K Gifford-McMahon cryocooler (after Green, 2001).

^{3}He-rich phase and a ^{4}He-poor phase. Lambda plates can reduce the temperature to 2.2 K. In all the above cases, the cooling systems are very expensive, require high maintenance and have very low running efficiency. But this is not an issue because one must have such low temperatures to operate these ultra-sensitive devices.

MRI magnets

Preventing loss of helium coolant in superconducting devices is an important issue in the case of MRI magnets. Implementation of zero-boil-off technology provides a viable solution as such, because helium losses during MRI quenches may reach unacceptable cost levels and affect the servicing of the medical diagnostic devices. Current Nb-based superconductors, such as NbTi and Nb_3Sn, require cooling by liquid helium at 4.2 K, or in some cases even cooling by supercritical helium at 1.8 K, to achieve the required superconducting current in the electromagnet's winding to obtain the high magnetic induction, *B*, generated by the magnet. The introduction, in 2001, of the relatively simple superconducting conductor MgB_2, characterised by T_c 40 K, raised the expectation that such superconductors would be used at temperatures of 14–20 K (Glowacki *et al.*, 2003). Such an increase of the temperatures involved would permit the use of more efficient cryocoolers but, above all, will enable the use of indirect liquid hydrogen cooling (i-LH_2) for the MRI magnet systems. An outline of the current MRI market trend and future directions is discussed in Chapter 12.

Low-resolution, 0.3 T permanent-magnet MRI systems are selling well in China. Millions of poor farmers cannot afford to go to major cities for check-ups on MRI systems using liquid helium. A cryogen-free, or if interfaced with a hydrogen economy, an i-LH_2-cooled 1.5 T MgB_2 MRI system, could have a large social impact. See the later section on "Indirect cooling of magnets by liquid hydrogen" for more details on i-LH_2.

Magnetic separation

In 1986, Eriez Magnetic Inc. and J. M. Huber Corporation of Erens, Georgia, USA produced a superconducting separator generating a field of 2 T at the diameter of the canister of 2.1 m. It successfully operated on kaolin clay using ferromagnetic stainless steel wool. The LHe requirement was 1,000 litres of liquid storage on the site. For mineral processing, the switch-on/switch-off system at high magnetic induction was not economically viable. However, Oxford Instruments has successfully developed magnetic separation systems based on cryocoolers (see Chapter 13).

Induction heating

In large-scale aluminium, copper and brass extrusion works, processing speed and quality depend on the homogeneous distribution of heat within billets. Magnetic

Billet Heaters based on the superconducting $(Bi,Pb)_2Sr_2Ca_2Cu_3O_x$, (Bi2223) electromagnets operating at 30 K, manufactured by Zenergy Power and Bültmann achieve better results than conventional induction or gas heating (and achieve 50 per cent power savings, a 45-second heating time, and a 25 per cent productivity increase). The magnetic heating process delivers homogeneous radial temperature distribution and precisely reproducible axial temperature tapers in every type of billet (www.zenergypower.com), as shown in Figure 16.7.

Thanks to the use of HTS technology, the operating power (including that of the cryocoolers) can be dramatically reduced. This would not the case if LTSs cooled by LHe were to be used. Use of LTSs would require much better thermal insulation between the hot billet and superconducting winding, as well as increased quantities of superconductor, which would raise costs dramatically.

The crossover in $B_{irr}(T)$ for the $(Bi,Pb)_2Sr_2Ca_2Cu_3O_x$ conductor and MgB_2 conductor at 20 K (see Figure 16.2) brings with it the new possibility that simple and cheaper MgB_2 conductors could be used in the construction of second-generation magnetic heaters, cooled, indirectly, by LH_2.

Energy storage

Superconducting magnetic energy storage (SMES), supercapacitors and flywheels are possible options for relatively large and fast-accessibility high-power energy storage.

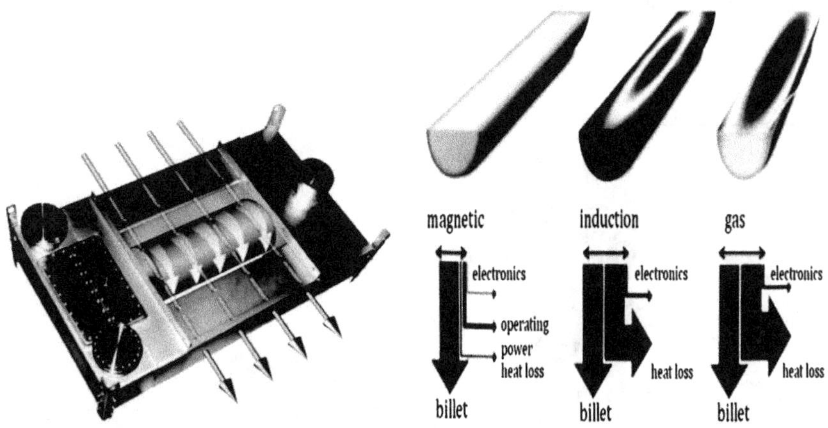

(a) (b)

Figure 16.7 Superconducting electromagnet based billet heating: (a) The 30 K DC superconducting Bi2223 coil inside a cryostat cooled by cryocoolers creates a steady magnetic field in which rotating the billet induces uniform heating by induction; (b) comparison of magnetic heating of the billet rotated in a magnetic field (energy efficient, fast and uniform heating) with inefficient and non-uniform induction heating by high frequency induction coils; similarly for gas heating. (Images are courtesy of Dr M. Backer from Zenergy Power Ltd).

SMES

In circumstances when a high electric power is needed relatively often for a short time, then superconducting magnetic energy storage (SMES) can be considered. In applications such as particle accelerators and in fault situations in the utility grid, high power pulses are needed at frequencies in the range of 0.1–1 Hz to compensate for fast alterations in the load and therefore require large energy bursts (Seeber, 1998). If these bursts are taken from, and fed back to, the utility grid its voltage is distorted and the quality of the electricity degrades. In addition, these bursts may decrease the operational lifetime of generators and electricity producers may be required to make extra arrangements. Thus, energy storage is an essential component for a cost-effective particle accelerator (Lyly *et al.*, 2010).

The major advantage of an SMES is the pre-existing cryogenic support provided by the cooling system of the superconducting magnets of the accelerator ring (Gehring *et al.*, 2006). However, the complex problem of AC losses still has to be controlled (Lyly *et al.*, 2010).

Currently, the feasibility of high-energy pulse SMES has been demonstrated for NbTi conductors cooled by LHe. High-temperature, multi-filamentary superconductors such as Bi2223 and YBCO are not yet mechanically feasible for large AC applications (Larbalestier *et al.*, 2001). There is an expectation that MgB_2 may replace NbTi conductors in applications at LH_2 temperatures.

The general design of an SMES is very similar for all conductor types (LTS, MTS and HTS), as shown in Figure 16.8. The only difference is in the cooling system and the readiness for DC or AC operation.

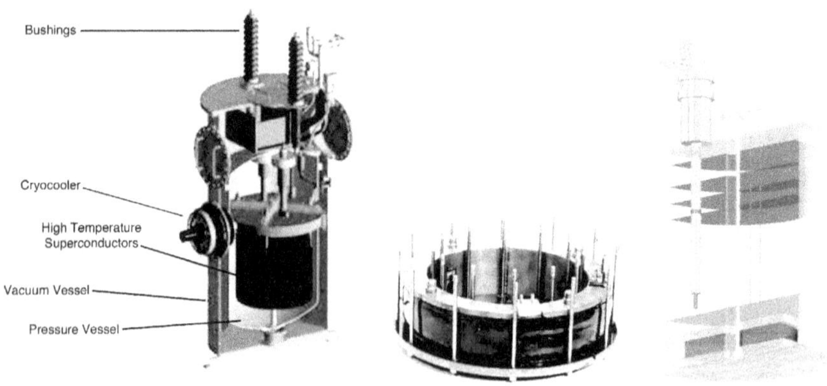

(a) (b) (c)

Figure 16.8 (a) Schematic of the SMES system based on a high-temperature superconducting coil; (b) $(Bi,Pb)_2Sr_2Ca_2Cu_3O_x$ 6 pancake coil with 10 kJ energy stored at 20 K; (c) cross-section of the cryostat system of an SMES, with a cryocooler providing cryogenic temperature of 4.2 K to the LTS superconducting winding at the bottom of the cryostat (Hiltunen *et al.*, 2008).

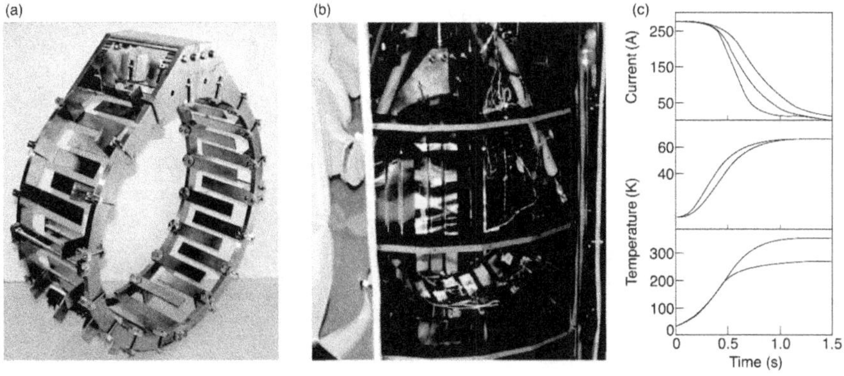

(a) (b) (c)

Figure 16.9 Nb_3Sn SMES system operating at 10 K: (a) Cu stabilisation for the cryogen-free, conduction-cooled SMES coil; (b) View of the fully assembled 200 kJ SMES system, nominal current of the coil was 275 A at 10 K; (c) Results of measured current decay, and quench analysis of thermal interface temperature. The hot-spot temperature reaches room temperature (Korpela *et al.*, 2003).

Flywheels

Superconducting magnetic bearings enable low-loss, contactless rotation with a form of levitation that is inherently stable. This gives them significant potential in high-speed applications, such as flywheel energy storage and turbomachinery. The most popular superconducting bearing configuration is the bulk superconducting/ferromagnetic configuration presented by ATZ, as shown in Figure 16.10. One of the major factors limiting levitation force for such superconducting magnetic bearings is the maximum possible *remanence* induction, B_r, of 1.4 T known to exist in rare earth permanent magnets. Therefore, considering the $B_{irr}(T)$ data for YBCO presented in Figure 16.2b, it is unnecessary to cool the superconducting stator to a temperature below LN_2, thus eliminating the need for cryogenic helium coolant.

Another solution to improve the performance of superconducting bearing is to use an NbTi superconducting magnet as the stator, and YBCO for the rotor. The design consists of a cryocooled, low-temperature superconducting stator coil of NbTi generating field of 5 T (4 K) with a Gd-Ba-Cu-O rotor cooled to 65 K (low-pressure liquid nitrogen cools the high-temperature superconducting (HTS) bulks (GdBCO)). The coil consists of two, face-to-face configuration (N-N) windings inducing a strong magnetic flux gradient. Such a solution requires cooling of the superconducting LTS coil to 4 K and makes the design quite complex and impractical.

Probably the most advanced high-energy, high-strength and stability bearing is the bulk, fully superconducting bearing, as shown in Figure 16.11.

Figure 16.11 shows the novel concept of a magnetic bearing, which uses the difference in irreversibility field of two superconducting components, to allow

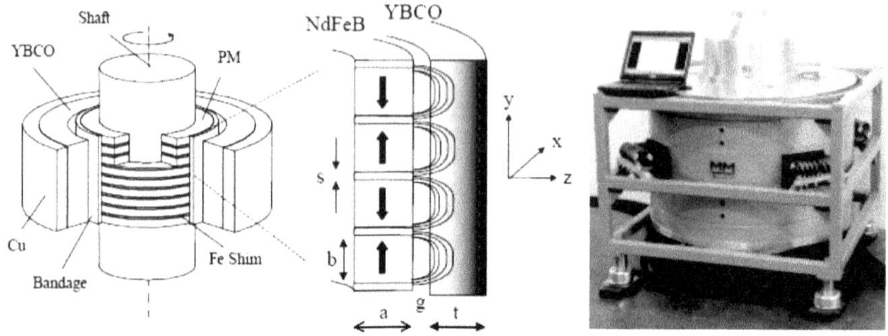

(a) (b)

Figure 16.10 (a) Schematic of diagram of the working *principium* of the superconducting/ferromagnetic bearing; (b) Image of the complete flywheel produced by ATZ, Germany. (Courtesy: ATZ).

one superconducting component to be field-cooled (that is, cooled in situ) in the magnetic field originating from the other component. The latter can first be magnetised at a higher temperature and can exploit a liquid hydrogen indirect cooling system (Clarke and Glowacki, 2010; Glowacki, 2010; McDonald *et al.*, 2002).

Magnetised (RE)BCO bulks with high trapped fields can be used as one of the components instead of permanent magnets, giving a significant increase in the levitation force density that can be achieved between the two components. Also, there has been progress in the manufacturing technology of large bulk superconducting plates or tubes of MgB_2 (T_c ~40 K) to overcome the granularity problems of HTS (see Figure 16.12). These have become very important as superconducting bearings materials for flywheels. Considering that the temperature provided by liquid hydrogen will be at the level of 15–20 K, the use of the i-LH_2 concept is strongly justified in such cases (Giunchi *et al.*, 2006).

Pulsed field magnetisation of the pellets is essential to create high field gradients using MPSC (multi-pulse technique with stepwise cooling). The new design has the potential to achieve levitation force densities over 100 N/cm^2. The increased force density that a superconducting bearing offers allows a rotor of a few tonnes to be levitated with a relatively small bearing. Conventional superconducting bearings, using permanent magnets as a field source, give a limited force density. As a result, doubling the rotor weight requires doubling the amount of superconductor used in the bearing.

As well as offering the potential to levitate very high loads, a bulk-bulk bearing can offer very high stiffness whilst also being compact. This gives the design potential for turbomachinery and general high-speed bearings where wear and low life expectancy are a problem with existing ball bearings. The new design can also accommodate high g-force environments unlike existing superconducting bearings. For cases where bearings are used in a cryogenic context, a bulk-bulk bearing could offer a particular economic advantage. These include cryogenic turbopumps for liquid hydrogen and other cryogens (Clarke and Glowacki, 2010). Space applications are also a natural choice, where increased rotational speed for controlling gyroscopes and stabilisation wheels can allow the weight of the rotating mass to be reduced (Patel *et al.*, 2011). There is also another context, in which a decentralised economy can be envisaged. This will be discussed in the next section of this chapter.

Indirect cooling of magnets by liquid hydrogen, i-LH_2

The motivation to cool any magnet cryogenically, DC or pulsed, superconducting or not, is to improve the electrical conductivity of its conductor. Considering

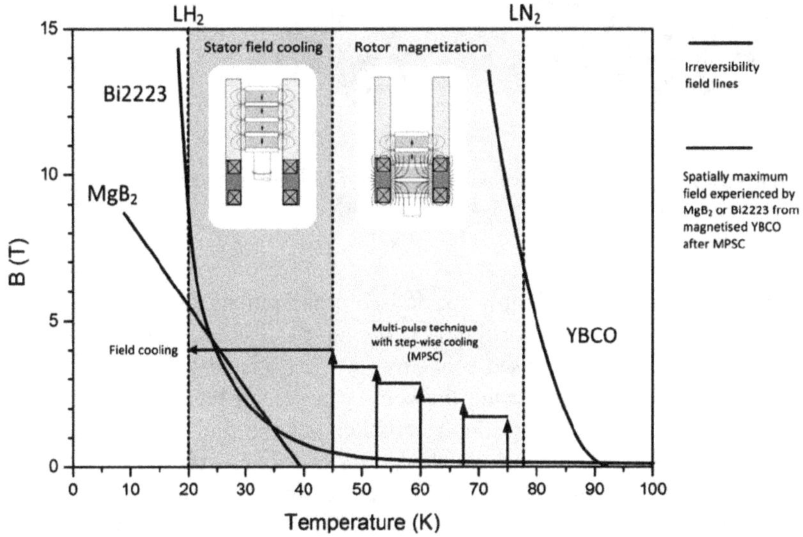

Figure 16.11 Typical irreversibility field lines for bulk YBCO, MgB_2, and $Bi_2Sr_2Ca_2Cu_3O_{10-x}$ (Bi-2223). B_{irr} for MgB_2 is defined as the field for which there is loss for macroscopic currents in a polycrystalline bulk. A large difference in the irreversibility fields for YBCO and MgB_2/Bi-2223 allows field cooling of the stator component of the bearing by lowering the temperature *after* the YBCO rotor component has been magnetised. The MPSC (multi-pulse technique with stepwise cooling) technique allows flux to be built up inside bulk YBCO over a large temperature range, before in-field (*in-situ*) cooling the MgB_2 or Bi-2223 down to 20 K. The insets schematically illustrate a possible configuration of a superconducting magnetic bearing in which the field source on the rotor is made up of cylindrical magnetised bulks of superconductor YBCO. The stator component is a hollow cylinder of superconductor MgB_2 or Bi2223. The rotor bulks are magnetised by displacing the whole rotor to allow alignment with a magnetising coil on the stator. Opposing field magnetisation is shown in which a bulk pair is simultaneously magnetised with opposite polarity by a coil split into two parts wound with opposite sense (Patel *et al.*, 2011).

Figure 16.12 Hollow cylinder and disk made of MgB_2 (diameter 43 mm) (courtesy: Prof. G. Giunchi, Edison s.p.a., Italy).

liquid hydrogen safety, only highly specialised organisations and companies can handle direct cooling. But indirect liquid hydrogen cooling, i-LH_2, can be seen as a viable option for a decentralised economy. In i-LH_2 installations a helium gas exchanger can be used, transferring the cooling power of the hydrogen bath at ~20 K to the desired cryomagnetic installation (Clarke and Glowacki, 2010; McDonald *et al.*, 2002). Economical calculations of cooling efficiency for a large 15 T electromagnetic non-superconducting device were supported by a detailed study of the purpose-built installation in the USA. The considered schemes were LHe, LH_2 and LN_2. This showed very clearly that direct cooling by helium or neon is between 70 and 100 times more expensive than indirect cooling by LH_2 (MacDonald *et al.*, 2010). The proposed schematic of the installation is shown in Figure 16.13. Taking into account the price of 1 litre of LH_2 at ~£0.5 (US$0.65), LHe at ~£3 (US$4.50) and Ne at ~£150 (US$225), and also the fact that hydrogen is the only cryogen apart from LNG that is an energy carrier, the choice of LH_2 as a cryogenic cooling medium for cryomagnetic applications is rather apparent.

Electromagnets of magnetic confined-fusion reactors

Chapter 15 gives an introduction to the requirements of magnetically confined fusion. It is projected that a DEMO fusion reactor and future reactors will require an alternative approach where medium-temperature and high-temperature superconductors will be used to provide high magnetic induction at elevated cryogenic temperatures (Glowacki and Nuttall, 2008; Clarke *et al.*, 2010; Glowacki and

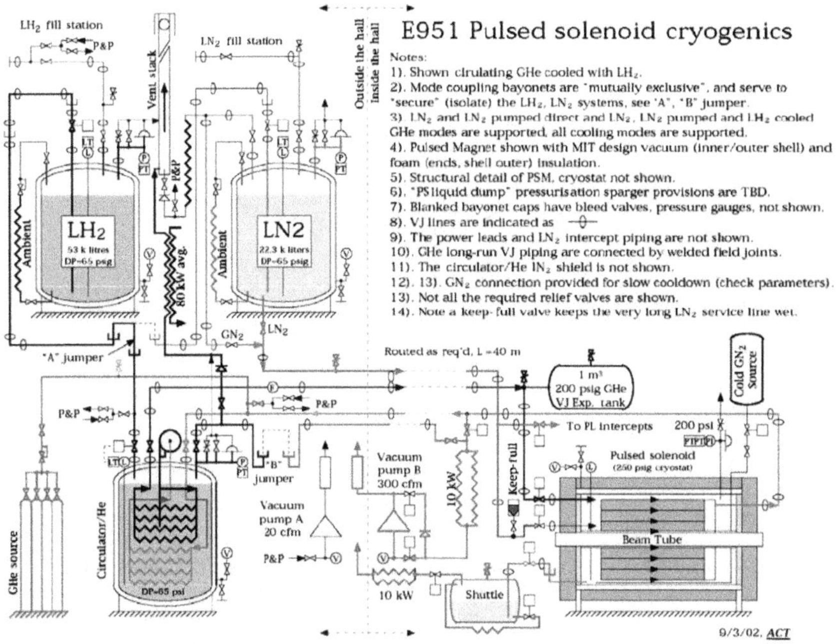

Figure 16.13 A process flow diagram for cooling of a pulsed solenoid magnet (normal conductor) via consumption of liquid hydrogen. The magnet itself would actually be cooled by helium gas that is cooled in a heat exchanger filled with the boiling liquid hydrogen. The heat exchanger would be located outside the experimental hall, which is schematically marked as a vertical broken line (McDonald *et al.*, 2002).

Nuttall, 2010). As shown in Figure 16.2b, some of the high-temperature superconductors, due to their specific temperature-dependent magnetic properties can be only effectively be used at LH_2 (20 K) or LNe (25 K) temperatures. For example, a preliminary design of a DEMO CICC, made with Bi-2212 round strands and operating at 20 K and 13 T, leads to a value of J_{cable} not very different from a Nb_3Sn CICC cable in the range of 54 amps per square millimetre (Duchateau and Hertout, 2008).

Indicative, early-stage design directives for the future DEMO fusion reactor require toroidal and poloidal electromagnets to be made respectively from $YBa_2Cu_3O_7$ and also MgB_2 conductors to achieve a desired magnetic flux density of 16 T in the plasma confinement (Fiet, 2008). However, there are approaches by Japanese researchers to develop advanced Nb_3Al conductors which will withstand the stresses and will provide the necessary field in the toroidal field coils (Figure 16.4). Unfortunately, such a quality research outcome may be expensive if one considers that there will be shortages of helium at a price acceptable for the DEMO operators (see Figure 16.5).

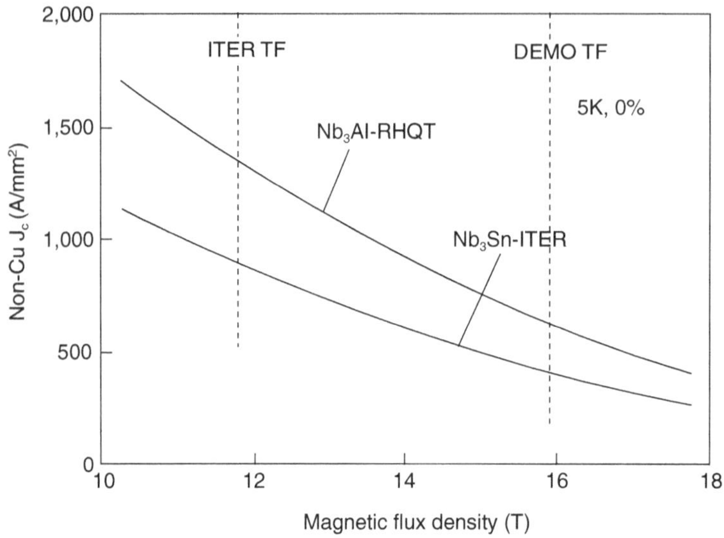

Figure 16.14 Comparison of performance of the Nb_3Sn and Nb_3Al strands for the TF coils for ITER and DEMO reactors. The critical current density is required to be 450 A at 15.9 T in the TF coil (temperature 5 K, strain 0 per cent) of the fusion DEMO reactor currently being designed by Japan Atomic Energy Agency (Hemmi *et al.*, 2009).

Especially for large installations, a predicted shortage of helium, and an already increased price, creates a necessity to develop new cooling systems. Little attention has been paid to the hydrogen option, since direct liquid hydrogen cooling would be very challenging. However, by adapting the already established concept of an intermediate low-loss helium loop, the substantial benefits of using 20 K cooling could be harnessed without encountering the direct safety impacts of hydrogen. As a coolant, 20 K (or possibly 15 K) helium gas will have different characteristics than a boiling cryogen. There will be no critical heat flux issue but local heat transfer coefficients may be lower. If microchannels are engineered into the system substrates then good thermal coupling could be achieved, perhaps surpassing the performance of conventional liquid helium cooling. Clarke and colleagues (2010) compare and discuss the technology implications of helium-cooled indirect hydrogen (i-LH_2) versus direct helium and nitrogen cooling, and identify the related research needs. Figure 16.15 shows a possible concept.

Hydrogen generation

Demand for hydrogen is growing every year and in just the USA, for example, the actual mass growth rate is predicted to increase dramatically from $n=2$ to $n=6$ (see Figure 16.16).

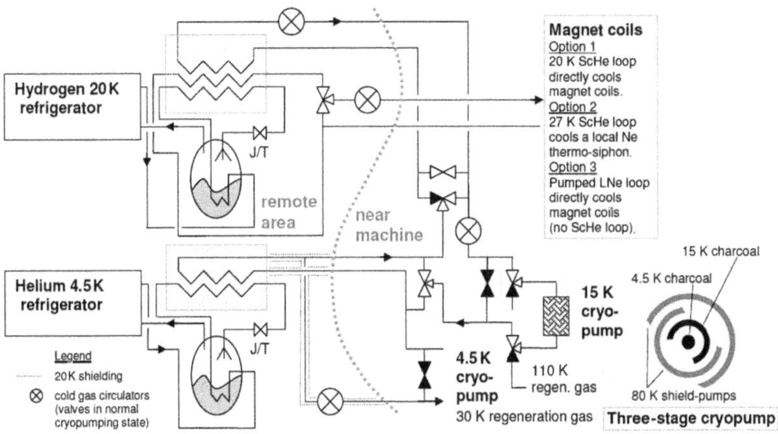

Figure 16.15 Combined use of hydrogen and helium refrigerators for three-stage cryopumps and HTS magnets (Clarke and Glowacki, 2010).

Steam reforming of natural gas is one of the major methods of hydrogen manufacture. However, this is of no help to a hydrogen economy, which must derive hydrogen from sources other than the fossil fuels it is intended to replace. The two directions considered are true renewables (solar, wind, tides and waves) or nuclear reactions. This is the dilemma: a decentralised or centralised hydrogen economy. While nuclear-generated electricity could be used for high-temperature electrolysis, nuclear heat can also be directly applied to split hydrogen from water. High-temperature (~1,000 °C) gas-cooled nuclear reactors (Chapter 14) have the potential to split hydrogen from water by thermochemical means using nuclear heat. Research into high-temperature nuclear reactors may eventually lead to a hydrogen supply that is cost-competitive with natural gas steam reforming. General Atomics predicts that hydrogen produced in a high-temperature gas-cooled reactor would cost US$1.53/kg (~£1/kg) (Schultz, 2003). Even taking into account that 30 per cent of the energy is used for the liquefaction of hydrogen, the price of the liquid hydrogen could be as low as US$0.153/litre-$H_2$ (£0.10/litre). The current price of liquid hydrogen obtained by steam reforming of natural gas is five times higher. There is a Scandinavian hydrogen network programme to integrate distribution and production of the hydrogen across Denmark, Norway and Sweden (Denmark Hydrogen; Norway Hydrogen; Sweden Hydrogen). In the UK there are also proposed hydrogen networks such as the "Scottish Hydrogen Corridor" along the A60 route (150 miles long between Aberdeen and Inverness), and along the M4 motorway as "CymruH_2Wales" that would join Wales with London.

Centralised H_2 economy – superconducting cable

In 1967 the British company BICC Ltd designed and built a laboratory-scale superconducting 3-metre link containing a coaxial pair of tubular niobium

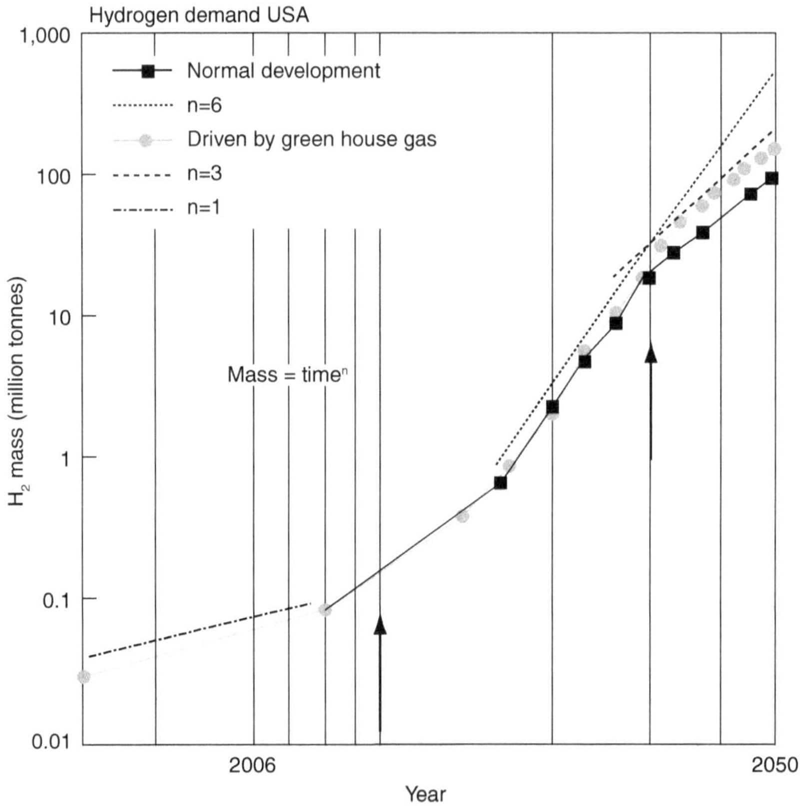

Figure 16.16 Predicted mass production demanded by the USA economy versus time. It is evident that the differences for two scenarios: *normal development* and demand driven by *greenhouse gas* are almost identical until 2050. The n value of the calculated mass=time^n may increase from n=2 to n=6 over next ten years.

conductors in a cryogenic envelope cooled by LHe, to test the feasibility of superconducting AC transmission (Edwards, 1968). In 1976, Siemens (Erlangen, Germany) built a 35 m test facility for a one-phase superconducting semi-flexible Nb-based AC cable. As well as the ordinary laboratory-scale experiments which have been performed to investigate various cable components, larger test facilities have also been installed at European laboratories in order to study size effects and as a first step to demonstrating the interaction of components and feasibility and performance of superconducting cables (Bogner, 1975).

The Nb_3Sn conductors for transmission cables were successfully tested at Kryzhanovskij Power Engineering Institute, KPEI, in Moscow, on the short-length demonstrator. A 100-metre long three-phase DC cable with tubular conductors was constructed for KPEI, Moscow. The conductors had an internal diffusion Nb_3Sn layer but the design suffered inherent mechanical problems due to its rigid structure. Complexity of the helium supply over the long distances and demanding cryogenics put all hopes for superconducting transmission cables on hold. The only fluid permitting use of LTS at temperature levels below 10 K is helium. For large-scale systems the low-temperature environment can be realised by a forced flow of supercritical helium through cooling channels along the conductors. The stable performance of a superconducting magnet can be disturbed by small heat releases in the conductors caused by magnetic or mechanical instabilities. These instabilities are transient effects measured in milliseconds, and so the transient heat transfer to the cooling fluid is important in ensuring the stability of superconducting systems (Bloem, 1986).

In the 1970s, there was intensive research conducted worldwide on the cryoresistive cables composed of pure aluminium or copper conductors cooled by LN_2 for delivery of the high current demand over the short distances. This rather simple technology was well developed but did not provide the answer for long-distance energy transfer, which was of great interest for large, developed countries such as Russia and the USA.

The discovery of high-temperature superconductors such as YBaCuO revitalised transmission research projects based on the experience of the LN_2 cryoresistive cables.

Initially cables had to be constructed on the basis of the Bi2223 conductor, which was characterised by a low irreversibility line (see Figure 16.2b), but due to the low magnetic induction generated by the cable this conductor could be used for energy transmission using LN_2 at temperatures ranging from 65–80 K. However, a model of the LN_2 cooling/pumping station infrastructure was

Figure 16.17 Three-phase HTS superconducting cable, courtesy of Siemens, Erlangen, Germany.

developed for DC and AC cables, but the actual cost of the fully stabilised, twisted AC $(Bi,Pb)_2Sr_2Ca_2Cu_3O_x$ conductor could not be lowered due to the materials, energy and labour-intensive processing. The only possible financial justification could be to lower the operational temperature of such a conductor to that of liquid neon (25 K) or hydrogen (20 K). On the other hand, at a temperature of 20 K, as presented in Figure 16.2a, there is a crossover in $B_{irr}(T)$ for the $(Bi,Pb)_2Sr_2Ca_2Cu_3O_x$ conductor and MgB_2 conductor, indicating that simpler and cheaper conductors could be used to construct a superconducting transmission line cooled by LH_2.

Recently, Grant and Hassenzahl (Grant, 2010; Hassenzahl *et al.*, 2009) have addressed several opportunities to exploit MgB_2 conductor developments for transformer and electric cable applications. They have also discussed the possibility of a novel and visionary power delivery system centred on an MgB_2-based DC cable cooled by gaseous or liquid hydrogen, supplying both electrical and chemical energy to the end user. This concept of the supercable with dual delivery purpose – hydrogen and electric power – was considered in detail by Grant (2004) and is presented in Figure 16.18.

Another recent study of the applicability of a 1 GW-class hybrid energy transfer line of hydrogen and electricity has been conducted. Details are given by Yamada and colleagues (Yamada *et al.*, 2010), as shown in Figure 16.19. Current plans are based on using high-temperature fission reactors to produce H_2, with a liquefier at a nearby liquefaction plant, and transfer of LH_2 by superconducting DC cable made of MgB_2 through the Mutsuogawara shipping port for wide distribution. The target distance of hybrid energy transfer lines is

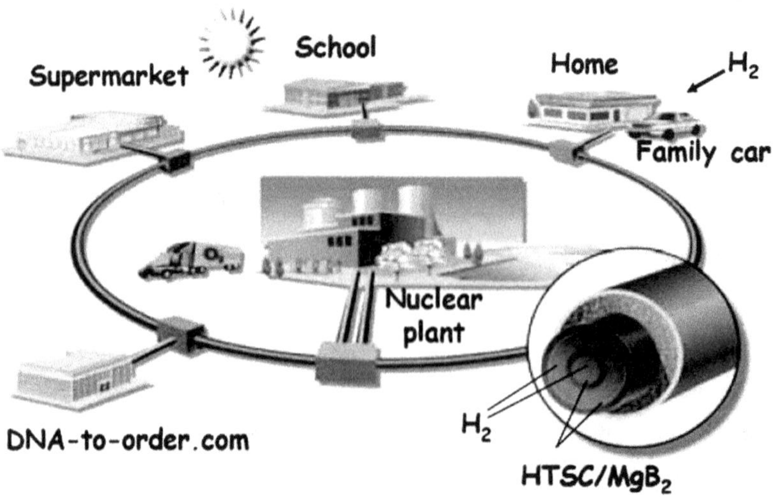

Figure 16.18 Vision concept of an urban community whose complete energy infrastructure comprises electricity and hydrogen generated by nuclear fission and solar roof power distributed through a SuperCable ring bus (Grant, 2004).

1,000 km. Hydrogen refrigeration stations will be placed every 10 km of the unit section. Capacity of the liquid hydrogen transportation is 100 tonnes per day, while a typical hydrogen LH_2 plant delivers 3 tonnes per day. There is a future plan in which cogeneration of H_2 will be made by a fusion reactor. Additionally hydrogen liquefaction can be integrated with gasification, where precooling of the H_2 can be conducted at the expense of gasification of the LNG.

Decentralised H_2 economy

The concept of small-volume, decentralised liquid hydrogen generation devices (Glowacki *et al.*, 2010) faces major engineering and materials challenges relating to process efficiency, where even with the largest plants (10,000 kg/h) need to use about 30 per cent of the HHV (higher heating value) energy for the liquefaction process (see Figure 16.20). The solution to this challenging and complex problem may only come from a new design of the compact liquefaction system where an initial cooling stage of the H_2 gas will come from the mixed-gas portable coolers operating at the 60–80 K temperature range (Gong *et al.*, 2010), but the final liquefaction stage will be reinforced with magnetic refrigeration (Kamiya *et al.*, 2007) and expansion turbines (Clarke and Gardner, 1976), which can dramatically improve the efficiency of decentralised small liquefiers.

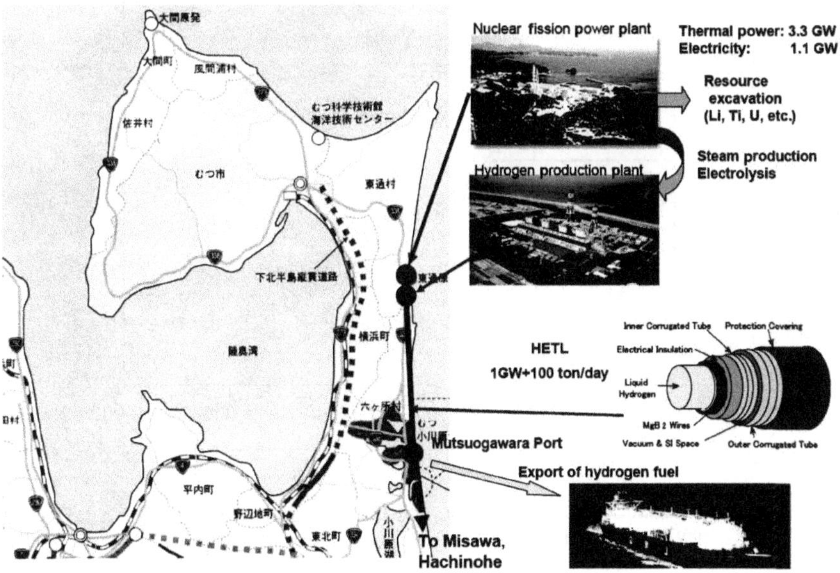

Figure 16.19 Schematic representation of a hydrogen generation, liquefaction, transfer and distribution system using fission and liquefaction. The rated current and maximum voltage of the DC 1 GW power line are 10 kA and 100 kV, respectively. LH_2 will be exported via sea tankers for distribution (courtesy: Yamada *et al.*, 2010).

Ceramic polycrystalline magnetic refrigerators such as the dysprosium-gadolinium-aluminium-garnet can achieve condensation at the level of 90 per cent Carnot efficiency, and liquefaction efficiency reaching 50 per cent of the Carnot efficiency (Kamiya *et al.*, 2007). There is also an urgent need for research on more effective hydrogen generation, purification, compression and ortho-para conversion to improve liquefaction efficiency and long-term storage.

The implementation of smaller hydrogen cryomagnetic installations such as envisaged in decentralised hydrogen energy generation, storage and use may be achieved with a superconducting energy storage device being indirectly cooled by hydrogen in an i-LH_2 loop, and operated as outlined in Figures 16. 21a and 16.21b (Glowacki *et al.*, 2010).

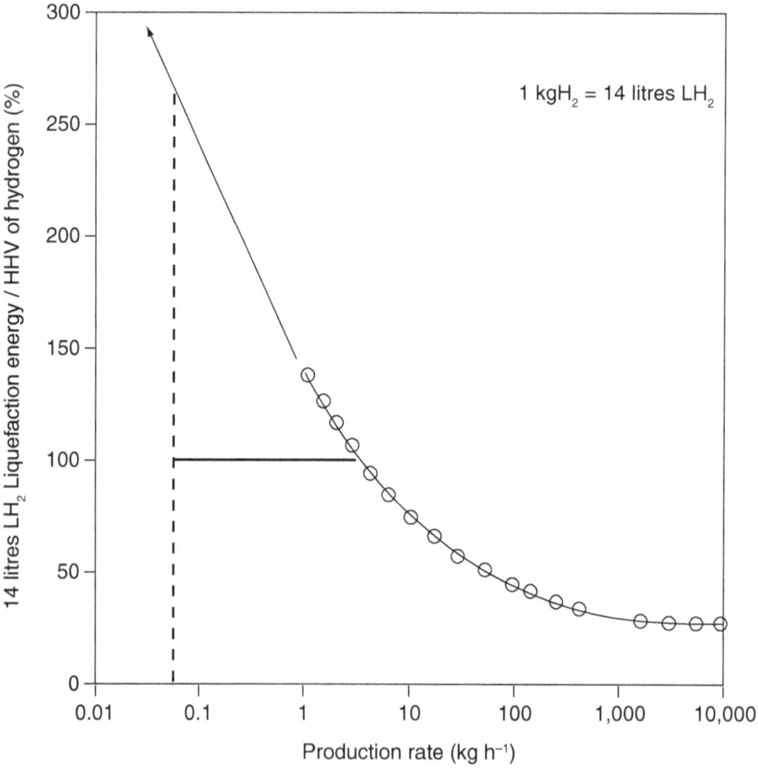

Figure 16.20 For small, basic liquefaction plants the energy needed to liquefy hydrogen may exceed the HHV of the gas. But even with the largest plants (10,000 kg/h), about 30 per cent of the higher heating value (HHV) energy is needed for the liquefaction process. The total energy needed to generate and compress hydrogen at filling stations exceeds the HHV energy of the delivered hydrogen by a factor of at least 1.5. The vertical broken line marks a very low production of 1 litre LH_2/h. It is obvious that the standard H_2 liquefaction process requires dramatic efficiency improvement if it is going to provide energy carrier and coolant for superconducting devices.

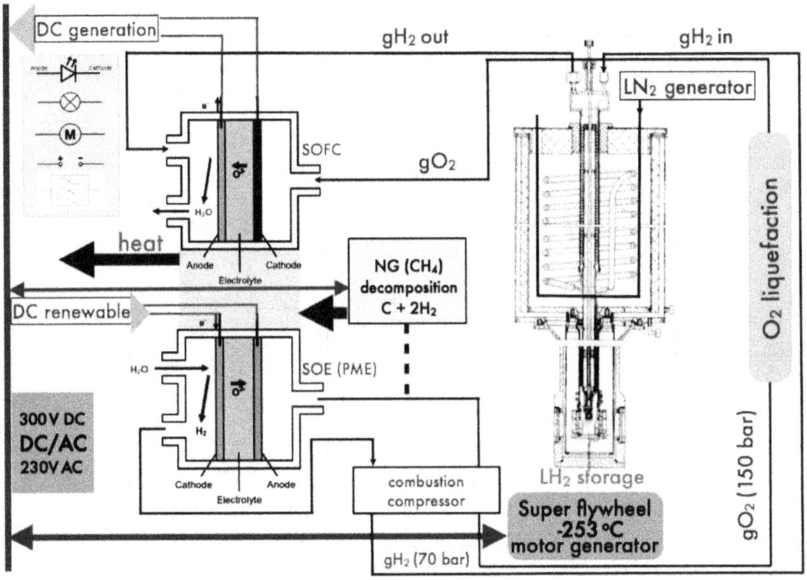

(a)

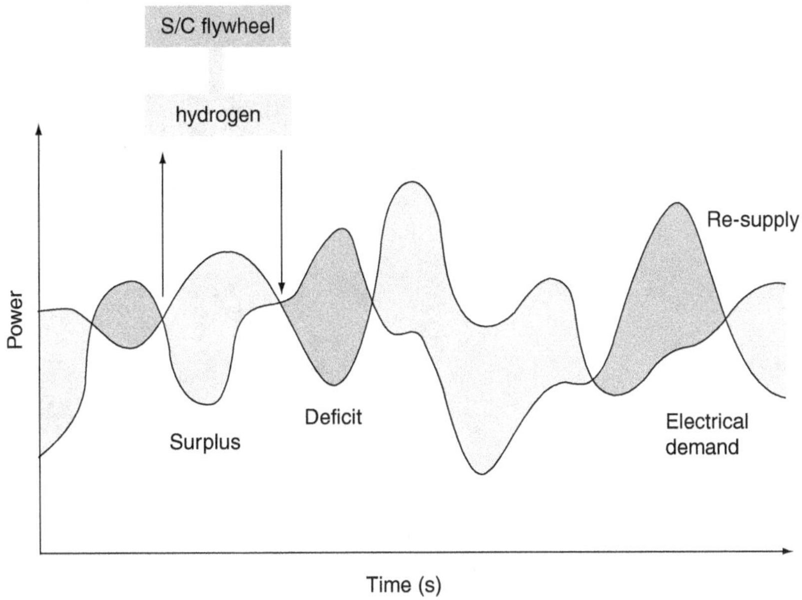

(b)

Figure 16.21 (a) Simplified layout of the decentralised liquid electricity-heat-hydrogen cycle (Glowacki *et al.*, 2010). This clearly involves integration of hydrogen generation, storage and usage with the cooling of superconducting devices such as flywheels (see Figures 16.10 and 16.11). Balanced CHP and integration with SmartGrid is possible. (b) Schematic of the energy management process based on a hydrogen cycle.

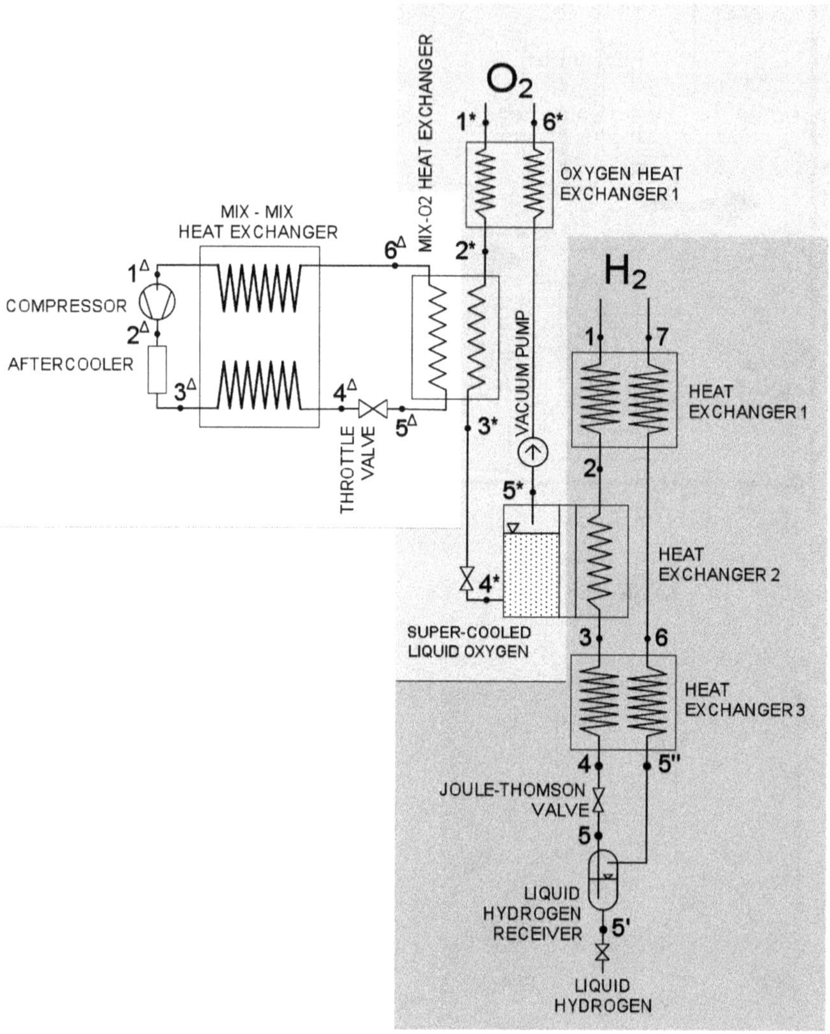

(a)

Figure 16.22 Hybrid O_2/H_2 liquefier: (a) Schematic of the novel design of oxygen and hydrogen cycle characterised by efficiency of 40 per cent liquefaction of hydrogen at rate of 1 ml of LH_2/min. The O_2/H_2 liquefier is combined at the initial cooling stage with a novel PEN J-T cycle marked on the left-hand side of diagram (Glowacki 2010); (b) Temperature-entropy phase diagram for oxygen liquefaction cycle – point 5* represents LO_2 under low pressure at a temperature of 55 K; (c) Temperature-entropy phase diagram for hydrogen liquefaction cycle.

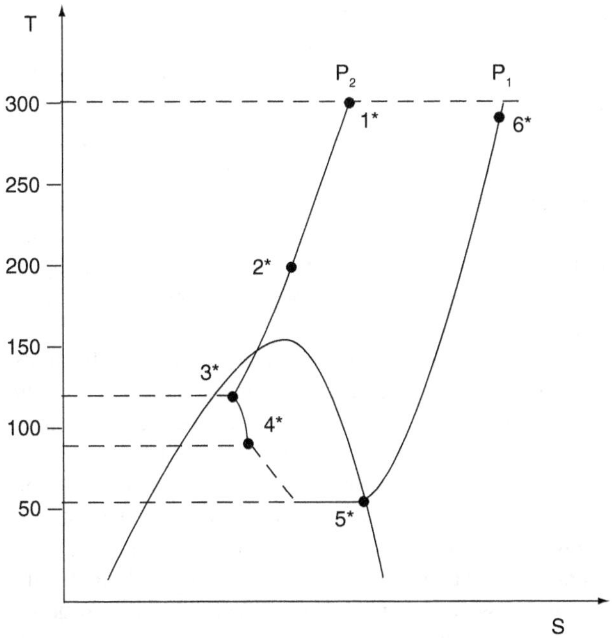

(b)

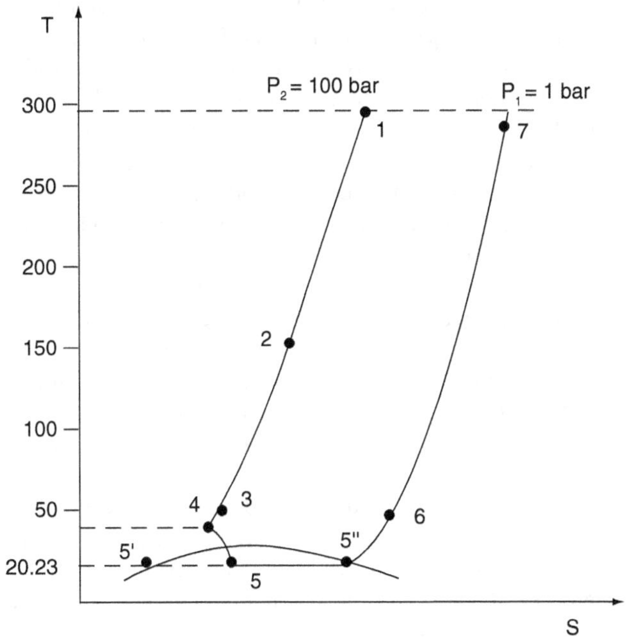

(c)

Figure 16.22 continued

Considering the decentralised hydrogen economy in conjunction with H_2O splitting it is important to take note that LO_2 and LH_2 can be used in their cryogenic form. Oxygen has a melting point of 54.1 K and hydrogen boils at 20.6 K under normal pressure. It is suggested that liquid oxygen might be used, under partial vacuum, at its lowest temperatures to act as a precoolant for the hydrogen gas.

To improve safety during liquefaction of hydrogen a novel composite solid nitrogen reservoir will be researched to act as a heat exchanger between liquid O_2 and gaseous H_2. Cryogenic LH_2 would be stored in a cryogenic, superinsulated dewar integrated with the liquefier, as presented in Figure 16.22. Additionally, a thermal link from the LH_2 energy storage to the superconducting flywheel is a viable option, as in Figure 16.21.

Many of the current and future energy and electrotechnology applications require the employment of medium and high-temperature superconductors working at cryogenic temperatures of approximately 20 K to achieve their desired efficiency. If such cryogenic temperatures needed to be created exclusively to enable superconducting devices, with no connection to hydrogen economy, the cost would be high – but much less than for 4K. For example, to generate a cooling power of 1 W at a temperature of 20 K requires ~100 W of power for the cryogenic cryocooler at room temperature. The overall efficiency of the cooling will improve dramatically if the liquid hydrogen is produced as an energy carrier and can have an additional use as a cryogenic coolant for superconductors deployed in the energy cycle.

If one envisages that liquid hydrogen integrated with superconducting storage energy systems may act as an energy vector, buffering the energy demand from renewable sources (as in Figure 16.21), one may also consider a new approach where electrothermal integration of a solid oxide electrolyser cell with solid oxide fuel cell can be viable option. A simplified approach is to use LH_2 and SOFC in conjunction with SMES, as presented in the literature, to provide an energy back-up for buildings and hospitals (Hirabayashi *et al.*, 2006, 2008).

These concepts do not consider the combined heat and power (CHP) cycle where hydrogen liquefaction is also involved. Noting the fact that solid oxide cell function (presented in Figure 16.21) may work either as a solid oxide electroyser cell or solid oxide fuel cell, the novel hybrid design of the multifuncional stack can be employed enabling an electrothermal balace as needed for a decentralised energy cycle. High-temperature SOFC benefits in the form of a 30 per cent improved performance if pure hydrogen is used rather than methane. In the more general picture, hydrogen from the storage can be used by polymeric fuel cells that required ultra-pure hydrogen gas, which in case of the cryogenic liquid hydrogen storage is simply sourced from LH_2 evaporation.

Another interesting and important issue about the proposed LH_2 cycle is that in most cases the input from renewable energy sources to support high-temperature electrolysers comes in the form of DC current and DC voltage. The same holds for the generation of the electricity from high-temperature solid oxide fuel cells. Therefore, one may recognise that an integrated LH_2 energy cycle that merges the lower-cost H_2 liquefaction and storage with a DC energy cycle becomes a distributed energy generic cluster for sustainable cities and industrial applications.

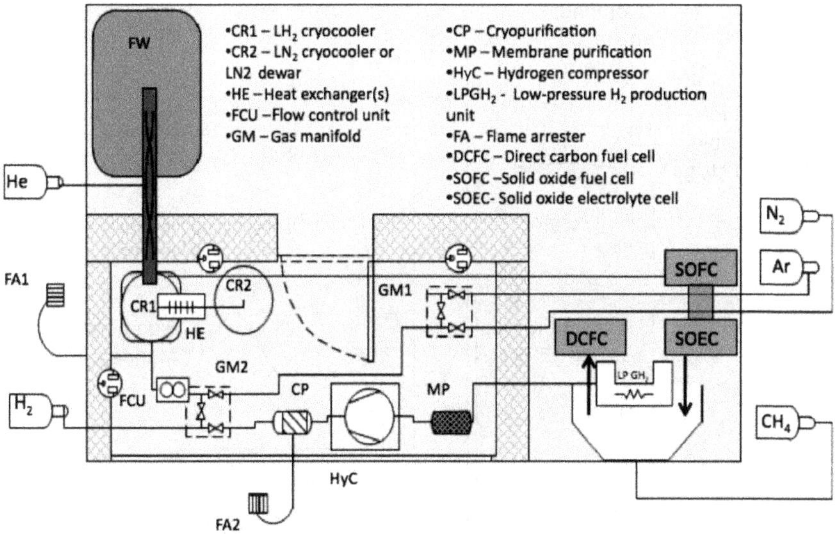

Figure 16.23 Top view of the new hydrogen laboratory for hydrogen generation (CH_4 microwave decomposition, or alternatively electrolysis, liquefaction and storage CR1, as described in Figure 16.21). FW represents a flywheel which is cooled by a helium gas cryofan from the LH_2 container (as presented in Figure 16.24). Hydrogen can be used in FCs or for transport.

The overall efficiency of CH_4 microwave plasma-assisted decomposition, presented schematically in Figure 16.23, is very high, and subsequently the gross cost of 1 kg of normal-pressure H_2 is US$1.5 (£1), which is very low in comparison with other methods such as electrolysis and steam reforming, and additionally does not produce CO_2 (Jasinski *et al.*, 2008).

Conclusions

The steady increase in the price of helium is already affecting many cryogenic-based businesses, forcing them to divert to new, more efficient technologies and even less economical cryogen-free cryogenics (see Chapter 11). There are cryomagnetic applications, which will require dynamic and efficient cooling, and one of the solutions will be indirect cooling by hydrogen, i-LH_2, which will use compressed helium in a closed cycle. This is acceptable if the helium leak rate is small.

The i-LH_2 solution demonstrates, clearly, that cooling directly with helium or neon is 70 times to 100 times more expensive than indirect cooling by LH_2. This presents opportunities for the superconductivity community and the hydrogen economy.

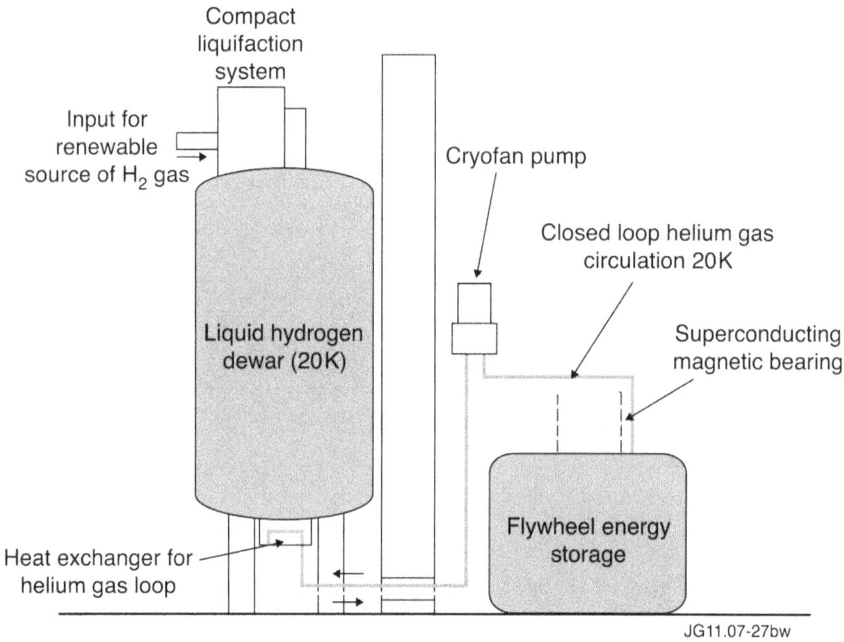

Figure 16.24 Indirect cooling of a fully superconducting MgB_2/YBaCuO bearing using a liquid hydrogen dewar and gaseous helium cryofan heat exchanger (Patel *et al.*, 2010). Helium gas is pumped around a closed loop with one end cooled to 20 K via a heat exchanger connected to a liquid hydrogen dewar. This set-up is at the design and construction stage and is going to be part of a liquid hydrogen laboratory in the new Department of Materials Science and Metallurgy building, University of Cambridge (completion in 2012).

Use of SmartGrid and CHP approaches might be extended to formulate decentralised and centralised hydrogen-based economies for the benefit of the public. There are obviously challenges, however. For example, a solution to the complex problem of lower liquefaction efficiency of smaller liquefaction rate may only come from a new design of the compact liquefaction system, where an initial cooling stage of the H_2 gas will come from the mixed-gas portable coolers operating at the 60–80 K temperature range. The final liquefaction stage would be reinforced with magnetic refrigeration and expansion turbines that could dramatically improve the efficiency of the decentralised small liquefiers. There is also an urgent need for research on more effective hydrogen generation, purification, compression and ortho-para conversion to improve liquefaction efficiency and long-term storage.

It is proposed that a sustainable LH_2 energy cycle be developed, exploring the interconnection between SOEC → LH_2 → SOFC and superconducting energy storage devices. This requires intensive research to optimise rapid development of the hybrid solid oxide cells, with interchangeable functions and integrated electrothermal management to secure real opportunities for implementation of the sustainable LH_2 energy cycle. Integration of a well-balanced hydrogen generation, compression, storage and usage with renewable DC energy sources can become the distributed energy generic cluster for sustainable cities and various decentralised industrial applications.

Acknowledgements

This work was carried out with the support of: SUPERGEN 14 project "H-delivery"; EU F7 "EFECTS"; Engineering and Physical Sciences Research Council; Science and Technology Facilities Council; and Siemens Magnet Technology UK. The authors wish to acknowledge Dr A. Piotrowskiej-Hajnus and Dr R. Tomov and also R. Clarke from the Culham Centre for Fusion Energy for their valuable discussions.

References

Ballarino, A and Carrillo, A. (2008) "Current Leads", Lectures on Superconductivity, ed. Glowacki, B. A. Applied Superconductivity and Cryoscience Group, University of Cambridge, 7 October, www.msm.cam.ac.uk/ascg/lectures/applications/currentleads.php.

Bloem, W. B. (1986) "Transient heat transfer to supercritical helium at low temperatures", PhD thesis, Technische Universiteit Delft, The Netherlands.

Bogner, G. (1975) "Cryopower transmission studies in Europe", *Cryogenics*, Vol. 15(2), February: 79.

Cai, Z, Clarke, R. H., Glowacki, B. A., Nuttall, W. J. and Ward, N. (2010) "Ongoing ascent to the helium production plateau – Insights from system dynamics", *Resources Policy*, 35: 77–89.

Chen, Y., Selvamanickam,V., Zhang, Y., Zuev, Y., Cantoni, C., Specht, E., Parans Paranthaman, M., Aytug, T., Goyal, A. and Lee, D. (2009) "Enhanced flux pinning by BaZrO3 and Gd,Y … 2O3 nanostructures in metal organic chemical vapor deposited GdYBCO high temperature superconductor tapes", *Applied Physics Letters*, 94, 062513.

Clarke, E. and Gardner, J. B. (1976) "Refrigeration with expansion turbines", *Contemporary Physics*, Vol. 17(6): 507–28.

Clarke, R. H. and Glowacki, B. A. (2010) "Indirect hydrogen versus helium or nitrogen cooling for fusion cryogenic and magnet systems", *Proceedings of ICEC 23 – ICMC, 19–23 July*, Oficyna Wydawnicza Politechniki Wrocławskiej, Wrocław, Poland.

Cryomech website, Syracuse, New York, www.cryomech.com (last accessed January 2012).

Denmark Hydrogen www.hydrogencarsnow.com/denmark-hydrogen-link.htm.

Duchateau, J. L. and Hertout, P. (2008) "Which superconducting material for the toroidal field system of the fusion DEMO reactor?", *Journal of Physics: Conference Series*, 97, 012038.

Edwards, D. R. (1968) Electrical and electronics technician engineer, March.

Fiet, W. (2011) "HTS4Fusion (an European Fusion Development Agency Programme)",

ITER Newsline, Vol. 179, 31 May, http://www.itep.kit.edu/hts4fusion2011 (accessed January 2012).

Gehring R, Juengst, K.-P., Kuperman, G., Bordry, F., Burnet, J.-P. and Voelker, F. (2006) "A SMES-Based Power Supply for Accelerator Magnets", *IEEE Transactions on Applied Superconductivity*, 16: 594.

Giunchi, G., Ripamonti, G., Perini, E., Ginocchio, S., Bassani, E. and Cavallin, T. (2006) "Highly densified MgB2 bulks by reactive Mg liquid infiltration", *Advances in Science and Technology*, 47: 7–16.

Glowacki, B. A. (2010) "Liquid Hydrogen as a energy vector and coolant for superconducting devices", *Biuletyn Polskiego Stowarzyszenia Wodoru i Ogniw Paliwowych*, 5: 13–21.

Glowacki, B. A., Majoros, M., Vickers, M., Eisterer, M., Toenies, S., Weber, H. W., Fukutomi, M., Komori, K. and Togano, K. (2003) "Composite Cu/Fe/MgB2 superconducting wires and MgB2/YSZ/Hastelloy coated conductors for ac and dc applications", *Superconductor Science and Technology*, 16: 297–305.

Glowacki, B. A. and Nuttall W. J. (2008) "Assessment of liquid hydrogen cooled MgB2 conductors for magnetically confined fusion", *Journal of Physics: Conference Series*, 97, 012333.

Glowacki, B. A. and Nuttall, W. J. (2010) "Hydrogen as a fuel and as a coolant – from the superconductivity perspective", *Journal of Energy Science*, 1: 15–29.

Glowacki, B. A., Tomov, R., Chorowski, M., Piotrowska-Hajnus, A., Kraus, M., Nowak, R. and Jewulski, J. (2010) "Sustainable LH2 energy cycle", *Journal of Energy Science*, 1: 29–35.

Gong, M., Wu, J. and Qinguo Hu, Q. (2010) "Development of a –186°C cryofreezer based on a dual mixed-gases Joule-Thomson refrigeration cycle", presented at ICEC 23 – ICMC 2010, 19–23 July, Wrocław, Poland.

Grant, P. M. (2002) "Potential Electric Power Applications for Magnesium Diboride", *Materials Research Society Symposium Proceedings*, Vol. 689(3).

Grant, P. M. (2004) "The SuperCable: Dual Delivery of Hydrogen and Electric Power", presented at the IEEE PES Meeting, New York, October, www.w2agz.com/BD_WROC10.htm.

Grant, P. M. (2010) "Potential Electric Power Applications for Magnesium Diboride", presented at ICEC 23 – ICMC 2010, 19–23 July 2010, Wrocław, Poland.

Green, M. A. (2001) "The effect of low temperature cryocoolers on the development of low temperature superconducting magnets", *IEEE Transactions on Applied Superconductivity*, 11(1): 2615.

Hassenzahl, W. V., Eckroad, S. E. C., Grant, P. M., Gregory, B. and Nilsson, S. (2009) "A High-Power Superconducting DC Cable", IEEE Transactions on Applied Superconductivity", Vol. 19(3), June.

Hemmi, T., Koizumi, N., Nunoya, Y., Okui, Y., Matsui, K., Nabara, Y., Isono, T., Takahashi, Y., Okuno, K., Banno, N., Kikuchi, A., Iijima, Y. and Takeuchi, T. (2009) "Characterization of Nb3Al strands subjected to an axial-strain for a fusion DEMO reactor", *IEEE Transactions on Applied Superconductivity*, 19: 1540.

Hiltunen, L., Martini, L. and Mikkonen, R. (2008) "SMES and Flywheels", Lectures on Superconductivity, ed. Glowacki, B. A. Applied Superconductivity and Cryoscience Group, University of Cambridge, 7 October, www.msm.cam.ac.uk/ascg/lectures/applications/currentleads.php.

Hirabayashi, H., Makida, Y., Nomura, S. and Shintomi, T. (2006) "Feasibility of hydrogen cooled superconducting magnets", *IEEE Transaction on Applied Superconductivity*, Vol. 16(2): 1435.

Hirabayashi, H., Makida, Y., Nomura, S. and Shintomi, T. (2008) "Liquid hydrogen cooled superconducting magnet and energy storage," *IEEE Transaction on Applied Superconductivity*, Vol. 18(2): 766.

ITER website www.iter.org/mach/Pages/Magnets.aspx.

Jasinski, M., Dors, M. and Mizeraczyk, J. (2008) "Production of hydrogen via methane reforming using atmospheric pressure microwave plasma", *Journal of Power Sources*, 181: 41–5.

Kamiya, K., Takahashi, H., Numazawa, T., Nozawa, H. and Yanagitani, T. (2007) "Hydrogen liquefaction by magnetic refrigeration", in S. D. Miller and R. G. Ross Jr (eds) *Proceedings of the Fourteenth Conference on Cryocoolers*, Boulder, CO: 637–44.

Korpela, A., Lehtonen, J., Mikkonen, R. and Perala, R. (2003) "Quench in a conduction-cooled Nb3Sn SMES magnet", *Superconductor Science and Technology*, 16: 1262–7.

Larbalestier, D., Gurevich, A., Feldmann, D. M. and Polyanskii, A. A. (2001) *Nature*, 414: 368.

Lyly, M., Hiltunen, I., Jarvela, J., Korpela, A., Lehti, L., Stenvall, A. and Mikkonen, R. (2010) "AC-loss considerations of a pulse SMES for an accelerator", *Journal of Physics: Conference Series*, 234, 032036.

McDonald, K. T., Iarocci, M., Kirk, H. G., Mulholland, G. T., Titus, P. H. and Weggel, R. J. (2002) "Use of He gas cooled by liquid hydrogen with a 15-T pulsed copper solenoid magnet", Technical/Economic Report, 15 August: 1–5 E951 also published in *Proceedings of ICEC 23 – ICMC 2010*, 19–23 July 2010, Oficyna Wydawnicza Politechniki Wrocławskiej, Wrocław, Poland).

Norway Hydrogen www.hydrogencarsnow.com/norway-hynor-project.htm.

Patel, A., Palka, R. and Glowacki, B. A. (2011) "New fully superconducting bearing concept using the difference in irreversibility field of two superconducting components", *Superconductor Science and Technology*, 24, 015009.

Schultz, K. R. (2003) "Use of the Modular Helium Reactor for Hydrogen Production", GA-A24428, September, University of Illinois.

Seeber, B. (1998) *Handbook of Applied Superconductivity*, Institute of Physics Publishing, London: Ch. H7.1.

SmartGrid (2010) Conference website, Brussels, 2–3 December.

Sweden Hydrogen www.hydrogencarsnow.com/hydrogen-sweden.htm.

Yamada, S. S., Hishinuma, Y., Uede, T., Schippl, K., Yanagi, N., Mito, T. and Sato, M. (2010) "Conceptual design of 1 GW class hybrid energy transfer line of hydrogen and electricity", *Journal of Physics: Conference Series*, 234, 032064.

Wang, C. (2009) "Small scale helium liquefaction systems", *Journal of Physics: Conference Series*, 150, 012053.

Withington, S. and Goldie, D. (2008) "Astrophysics detectors", *Lectures on Superconductivity* Ed. Glowacki, B. A., Applied Superconductivity and Cryoscience Group, University of Cambridge (7 October), www.msm.cam.ac.uk/ascg/lectures/electronics/.

17 Is there a helium problem?

Ways forward

Ralph Scurlock and Art Francis

Introduction

If there is a helium problem, what should be done about it? Has the US National Council of Sciences review fixed the problem, and what about issues faced by those outside the United States? Should the rest of the world follow the lead set by the US policy work? This chapter includes a serious discussion of "venting" and "moral" issues (e.g. grades of user).

Is there a helium problem?

From the point of view (RGS) of a large university user (relatively small compared with the largest users) of the order of 100,000 cf/year, the answer is, "Yes, without any doubt."

From the point of view of a US gas specialist (Art Francis), the federal control of helium as a natural resource has collapsed in the face of competition from private industry and extensive legal wrangling. The federal store of helium is being sold off, and after 2015 the US will become an importer of helium. Nothing is being done to counter the ensuing chaos and helium famine. Yes, there is a helium problem; indeed, there may be no helium for many users in the future if international action is not taken urgently.

Today in 2012, we are experiencing the results of the rather chaotic mismatch between rapidly rising but highly variable demand for helium, and the available but highly variable supply, resulting in an accompanying continuous rise in price. Without access to a substantial buffer store, suppliers cannot guarantee to meet the needs of helium users, who have to maintain a larger and larger stock of helium gas to meet research and training requirements.

The US helium conservation programme was terminated in 1996, and supply has been placed in the hands of private companies, leaving users such as ourselves to seek what is available to buy, at premium prices. As for the future, the outlook for 2060 is grim in the face of an expected global famine of helium, with any available scarce stocks inevitably fetching prices many times the price we are paying today.

Something has to be done urgently, on a scale far wider, and also more effective, than the US achieved up to 1996. We need to consider and pioneer the setting up and funding of a body to look after a diminishing earth resource before it is too late to do so – possibly under the authority of the United Nations. This body would have to coordinate government administrations, including the US Bureau of Land Management, and private supply companies in the energy, oil and gas industries, with the variable demands of a growing number of helium users. The initiative, for monitoring and advising on the supply, storage, and use of helium, would include the need to expand the role, developed by the US administration during the past 90 years, into an international one.

Non-US helium users have been enjoying freely available access to helium provided by the US administration up to 1996, and this has encouraged a culture of taking everything for granted. This free ride has come to an end with the 1996 Helium Act, and a new worldwide body needs rather urgently to be created. The principle of conservation and storage of helium extracted from natural gas must be re-established, preferably with extraction and refining plant positioned adjacent to newly discovered sources.

This new body should include:

1 an advisory committee or council, to agree policy and provide a monitoring role;
2 an executive committee to exercise planning, control, conservation, storage and marketing of helium as a diminishing earth resource.

If helium is wasted by venting into the atmosphere before it can be used, then re-extraction of this vented helium will be extremely expensive (as described in Chapter 8), and will produce a large carbon footprint. The outcome will be a limited supply of helium gas. What then?

Let us consider the situation in China in 1979, at the end of the Cultural Revolution. The only source of helium in China was from atmospheric air, via air separation plants; there was no LNG or known helium-bearing natural gas in China at that time. The amount of helium gas or liquid available for research was extremely limited, and this shortage stifled Chinese development of all the exciting cryogenic innovations and applications being pursued in the Western world. For example, while Chinese work on LTS superconductors and applications such as superconducting generators was of an extraordinarily high quality, its progress was limited by the tens of hours per year for which liquid helium was available, compared with thousands of hours per year in my own laboratories in Southampton. To economise on the use of liquid helium, cryostats were portable, and were wheeled to a central helium liquefier for filling with liquid, and then taken back to the experimental laboratory. The Chinese were doing their best with the available materials, but it was the shortage, not the cost, of helium produced by extraction from the air that was the main problem for them.

Such a shortage will be the determining factor for all helium users, worldwide, once natural gas sources are exhausted. There would then be short-term helium famines to contend with, followed by longer and more permanent famines; only small stocks of helium would be for sale, at inflated market prices, way above extraction, storage and distribution costs.

Discussion of possible ways forward

The 2010 NRC report

What can be done to delay the onset of this famine? There are a number of possible ways forward to help delay the helium famine for 20 years, or perhaps longer; but none to stop the eventual famine.

The most recent 2010 US National Research Council report, "Selling the Nation's Helium Reserve", fails to address this a worldwide problem. Indeed, the recommendations in the report are almost exclusively concerned with US users of helium and their division into alpha users and beta users, in the face of a diminishing domestic supply, together with the anticipation that the USA will become an importer of helium after 2015. In other words, the report supports a protectionist policy, to be adopted in 2015 or whenever the remaining federal reserve of conserved helium has been sold. The proposed division between alpha and beta users with different priorities for helium use was previously discussed and thrown out in horror at the Cambridge seminar on the "Future of Helium", in March 2009. In despair, let us look elsewhere for guidance in determining the way forward.

Towards an optimum programme, written by Art Francis, Praxair Inc., 1996

As a start, let us look at the following extended section of analysis, which includes a set of nine points to be considered in setting up an international conservation programme. These were first proposed by Art Francis, Praxair, in his 1996 manuscript "Helium at the Crossroads", and provide the inspiration needed to go forward.

Realities of future conservation

In order for future helium conservation to be successful, it will be necessary for the private sector and governments to share involvement and include the participation of foreign governments and enterprises. To achieve an optimum integration of effort among these disparate groups, it is necessary to understand the differences in their likely motivations. Private enterprise is by nature driven by the desire for profit. Government activities are carried on to promote the general welfare and to strengthen the state. The funds that governments use in pursuit of objectives are obtained from taxes or bonds. Only a few of the activities of government result in any directly related revenues. Very few activities are established on a pay-as-you-go basis. Consequently, government programme managers pay little attention to the economic drives that guide private enterprise. To coordinate parallel government and private sector helium activities, the government must be sensitive to economic realities.

One of these realities is that the value of stored crude helium will always be greatest when used to sustain helium-dependent applications in the event that the regular supply fails in some way. Helium activities by government would serve best if they were directed toward such emergency situations. If the US Bureau of Mines helium activity is also engaged in providing the normal supply to some or all federal agencies, then that supply should come from helium extracted from the current production of fuel gases.

In carrying out either normal or emergency supply functions, the government activity should be managed in such a way as to pay its own way, or at least provide sufficient income to cover all the relevant cost of the enterprise, including return on invested (borrowed) capital to justify further investment as necessary. This should be done, not because government needs to make a profit or meet a mandated self-amortisation goal, but rather to cause the government-operated enterprise to function by the same forces that govern the private sector. When this concept is ignored, one sector will enlarge the scope of its activities and encroach on those activities more logically carried out by the other sector.

Another basic reality that should be understood is that definite programmes and detailed statutory laws can be successful only when the subject is well understood and relationships remain relatively constant. For example, failures of the US Helium Act of 1960 and the resulting programme have developed from lack of understanding and experience. Goals were set too rigidly. No effective mechanisms were provided for modifying objectives over a 30-year period of great changes, often unforeseeable. Discoveries can rarely be predicted. Helium demand, supply and application are still very strongly influenced by discovery and innovation. As a consequence, statutory law should establish authority and responsibility to pursue a desired direction. In so far as possible, helium conservation should be a voluntary reaction to the recognition of new developments. Ideally, shared knowledge should stimulate all the interested parties to carry out necessary modifications to ongoing activities. Government intervention ought to be directed towards encouraging conservation and discouraging waste.

Except when a crisis is fully recognised, the process by which change is brought about in a democracy is often ponderous and slow in reacting to timely developments. Furthermore, once an idea is finally adopted by a government, it is unlikely to continue in practice if the actual application shows little or no advantage. In the case of helium, it is likely that most innovations in supply and use will occur in the private sector. A new idea does not need to be accepted by all concerned parties or even a majority. If the management of only one enterprise tries an innovation and if it demonstrates some clear advantage, then its application will spread. If the innovation provides no advantage, it will be abandoned.

Basic facts

To optimise future helium use, some basic facts should be realised:

1 Helium contained in gases being produced for fuel use will be lost if the helium is not extracted for current use or storage.
2 Helium currently in storage, and helium in gases not currently being produced, should remain available for future use.

The prime objective of helium conservation should be to minimise the use of helium taken from storage and/or natural gas not already being produced for fuel.

The supply of helium is dependent on a sufficient flow of helium contained in fuel gas into extraction plants. Because the rate of this flow is established by gas industry needs, and not by the needs of the helium industry, the production volume of crude helium will not synchronise with helium demand. In order to encourage the growth of helium use, it is necessary to utilise a large crude helium storage system in a manner analogous to the role of a flywheel on an engine. The storage system insulates helium users from the gas industry's influence on crude helium extraction. In this way, the refined helium supply and demand will be kept in balance with consumer demand at reasonably stable prices.

Pressures on the private sector to minimise the investment of capital for inventory will limit the quantity of privately owned helium in storage, especially for long holding times. Holding previously extracted crude helium in storage, especially for long periods of time, is more properly an activity of governments. This is appropriate for in-ground reserves of non-fuel helium-bearing gases as well. Preserving these helium deposits will probably require purchase of such fields by the government. Exploitation of these deposits should be prevented, even if higher helium prices might make such exploitation economically attractive to a private owner.

An important fact that should be faced squarely is that, while a large portion of known world helium reserves are located in the United States, half the projected world demand for helium in the next 50 years will occur outside the US. Much of this demand should be supplied from helium-bearing natural gas deposits located outside the US. As in the US, these gases will be produced primarily for fuel use. A flywheel of crude helium storage will be needed to balance helium supply and demand.

The opportunity to create such storage, either in consuming or producing locations outside the US, does not exist at present (1996). For example, the US storage system provides the only back-up for helium production in Poland, a major foreign source of helium. Up to the present, no foreign government has made any effort to establish a conservation storage programme of its own, or to participate in the US programme in any way. Frankly speaking, foreign helium consumers have had a free ride at US expense.

Worldwide conservation

To optimise the use of world helium resources, the United States conservation programme should be recognised as the historic starting point for worldwide helium conservation. Foreign governments should be encouraged to support the US programme. This support should include investment in stored crude helium. Information concerning gas deposits containing exploitable helium and/or information about expected consumption by current and projected helium applications should be supplied to the Bureau of Land Management.

It is basically unsound to treat helium conservation as a purely domestic problem. The US appears to possess a disproportionately large share of world helium resources, but many other nations have large portions of other resources. In so far as possible, all resources should be managed to establish optimum worldwide use patterns, free of national self-interest.

In the past, the exclusion of private sector helium activities in the US helium conservation plan caused that plan to become derailed. Subsequent actions by all parties did not serve the helium conservation concept well. It is logical to assume that the 1996 failure to include foreign helium activities could also have unforeseen negative consequences. The history of helium conservation suggests that the problems encountered by the US leading to termination of its conservation programme *are not the critical issues.*

Points to be included in a Worldwide Conservation Agreement

The primary objective of all concerned parties should be to reach an effective agreement that defines the goals of helium conservation and the respective roles of the US government, foreign governments and the private sector worldwide. The agreement should include the following points:

1 Current world needs should be met by extracting helium from natural gases exploited for fuel use.
2 Helium contained in non-fuel gases should remain in place, not to be exploited until supplies of helium-bearing fuel gases are fully depleted and helium supply has substantially decreased due to that depletion.
3 Helium extracted from fuel gases and not immediately required for current consumption ought not to be vented. It should be stored at minimum practical cost. In locations where storage is not available, other arrangements should be made to avoid venting any helium produced in excess of immediate need.
4 Opportunities should be sought to establish low-cost storage at locations of major helium extraction operations.
5 Crude helium from storage should be made available to supplement normal supply from extraction plants, when the production of helium-bearing fuel gas is insufficient to meet the needs of helium consumers. Use of stored helium should command premium prices that would provide appropriate economic incentives to encourage expanded extraction from helium-bearing fuel gases.

6 All sources of economically extractable helium-bearing fuel gases and helium-consuming applications should be identified and continuously monitored.
7 Forecasts of supply and demand, and inventory changes, should be made at regular intervals and publicly reported.
8 Governments involved in helium conservation should utilise economic mechanisms such as prices charged, and/or paid, for helium, as well as taxes or subsidies relating to helium, in such a way as to encourage the private sector to accomplish the previously stated objectives.
9 These objectives should be pursued by means of information exchange and discussion, with minimum reliance on formal agreements, contracts and statutory law.

Responsibility of the private sector

The private sector can most effectively introduce new technology to improve the efficiency of helium production, distribution and application. Private enterprise can also more effectively integrate helium with the supply and marketing activities of related commodities. Therefore, private enterprise should have the primary responsibility for normal worldwide helium supply.

Practical ways forward for both non-US and US users

Bearing in mind the points raised in the previous section (i.e. "Points to be included in a Worldwide Conservation Agreement") by Art Francis, and applying some common sense, there are a number of possible ways forward, including the following.

Local storage by individual users

The obvious initial way forward is for users to build up their own personal stock of compressed helium gas, in anticipation of a famine. However, if this policy were to be adopted by everyone, it would merely accelerate the creation of such a famine.

Supranational storage and conservation

For helium-bearing natural gas being found outside the USA, it would be sensible to introduce local national/supranational storage in-ground, in, say, Algeria and North Africa, Qatar and the Arabian Gulf, Poland and Eastern Europe, Russia and neighbouring states, Nigeria and Central Africa, Venezuela and South America, Australia and the North-West Shelf, so that all the helium could be extracted, rather than be vented to the atmosphere. The volume exceeding local demands could thus be stored for future use as a marketable product. This local storage would then act as a buffer in smoothing out the mismatch between local demand and supply without wasting helium by venting. The economics and operation of this action need to be studied urgently because of the intrinsic ten-year timescale needed from a policy decision being made to the actual commissioning of such plant.

At the same time, it should be noted that US helium refiners need a large amount of replacement capital for their ageing cryogenic plant in the next ten years, just to maintain their present refining capacity, which anyway is insufficient to meet future demand.

Improvement in extraction efficiency

There are a number of ways of increasing extraction efficiency, including a need for the following.

1 Cease venting helium into the atmosphere, by extracting close to 100 per cent of helium from all natural gas fuel sources, via cryogenic separation, prior to selling on the natural gas for use as a fuel. Selling fuel gas containing it must no longer waste helium.
2 Cease the mining of helium-bearing natural gas and oil shales, if extracted helium cannot be marketed or alternatively stored in-ground in depleted gas fields. The helium would then remain in store in the unused gas field, until such time as needed.
3 Prioritise LNG plants that can separate helium continuously, at nominal additional cost, by providing them with additional capital, as primary producers.

Future planning

There are a number of steps to be included in planning for the future.

1 To set up local production, storage and supply networks, perhaps modelled on the now terminated US conservation programme, but financially independent of any US federal programme.
2 To have some agreed planning and control, ensuring that extraction capacity is sufficient for a large excess of extracted helium to be made available for conservation in local underground storages. Thus excess unplanned demand spikes can be met by withdrawal from the same underground storage.
3 To consider setting up the necessary funding. Helium gas should be regarded like other world derivatives as a sound asset and a good investment, with a value after storage and refining that increases with time above inflation. Sales of helium from storage would include the gross cost of storage, as well as the initial purchase price. Ownership would then remain totally within a consortium of storage and refining companies, via a legal protocol of, say, an International Helium Committee, thereby avoiding the legal tangles that contributed to the termination of the US conservation programme.

Improvements and changes in technology

There are a number of ways of improving current technology, including the following.

1 Encouraging green cryogenics, and a green culture by all users, so as to make much more efficient use of helium. The Institute of Cryogenics, University of Southampton, practised such a culture over many years, during which the number of liquid helium users – students, research fellows, staff and technicians – rose dramatically. At the same time the equipment and cryostats grew increasingly large. However, the volume of grade-A helium gas required as a replacement for gas lost and not returned for reliquefaction remained relatively constant. This was because of the efficient cooldown of systems with liquid helium, and the use of low-loss transfer lines with gas-tight seals. This even held good for high-speed rotating cryostats and machines, enabling all operators to learn how to reduce gas losses. As the volume of liquid helium used grew, so the gas lost per cycle of usage fell progressively from 10 per cent down towards 1 per cent, while the replacement volume of gas required each year remained constant, and firmly below 10 per cent of the stock held.

2 Substitutes for helium need to be developed. Other inert gases are more readily available from the atmosphere, and are far cheaper to produce. Surely other inert gases, as well as hydrogen and nitrogen, could be used instead of pure helium for many of the applications that only arose, in the first place, because helium had been so readily available and so cheap?

3 Closed-cycle cryocoolers are rapidly coming into use for cooling down to LHe temperatures on a small scale with cooling powers in the range of 1.0–1,000 W. While cryocoolers are extremely handy for dry-cooling small systems in the absence of any additional cooling media or equipment, they can also be adapted as boil-off helium condensers, or as small liquefiers/reliquefiers. Cryomech Inc. have redesigned the precooling section of their condensers and obtained a 100 per cent improvement in liquid helium produced with no increase in compressor power. This simple modification, doubling the output, indicates what improvements can be achieved in cryocooler technology.

4 The operation of large S/C magnets, using low-T_c superconductors like NbTi and Nb_3Sn, generally requires liquid helium coolant to maintain thermal stability when carrying high currents in high fields. Changes in technology, as the latest ceramic superconductors come into high-current, high-field operation, suggest that liquid helium coolant will eventually be replaced by liquid nitrogen or liquid air. Otherwise, major users of liquid helium such as the LHC, and fusion reactors, will run into the helium famine problem quite soon.

5 While grade-A 99.996 per cent pure helium gas is needed in cryogenic systems and cryocoolers operating at 4 K with liquid helium and up to, say, 20 K with gaseous helium, high-purity gas is not necessarily needed for

many uses at higher temperatures. Lower grades of helium gas, with non-toxic inert gases such as neon and argon, could therefore be more widely marketed for these purposes.

Prioritisation of users and applications

When shortages really occur, an agreed set of rules, together with an authority in charge, would have to prioritise users. Possible divisions could include the following.

1 A division could be between grade-A helium users and non-grade-A users.
2 Grade-A users who really need LHe as a first priority, possibly on individual small scales of less than 100,000 cubic feet per year; eg. MRI systems in hospitals, university research departments, etc.
3 Large grade-A LHe users in peak situations; eg. rocketry using cryogenic LOX and LH_2, and particle accelerators like the Tevatron and the LHC.
4 Non-grade-A users on very large scales, such as deep-sea divers for breathing gases, and nuclear reactors, both fission and fusion, for maximum heat transfer efficiency, and airships.
5 Heliarc welding in the USA (replaceable by argon-arc welding outside the USA), and balloon gas for party balloons.

The simple division between alpha and beta users proposed by the US NRC 2010 report would probably be extremely controversial and divisive. Discussion of prioritisation should not be allowed to interfere with an international conservation programme.

Proposal for International Helium Council and World Helium Bank

To provide a global helium conservation programme, the formation of an International Helium Council and World Helium Bank needs to be actively considered. The idea of a World Helium Bank is not a new one. In the late 1980s, Ed Hamell of Los Alamos National Laboratory in the US surveyed and reported on the failing US conservation programme, and suggested that a much broader international programme needed to be founded, but nothing happened.

In the early 1990s, Art Francis of Praxair also suggested an international programme to take over the then terminating US conservation programme, but again nothing happened. These two farsighted men were whistling in the wind!

The proposed International Helium Council would need to adopt the US system of conservation in a modified form acceptable to all helium producers and users, by 2015.

The legal problems that brought down the US conservation programme, could possibly best be avoided by creating a World Helium Bank in preference to private deals between individual suppliers and large users. This bank would purchase all surplus helium not immediately sold after extraction, for conservation and storage as a derivative product. Helium taken from storage would then be sold to meet additional demand at a price determined by the market and its full storage costs, similar to other derivative products on the world markets.

A financial float might be needed from participating companies and countries. This could be raised via the purchase and sale of shares in the World Helium Bank.

There are number of options to explore, urgently, to satisfy the nine points raised by Art Francis.

Outcomes of a successful IHC and World Helium Bank

Looking ahead, perhaps we can come to terms with conserving a natural resource, with the right will. However, the 2009 Copenhagen Conference on Climate Change did not achieve much on a global scale, due to self-interest preventing a common view being attained.

Perhaps we can do better, and develop some optimism about the long-term future.

The operation of a World Helium Bank through an International Helium Council would be very welcome to all users. The real problem lies with setting up a global authority to bring together national and institutional interests, together with adequate finance in collaboration with private producers and storage companies, and with the necessary urgency.

The IHC needs to be set up by 2015, when the US federal reserve of helium will be approaching zero. It would take a World Helium Bank some ten years to begin a world conservation programme, which therefore should aim to be operational by 2025.

If this were to happen, the following outcomes could be achieved.

1 Continued stimulation and stabilisation of the cryogenic use of liquid helium temperatures in a growing array of devices, instruments, etc., for applications to benefit mankind.
2 Stimulation and provision of a future for further applications of helium gas, with the gas recycled in use.
3 It could, on the other hand, result in a reversion to the twentieth-century culture of the profligate use of helium gas, when it was vented into the atmosphere after use with little attempt to conserve or recycle it.

Strict control of helium as a natural resource, on a global basis, will have to be pioneered and exercised in the near future, and it will have to be achieved effectively for all to benefit.

18 The future of helium

Policy, molecules and machines

William J. Nuttall, Richard H. Clarke and Bartek A. Glowacki

Historical link

The preceding chapters of this book make clear two fundamental realities concerning the future for helium. First, helium is a by-product of natural gas production; and second, helium is a vital substance for key high-technology industries of the future. The long-term future need for helium is clear, but the long-term scenarios for supply are far less certain.

The earlier chapters have explained in some detail the various supply-side challenges and opportunities that lie ahead. Consideration has been given, not only to the United States but to other parts of the world that are emerging, or have the potential to emerge, as major helium suppliers in future. Today the vast majority of helium used, and 100 per cent of commercially traded helium, comes from the upstream part of the natural gas supply chain.

The twentieth century was a time of enormous technological change fuelled almost entirely by fossil fuels. Through that century the fuel of choice changed. In the earliest decades of the century coal powered trains and ships provided long distance transport. First in ships, and then in trains, coal was replaced by diesel oil. The origin of that process was arguably Winston Churchill's decision in 1911 as First Lord of the Admiralty to adapt the Royal Navy fleet to run on oil. One of several attendant advantages was that a flotilla's location would not be given away by a large plume of dark coal smoke, many miles high (Dahl, 2000). Churchill's decision had far-ranging geopolitical consequences, as Britain sought to secure reliable oil supplies, but it is the technical consequences that concern us here. The transition to oil spread from shipping to include trains as the century progressed. The growing popularity of the automobile through the twentieth century further enhanced the role of oil. For much of the century, the oil industry, from exploration through to refining, regarded natural gas as a potentially dangerous problem to be avoided. The discovery in an oilfield of large amounts of natural gas was a source of dismay. The gas would usually be flared as an inconvenience, although high-pressure (non-combusted) venting was also developed (OGP, 2000). In the later decades of the twentieth century, however, the international oil companies' attitude to natural gas changed dramatically. At first, natural gas replaced synthetic town gas for homes and businesses, and then it was increasingly used in electricity generation. As described in Chapter 5, the natural gas industry has itself evolved away from a purely pipeline based supply chain to see the growth of cryogenic liquefied natural gas shipments. Despite a growing recognition of the energy resource wastage and the climate damage associated with natural gas, the practice of flaring grew throughout the twentieth century, only estimated to have peaked globally in 2005 at 162 Bcm. Such flaring was roughly equivalent to 30 per cent of annual EU natural gas consumption (World Bank, 2009). Oilfield natural gas tends not to be especially rich in helium. The lesson learned from the natural gas story is that a commodity that was once deemed worthless can, in time, become extremely valuable, such that previous industry practice can start to seem wasteful in the extreme. Such a parallel could well hold true for helium in the future. As Chapter 10 describes, the role of 'venting' is crucial to the evolving dynamics of the helium market.

The twentieth century comprised successive waves of fossil fuel development. Today coal, oil and natural gas all remain important sources of the global energy supply, but they all face an uncertain future. More than anything, that uncertain future is determined by global climate policy.

Elsewhere, one of us has written on the topic of three competing timeframes for the future of fossil fuel energy supply (Nuttall and Manz, 2008). These three timeframes are the timescale of fossil fuel resource depletion, the timescale of global climate change, and the timescale of technological innovation. Nuttall and Manz (2008) have posited that energy policy has been complicated by implicit confusion between two timescales widely perceived to be very similar: the timescale of fossil fuel depletion and the timescale of climate change. It is far from clear which timescale will in fact shape our future. The residence time of carbon dioxide in the atmosphere is a complex matter. At the turn of the century it was estimated to lie somewhere in the range 5–200 years (IPCC, 2001). Carbon dioxide exchange between the atmosphere and the oceans ensures that the effects of increasing atmospheric carbon dioxide concentration endure longer than the period spent by individual molecules in the atmosphere. Hence, the International Panel on Climate Change (IPPC) advises that important parts of the carbon dioxide problem endure for decades after the atmospheric emission of carbon dioxide in fossil fuel combustion. Importantly, of course, it is directly the concentration of greenhouse gases in the atmosphere that determines global climate change, rather than the more indirect process of emissions. As such, media commentary on CO_2 emissions is commentary on a topic at some arm's length from the key issue – global warming. Importantly, the CO_2 residence time in the atmosphere is far longer than the timescale of energy and climate policy action. As such, it matters little *when* we make carbon dioxide emissions. The important question becomes *whether* we make carbon dioxide emissions. A future seeking to enhance energy security by delaying the date of fossil fuel resource depletion can be of only minor impact on the issue of global warming. A focus on atmospheric CO_2 concentrations over the long term motivates an interest in carbon dioxide capture and storage or the notion that the age of fossil fuels will end with much resource left untapped underground.

Such considerations allow us, broadly, to imagine two scenarios for the future, which might be summarised as follows:

Scenario 1 – fossil fuel depletion

Over the coming decades, the following will occur.

1. Fossil fuel production costs will rise as easy sources are used up.
2. CO_2 costs will rise (but such cost rises will be a marginal addition to the main effect).
3. Fossil fuels will deplete (and reserve replacement ratios will fall).
4. Initially, demand will continue to grow (essentially exogenously).
5. Prices will rise.
6. Demand will drop in response to rising prices.
7. Fossil fuels will become exhausted.
8. Demand will finally fall, primarily because molecules of fuel are not physically available.
9. A large proportion of fossil fuel carbon will have been transferred to the atmosphere.
10. To a first approximation, no matter how much depletion is slowed, long-run global warming is unchanged.

Much of the economics of resource depletion still follows an assumption of such a scenario. Harold Hotelling predicted in a most influential piece of work (Hotelling, 1931) that the prices of exhaustible resources, such as fossil fuels, should be expected in a competitive market to rise at the going (and exogenous) rate of interest. This has become known as Hotelling's rule (see Chapter 10). It was of course developed independent of any concern for environmental externalities. For a good introduction to such issues, see the recent book by Bergstrom and Randall (2010).

Scenario 2 – climate change-dominated future

Over the coming decades, an alternative scenario runs as follows.

1. Climate change becomes an ever greater problem and fear.
2. Fossil fuel production remains cheap.
3. Social cost of CO_2 rises and forward predictions look worse.
4. Reductions of fossil fuel use achieved via rising CO_2 prices and prohibitions.
5. Demand falls in response to retail price and use constraints.
6. While retail prices rise wholesale prices fall.

7 High-cost production shuts down.
8 Average and marginal production costs fall.
9 Much fossil fuel remains in the ground for eternity.
10 Fossil fuels are never depleted in the Hotelling sense.
11 Long-term average CO_2 concentrations are significantly lower than in the other scenario.

It is arguable that only in this latter scenario will the global climate be safely stabilised.

Such observations remind us of the need to reflect on the aphorism of former Saudi Oil Minister, Sheikh Ahmed Zaki Yamani:

> The Stone Age didn't end for lack of stone, and the oil age will end long before the world runs out of oil.

Implications for helium

Commentators have started to draw attention to the wider implications arising from the end of the age of fossil fuels for the future of our modern society. For example, Aric McBay (McBay, 2010) has warned of the possible difficulties ahead concerning the supply of plastics and fertilisers currently made as by-products from fossil fuels. As global demand for oil diminishes, the fractions emerging from the oil refining process can be adjusted to accommodate the differences in demand between, for instance, petroleum and plastics. However, it is important to note that such potential for adjustment is constrained. If the reduction in petroleum demand were to be dramatic then the consequence for the plastics industry could be profound.

It has been the purpose of this book to help elucidate a similar set of issues for the future of the global helium industry. At the end of the first decade of the twenty-first century it is far from clear whether the century ahead will see a sustained rise in the importance of natural gas, noting the rise of a liquified natural gas supply chain and innovations permitting production of natural gas from shale rock formations perhaps greatly expanding economically accessible resources. It is quite possible that the twenty-first century will be the age of natural gas. It is similarly probable, however, that another future lies ahead – a largely decarbonised world in which coal, oil and gas resources will stay forever in the ground, unused. While a global transition away from coal and oil towards natural gas would represent a major reduction in carbon dioxide emissions, it would not achieve the levels of reduction desired by policymakers for 2050 (80 per cent by EU and 83 per cent by USA), and any associated diminution of harmful climate change could be far from sufficient.

Whether the twenty-first century will be the age of natural gas or represent a genuine transition to a post-fossil fuel era remains to be seen, but this choice will have profound consequences for the long-term future of the helium supply industry.

Our high-technology future

While the issues surrounding the future for plastics and fertilisers raise interesting questions, the issues surrounding helium are even more complex. This is because helium can, at first impression, appear to be an essential commodity for a wide variety of future low-carbon technologies. For instance, as various chapters of this book have shown:

- Helium is needed for important forms of low-carbon electricity generation (from nuclear fission (Chapter 14) and from nuclear fusion (Chapter 15)).
- Helium is needed to cool today's commercially deployed superconductors with important applications in medicine (Chapter 12) and industry (Chapter 13). Most importantly, superconductors have a major role to play in future electricity transmission, distribution and storage (Chapter 16). These electricity industry innovations will be essential as the electricity system adjusts to accommodate an ever larger proportion of intermittent low-carbon renewable sources of power.
- Finally, helium is needed to develop telecommunication infrastructures further. A major demand for helium comes from the fibre optic industry (Chapter 9). Improved telecommunications have the potential greatly to reduce the need for business travel, just as conventional postal services have been in decline in developed countries.

In a long-term future of dramatic decarbonisation it is far from clear how the combined futures of the helium supply industry and the natural gas industry will come together to deliver the helium for the technologies that will be so desperately needed. Perhaps we shall see a dramatic reduction of helium and natural gas venting by the natural gas industry. Second, perhaps we shall see these high-technology futures delivered with helium substitutes. The possibility of liquid hydrogen superconductors was discussed in Chapter 16. Third, we might see helium sourced from helium-rich underground resources, independent of the natural gas supply chain. Such ideas have been discussed in Chapters 6 and 7.

The Danish physicist Niels Bohr famously remarked:

> *Prediction is very difficult, especially about the future.*

As we consider the future for helium, his remarks seem particularly prescient, but we have attempted to address such considerations in Chapter 10. We close this book with one overarching observation, the basis of which is that helium is a by-product. Helium that is not separated and used is not conserved. There may be helium users and entire sectors of helium users who believe in the need to conserve helium resources. They may imagine that high-technology moves to eliminate their demand for helium, perhaps via substitution with an alternative, will preserve precious helium for more important applications. Such optimism is misplaced. The helium not purchased and used will not remain underground to be extracted at a later date.

It will remain mixed with natural gas, which will be combusted and hence the helium vented to the atmosphere. Those sectors which reduce their demand for helium may reduce their short-term exposure to the risk of helium price rises, but they will do little or nothing to preserve this important but finite resource.

Those who seek to invest in machines rather than molecules may find that one day they are forced to consider the expensive air separation of helium to provide the limited amounts of gas they will always require for their essential purposes.

Helium is a by-product of a vast and an unsustainable natural gas industry and this is a large-scale reality that is unlikely to change for many decades. In particular, prospective large users of helium should consider the following two issues.

1 If there is a market failure in which natural gas-centred decision making dominates, even when helium extraction and separation would be economically attractive – e.g. because of a corporate tendency to focus on core-business competencies – then arguably, large helium users should consider purchasing helium-rich underground gas fields in order to protect their own interests. This would also buy time in which technological improvement might occur (Chapter 17).
2 In all scenarios, and most importantly, such large helium users should develop their core competencies so as to develop low-leakage systems (Chapter 15). Leakage will clearly become a most pressing issue in the decades ahead. There is time for slow, incremental improvement if that journey starts now.

References

Bergstrom, J. C. and Randall, A. (2010) *Resource Economics*, Edward Elgar, Cheltenham, UK

Dahl, E. J. (2000) 'Naval Innovation: From Coal to Oil', *Joint Forces Quarterly*, Winter 2000–1 (excerpt available at www.epmag.com/archives/digitalOilField/5911.htm, accessed 15 January 2011)

Hotelling, H. (1931) 'The Economics of Exhaustible Resources', *Journal of Political Economy*, 39: 137–75

International Panel on Climate Change (2001) 'Working Group I: The Scientific Basis', available at: www.ipcc.ch/ipccreports/tar/wg1/016.htm accessed 15 January 2011

McBay, A. (2006) *Peak Oil Survival: Preparation for Life After Gridcrash*, Globe Pequot, Guildford, CT, USA

Nuttall, W. J. and Manz, D. L. (2008) 'A New Energy Security Paradigm for The Twenty-First Century', *Technological Forecasting and Social Change*, Vol. 75(8), October: 1247–59

OGP, International Association of Oil and Gas Producers (2000) 'Flaring and Venting in the Oil and Gas Exploration and Production Industry', Report No. 2.79/288, January

World Bank (2009) 'World Bank, GGFR Partners Unlock Value of Wasted Gas', 14 December 14, available at http://web.worldbank.org/WBSITE/EXTERNAL/TOPICS/EXTSDNET/0,,contentMDK:22416844~menuPK:64885113~pagePK:64885161~piPK:64884432~theSitePK:5929282,00.html (accessed 15 January 2011)

Author index

Subject index

For Product Safety Concerns and Information please contact our EU representative GPSR@taylorandfrancis.com
Taylor & Francis Verlag GmbH, Kaufingerstraße 24, 80331 München, Germany

www.ingramcontent.com/pod-product-compliance
Lightning Source LLC
LaVergne TN
LVHW011109300926
849160LV00008B/316

* 9 7 8 1 1 3 8 7 7 4 8 6 5 *